FEASIBILITY AND INFEASIBILITY IN OPTIMIZATION:
Algorithms and Computational Methods

***Recent titles in the* INTERNATIONAL SERIES IN OPERATIONS RESEARCH & MANAGEMENT SCIENCE**
Frederick S. Hillier, Series Editor, *Stanford University*

Sethi, Yan & Zhang/ *INVENTORY AND SUPPLY CHAIN MANAGEMENT WITH FORECAST UPDATES*
Cox/ *QUANTITATIVE HEALTH RISK ANALYSIS METHODS: Modeling the Human Health Impacts of Antibiotics Used in Food Animals*
Ching & Ng/ *MARKOV CHAINS: Models, Algorithms and Applications*
Li & Sun/ *NONLINEAR INTEGER PROGRAMMING*
Kaliszewski/ *SOFT COMPUTING FOR COMPLEX MULTIPLE CRITERIA DECISION MAKING*
Bouyssou et al/ *EVALUATION AND DECISION MODELS WITH MULTIPLE CRITERIA: Stepping stones for the analyst*
Blecker & Friedrich/ *MASS CUSTOMIZATION: Challenges and Solutions*
Appa, Pitsoulis & Williams/ *HANDBOOK ON MODELLING FOR DISCRETE OPTIMIZATION*
Herrmann/ *HANDBOOK OF PRODUCTION SCHEDULING*
Axsäter/ *INVENTORY CONTROL, 2nd Ed.*
Hall/ *PATIENT FLOW: Reducing Delay in Healthcare Delivery*
Józefowska & Węglarz/ *PERSPECTIVES IN MODERN PROJECT SCHEDULING*
Tian & Zhang/ *VACATION QUEUEING MODELS: Theory and Applications*
Yan, Yin & Zhang/ *STOCHASTIC PROCESSES, OPTIMIZATION, AND CONTROL THEORY APPLICATIONS IN FINANCIAL ENGINEERING, QUEUEING NETWORKS, AND MANUFACTURING SYSTEMS*
Saaty & Vargas/ *DECISION MAKING WITH THE ANALYTIC NETWORK PROCESS: Economic, Political, Social & Technological Applications w. Benefits, Opportunities, Costs & Risks*
Yu/ *TECHNOLOGY PORTFOLIO PLANNING AND MANAGEMENT: Practical Concepts and Tools*
Kandiller/ *PRINCIPLES OF MATHEMATICS IN OPERATIONS RESEARCH*
Lee & Lee/ *BUILDING SUPPLY CHAIN EXCELLENCE IN EMERGING ECONOMIES*
Weintraub/ *MANAGEMENT OF NATURAL RESOURCES: A Handbook of Operations Research Models, Algorithms, and Implementations*
Hooker/ *INTEGRATED METHODS FOR OPTIMIZATION*
Dawande et al/ *THROUGHPUT OPTIMIZATION IN ROBOTIC CELLS*
Friesz/ *NETWORK SCIENCE, NONLINEAR SCIENCE and INFRASTRUCTURE SYSTEMS*
Cai, Sha & Wong/ *TIME-VARYING NETWORK OPTIMIZATION*
Mamon & Elliott/ *HIDDEN MARKOV MODELS IN FINANCE*
del Castillo/ *PROCESS OPTIMIZATION: A Statistical Approach*
Józefowska/*JUST-IN-TIME SCHEDULING: Models & Algorithms for Computer & Manufacturing Systems*
Yu, Wang & Lai/ *FOREIGN-EXCHANGE-RATE FORECASTING WITH ARTIFICIAL NEURAL NETWORKS*
Beyer et al/ *MARKOVIAN DEMAND INVENTORY MODELS*
Shi & Olafsson/ *NESTED PARTITIONS OPTIMIZATION: Methodology And Applications*
Samaniego/ *SYSTEM SIGNATURES AND THEIR APPLICATIONS IN ENGINEERING RELIABILITY*
Kleijnen/ *DESIGN AND ANALYSIS OF SIMULATION EXPERIMENTS*
Førsund/ *HYDROPOWER ECONOMICS*
Kogan & Tapiero/ *SUPPLY CHAIN GAMES: Operations Management and Risk Valuation*
Vanderbei/ *LINEAR PROGRAMMING: Foundations & Extensions, 3rd Edition*
Chhajed & Lowe/ *BUILDING INTUITION: Insights from Basic Operations Mgmt. Models and Principles*
Luenberger & Ye/ *LINEAR AND NONLINEAR PROGRAMMING, 3rd Edition*
Drew et al/ *COMPUTATIONAL PROBABILITY: Algorithms and Applications in the Mathematical Sciences*

**** A list of the early publications in the series is at the end of the book ****

FEASIBILITY AND INFEASIBILITY IN OPTIMIZATION:
Algorithms and Computational Methods

John W. Chinneck
Systems and Computer Engineering
Carleton University
Ottawa, Canada

John W. Chinneck
Carleton University
Ottawa, ON, Canada

Series Editor:
Fred Hillier
Stanford University
Stanford, CA, USA

ISBN-13: 978-1-4419-4519-8
ISBN-13: 978-0-387-74932-7 (e-book)

Printed on acid-free paper.

9 8 7 6 5 4 3 2 1

springer.com

For my parents, Mac and Shirley Chinneck, who fed that spark of curiosity with science books, telescopes and chemistry sets, even while I was busily dismantling useful household items, and for Linda and Annie, who make it all worthwhile.

For my parents, Max and [illegible], [illegible] bought me books, telescopes and chemistry sets even when I was busily dismantling [illegible] useful household items; and for [illegible], who make it all worthwhile.

Preface

Most applied optimization problems involve constraints: What is the maximum profit that a manufacturer can make given a limited number of machines and a limited labour force? What is the minimum amount of fuel that a fleet of trucks can consume while making a specified set of deliveries? What is the smallest amount of silicon needed to etch an electronic circuit while respecting limits on signal propagation time, inter-wire distance, etc.? Applications of constrained optimization are everywhere in industry, business, and government.

Of course, the solution returned by an optimization algorithm must also be feasible: we want the best possible value of the objective function that satisfies all constraints and variable bounds. Some optimization algorithms are not even able to proceed towards optimality until a feasible solution is available. In addition, the optimization question can be converted to a feasibility question, and vice versa. And what happens when an algorithm is unable to find a feasible solution? How do we know what went wrong? How do we repair the model? Questions of optimization, feasibility, and infeasibility are inextricably linked.

There has been a surge of important developments related to feasibility and infeasibility in optimization in the last two decades, a trend that continues to accelerate even today. New and more efficient methods for seeking feasibility in difficult optimization forms such as mixed-integer programs and nonlinear programs are emerging. The first effective algorithms for analyzing infeasible models have been discovered and implemented in commercial software. A community of researchers in constraint programming has begun to integrate their knowledge and approaches with the optimization community. Unanticipated spin-off applications of the new algorithms are being found. It's an exciting time.

The goal of this book is to summarize the state of the art in recent work at the interface of optimization and feasibility. It should serve as a useful reference for researchers, graduate students, and software developers working on optimization, feasibility, infeasibility, and related topics. Readers having a reasonable grounding in optimization (linear and nonlinear programming, mixed-integer programming, etc.) should have no difficulty following the material.

Lightweight coverage of topics in constraint programming, with an emphasis on constraint satisfaction problems, is included to illustrate the extensive overlap and convergence in the two literatures. An ideal version of the book would cover topics in constraint programming in the same depth as topics in optimization, but this is beyond the scope of this project: collecting and organizing the wealth of new developments relating to feasibility and infeasibility in optimization. I hope the resulting book is useful to both optimizers and constraint programmers, and that it helps accelerate the ongoing merger of the two communities merge into a stronger hybrid.

Acknowledgements

My graduate work was conducted during the late 1970s and early 1980s. Inspired by the energy crises of those times, I constructed network optimization models to minimize the use of energy in large industrial plants. Later I found the optimization modeling more interesting than the energy aspects of this work. I had noticed that some of the processing network models that I was using in the energy work suffered from an inability to carry flow in some of the arcs, a pathology later labeled *nonviability* (see Sec. 9.2). I developed algorithms to automatically identify and analyze this problem.

Enter Harvey Greenberg. At that time he was involved in a project to develop an *Intelligent Mathematical Programming System* (IMPS) (see e.g. Greenberg (1996b)), and consequently had an interest in algorithms for analyzing modeling errors of various types, such as nonviability. Harvey organized an extraordinary series of meetings on the IMPS topic for an eclectic group of researchers from academia and industry. Harvey invited me to one of these meetings and, as they say, the rest is history. Sitting in the bar one night after the IMPS meeting we had a discussion about whether or not you could isolate the cause of infeasibility in a linear program to an irreducible subset of the constraints defining the model. At the time, Harvey didn't think it could be done, but I did, so I bet him a beer that I could find a way to do so. As you will see in Part II of the book, I won that bet.

But there is a postscript to this story. I have now known Harvey for around twenty years, and we have gone on to make numerous one-beer bets on other issues in optimization. I have not won a single one of those subsequent bets, so I am currently several hundred beers in debt to him. But I have an even bigger debt than that. Harvey became my unofficial mentor, always ready to provide advice and suggestions and listen to my ideas. His influence on my work has been profound.

Harvey and Pascal Van Hentenryck both took the time to read an early draft of the book and provide advice and suggestions that greatly improved it. Both pointed out topics that should be greatly expanded upon, especially the material on constraint programming, but time is unfortunately limited, so a full treatment of that topic remains another project. And as clever as those two fellows are, I'm sure I've managed to hide a few errors in the manuscript that they did not find: those are mine alone.

Last but not least are the two incredible ladies in my life, my wife Linda and daughter Annie, who can finally look at this book and see what kept me glued to the computer for such long hours over the past year. Thanks for being there.

John W. Chinneck

Table of Contents

Dedication.....v
Preface.....vii
Table of Contents.....ix
List of Figures.....xiii
List of Tables.....xv
List of Algorithms.....xvii

Introduction.....xix
1. Preliminaries.....1
   1.1. The Optimization Model.....1
   1.2. Measuring Infeasibility.....2

Part I: Seeking Feasibility.....7
   Model Reformulation.....8

2. Seeking Feasibility in Linear Programs.....11
   2.1. The Phase 1 Algorithm.....11
   2.2. The Big-*M* Method.....13
   2.3. Phase 1 from Any Basis.....13
   2.4. Crash Start Heuristics.....15
   2.5. Crossover from an Infeasible Basis.....16
   2.6. Advanced Starts: Hot and Warm Starts.....17
   2.7. Seeking Feasibility and Optimality Simultaneously.....18
   2.8. Projection Methods.....19

3. Seeking Feasibility in Mixed-Integer Linear Programs.....23
   3.1. Pivot-and-Complement and Pivot-and-Shift Heuristics.....25
   3.2. The OCTANE Heuristic.....28
   3.3. The Feasibility Pump.....30
      3.3.1. The Feasibility Pump for Mixed-Integer Nonlinear Programs. 34
   3.4. Branching Variable Selection by Active Constraints Methods.....37
   3.5. Conflict Analysis.....42
   3.6. Market Split Problems.....43

4. A Brief Tour of Constraint Programming.....45
   4.1. Branching in the Satisfiability Problem.....49

5. Seeking Feasibility in Nonlinear Programs.. 51
5.1. Penalty Methods... 52
5.2. Determining the Characteristics of an NLP................................ 54
5.2.1. Convex Sampling Enclosures.. 58
5.2.2. Hit-and-Run Methods.. 59
5.2.3. Approximating Nonconvex Feasible Regions............................. 60
5.3. Bootstrapping in a Convex Constrained Region.............................. 60
5.4. Initial-Point Placement Heuristics... 63
5.5. Constraint Consensus Methods for Approximate Feasibility................. 65
5.6. Finding a Good Sampling Box for Multistart................................. 73
5.6.1. Tightening the Variable Bounds... 73
5.6.2. Best Heuristic Sampling Box.. 76
5.7. Multistart Methods.. 77
5.7.1. MSNLP Feasibility Mode... 80
5.7.2. Multistart Constraint Consensus.. 80
5.8. Bootstrapping Method of Debrosse and Westerberg........................... 85
5.9. Global Optimization.. 87

Part II: Analyzing Infeasibility.. 89

6. Isolating Infeasibility.. 93
6.1. General Logical Methods... 94
6.1.1. Logical Reduction of Models and Presolving............................ 95
6.1.2. The Deletion Filter... 97
6.1.3. The Additive Method.. 98
6.1.4. The Elastic Filter.. 101
6.1.5. Speed-ups: Treating Constraints in Groups............................. 104
6.1.6. Speed-ups: Combining the Additive Method and the Deletion Filter...... 109
6.1.7. Sampling Methods... 110
6.2. Methods Specific to Linear Programs....................................... 112
6.2.1. The Reciprocal Filter... 113
6.2.2. The Sensitivity Filter.. 114
6.2.3. Pivoting Methods... 116
6.2.4. Interior Point Methods... 118
6.2.5. Speed-ups: Combining Methods.. 118
6.2.6. Guiding the Isolation.. 120
6.2.7. Finding Useful Isolations.. 122
6.2.8. Analyzing Infeasible Network LPs...................................... 127
6.2.9. Software 128
6.3. Methods Specific to Mixed-Integer Linear Programming....................... 130
6.3.1. A Deletion Filter for MIPs... 133
6.3.2. Additive Methods for MIPs.. 134
6.3.3. An Additive/Deletion Method for MIPs................................. 137
6.3.4. Using the Information in the Initial Branch and Bound Tree.......... 138

6.3.5. Speed-ups....................140
6.3.6. Conclusions from Empirical Studies....................140
6.3.7. Software Survey....................143
6.4. Methods Specific to Nonlinear Programming....................143
6.4.1. Deletion Filtering....................144
6.4.1.1. Speeding the Isolation by Grouping Constraints 149
6.4.2. IIS Isolation by the method of Debrosse and Westerberg...... 150
6.4.3. Methods for Quadratic Programs....................151
6.4.4. Methods for Space-Covering Global Optimizers....................153
6.4.5. Software Survey. 154
6.5. Methods Specific to Constraint Programming....................154

7. Finding the Maximum Feasible Subset of Linear Constraints.................... 159
7.1. Exact Solutions.................... 161
7.1.1. An Exact Solution via MIP.................... 161
7.1.2. An Exact Formulation via Equilibrium Constraints.................... 162
7.2. IIS Enumeration and Covering.................... 164
7.3. Phase One Heuristics.................... 167
7.4. Chinneck's SINF-Reduction Heuristics.................... 169
7.5. Two-Phase Relaxation-Based Heuristic.................... 179
7.6. Randomized Thermal Relaxation Algorithms.................... 181
7.7. An Interior-Point Heuristic.................... 183
7.8. Working with IIS Covers....................184
7.8.1. Single Member IIS Covers.................... 185
7.8.2. Finding Specific IISs Based on IIS Covers.................... 186
7.9. The Minimum Number of Feasible Partitions Problem.................... 189
7.10. Partial Constraint Satisfaction in Constraint Programming....................193

8. Altering Constraints to Achieve Feasibility.................... 197
8.1. Shifting Constraints.................... 197
8.1.1. Using the Phase 1 Result.................... 198
8.1.2. Minimizing the l_1 Norm.................... 199
8.1.3. Least-Squares Methods.................... 199
8.1.4. Roodman's Bounds on Minimum Constraint Adjustments.....200
8.1.5. A Fuzzy Approach to Constraint Shifting.................... 202
8.1.6. A Goal Programming Approach to Constraint Shifting.................... 202
8.1.7. Constraint Shifting in Sequential Quadratic Programming..... 204
8.1.8. Violating a Limited Number of Constraints by a Limited Amount. 205
8.2. Adjusting the Constraint Matrix.................... 206
8.3. Related Research.................... 208

Part III: Applications 211

9. Other Model Analyses 213
   9.1. Analyzing Unbounded Linear Programs 213
   9.2. Analyzing the Viability of Network Models 213
   9.3. Analyzing Multiple-Objective Linear Programs 216
      9.3.1. Interaction Analysis of the Constraints 218
      9.3.2. Interaction Analysis of the Objectives 218
         9.3.2.1. Generating Different Interacting Sets of Objectives 220
         9.3.2.2. Which Objectives Conflict With a Particular Objective? 221
         9.3.2.3. Evaluating the Relative Amount of Objective Interference 221
      9.3.3. Summary of the Method 222
      9.3.4. Example 224

10. Data Analysis 227
   10.1. Classification and Neural Networks 227
   10.2. Data Depth 231
   10.3. Errors in Massive Data Sets 232

11. Miscellaneous Applications 235
   11.1. Radiation Treatment Planning 235
   11.2. Protein Folding 236
   11.3. Digital Video Broadcasting 237
   11.4. Automated Test Assembly 238
   11.5. Buffer Overrun Detection 239
   11.6. Customized Page Ranking 239
   11.7. Backtracking in Tree-Structured Search 240
   11.8. Piecewise Linear Model Estimation 242
   11.9. Finding Sparse Solutions to Systems of Linear Equations 243
   11.10. Various NP-Hard Problems 244

12. Epilogue 247

References 249
Index 265

List of Figures

- Fig. P1.1. Typical easy-hard-easy pattern for determination of feasibility status ... 7
- Fig. 2.1. Several steps in a cyclic orthogonal projection method 21
- Fig. 2.2. The consensus vector (solid) results from the component-wise averaging of the three feasibility vectors (dased) 21
- Fig. 3.1. Two-dimensional octagon around unit cube 29
- Fig. 3.2. Simplex iterations performance profiles with Cplex internal heuristics off (Patel and Chinneck 2006) 41
- Fig. 3.3. Simplex iterations performance profiles with all Cplex 9.0 heuristics turned on (Patel and Chinneck 2006) 42
- Fig. 5.1. Function shape is assessed via difference measurements along the line segment (Chinneck 2002) 56
- Fig. 5.2. Examples of convex and nonconvex region effects (Chinneck 2002) .. 57
- Fig. 5.3. Hit-and-run sampling in a convex enclosure (Chinneck 2002) 59
- Fig. 5.4. Example iteration of the Constraint Consensus method (Chinneck 2004) 66
- Fig. 5.5. Nonlinear interval analysis via sampling tightens the bounds on variable y. (a) inequality constraint. (b) equality constraint (Chinneck 2002) .. 74
- Fig. 5.6. Range cutting (Chinneck 2002) 75
- Fig. 5.7. Constraint Consensus invocations on 2D Branin function (MacLeod and Chinneck 2007) 82
- Fig. 5.8. Bin voting 82
- Fig. 5.9. Theorem 5.1 85
- Fig. 5.10. Theorem 5.3 86
- Fig. 6.1. Simple linear IIS 93
- Fig. 6.2. Identifying an IIS by sampling 110
- Fig. 6.3. Sampling indicates infeasibility relative to constraint 111
- Fig. 6.4. The sensitivity filter 115
- Fig. 6.5. The branch and bound solution fails to terminate (Guieu and Chinneck 1999) 131
- Fig. 6.6. An infeasible MIP with feasible LP relaxation (Guieu and Chinneck 1999) 132
- Fig. 6.7. Constraints in NLTEST1 148
- Fig. 6.8. Sum of the absolute constraint violations in NLTEST1 148
- Fig. 6.9. Sum of absolute constraint violations for NLTEST1 MIS1 149
- Fig. 6.10. Example of method of Debrosse and Westerberg 150

- Fig. 7.1. Example infeasibility 169
- Fig. 7.2. A pathological counter-example (Chinneck 1996c) 172
- Fig. 9.1. Example processing node with ratio equations 214
- Fig. 9.2. A nonviable pure network 214
- Fig. 9.3. Finding a minimal nonviability in a processing network model (Chinneck 1996b) 215
- Fig. 9.4. Examples of nonconflicting and conflicting objectives 219
- Fig. 10.1. Separating hyperplane 228
- Fig. 10.2. The grey point has data depth 2 231
- Fig. 11.1. A piecewise linear model consisting of three slabs 242

List of Tables

- Table 6.1. Characteristics of the netlib infeasible LPs 113
- Table 6.2. Summary of results for IS isolation methods for MIPs 141
- Table 6.3. Example IIS isolation by method of Debrosse and Westerberg (1973) 151
- Table 7.1. IIS cover cardinality on difficult LPs for two Phase 1 methods (Chinneck 1996c) 169
- Table 7.2. Comparison of algorithms on difficult infeasible LPs (Chinneck 2001a) 177
- Table 9.1. Normalized objective interference table 225
- Table 10.1. Classification data sets 228
- Table 10.2. Three algorithms for classification (Chinneck 2001a) 229
- Table 10.3. More MAX FS algorithms for classification (Amaldi et al. 2007) 230

List of Algorithms

- Alg. 3.1. General steps in the branch and bound method for solving MIPs 24
- Alg. 3.2. The pivot-and-complement heuristic for binary programs (Balas and Martin 1980) 27
- Alg. 3.3. The pivot-and-shift integer-feasibility seeking search phase (Balas et al. 2004) 28
- Alg. 3.4. Main steps in the OCTANE heuristic (Balas et al. 2001) 29
- Alg. 3.5. Simplified feasibility pump algorithm 31
- Alg. 3.6. The feasibility pump for binary MIPs (Fischetti et al. 2005) 32
- Alg. 3.7. The feasibility pump for general MIPs (Bertacco et al. 2005) 33
- Alg. 3.8. The feasibility pump for convex mixed-integer nonlinear programs .. 35
- Alg. 5.1. Bootstrapping procedure to achieve initial feasibility of a convex constrained region (Chinneck 2002) 62
- Alg. 5.2. The basic Constraint Consensus algorithm (Chinneck 2004) 67
- Alg. 5.3. Feasibility-distance based Constraint Consensus (FDnear, FDfar) (Ibrahim and Chinneck 2005) 68
- Alg. 5.4 Average direction-based (DBavg) Constraint Consensus (Ibrahim and Chinneck 2005) 69
- Alg. 5.5. Maximum direction-based (DBmax) Constraint Consensus (Ibrahim and Chinneck 2005) 70
- Alg. 5.6 Direction-based and bound-based (DBbnd) Constraint Consensus (Ibrahim and Chinneck 2005) 72
- Alg. 5.7. Bootstrapping method by Debrosse and Westerberg (1973) 88
- Alg. 6.1. The deletion filter 97
- Alg. 6.2. The additive method 98
- Alg. 6.3. The dynamic reordering additive method 100
- Alg. 6.4. The elastic filter 102
- Alg. 6.5. The depth first binary search filter 105
- Alg. 6.6. The generalized binary search filter 107
- Alg. 6.7. The additive/deletion method 109
- Alg. 6.8. The sensitivity filter 114
- Alg. 6.9. The deletion/sensitivity filter 119
- Alg. 6.10. The additive/sensitivity method 120
- Alg. 6.11. The *(IR-LC-BD)* deletion filter for MIPs 133

- Alg. 6.12. The basic additive method for MIPs 135
- Alg. 6.13. Dynamic reordering additive method for MIPs 136
- Alg. 6.14. Basic additive/deletion method for MIPs 137
- Alg. 6.15. Using information from the original branch and bound tree 139
- Alg. 6.16. The deletion filter for NLPs 146
- Alg. 6.17. Using the MIS 147
- Alg. 6.18. Finding the set of killing constraints 152
- Alg. 6.19. Deletion filtering for inequality-constrained QCQP 153
- Alg. 7.1. Minimum-weight IIS set covering algorithm (Parker and Ryan 1996) 164
- Alg. 7.2. The constraint frequency heuristic for the IIS cover (Tamiz et al. 1995) 166
- Alg. 7.3. Heuristic 1 for MIN IIS COVER (Chinneck 1996c) 171
- Alg. 7.4. Heuristic 2 for MIN IIS COVER (Chinneck 2001a) 176
- Alg. 7.5. Overall logic of the two-phase relaxation-based heuristic (Amaldi et al. 2007) 180
- Alg. 7.6. The IIS member labelling scheme 185
- Alg. 7.7. Finding one IIS for each member of the IIS set cover 187
- Alg. 7.8. Finding a single IIS given the complete set of IIS covers (Liffiton and Sakallah 2005) 188
- Alg. 7.9. Backtracking greedy algorithm for MIN PFS (Bemporad et al. 2005) 192
- Alg. 7.10. Partition refinement (Bemporad et al. 2005) 194
- Alg. 8.1. Adjusting the α and β conditions (Censor et al. 2006) 206
- Alg. 9.1. Analyzing MOLPs using IIS isolation algorithms 223

Introduction

To be, or not to be: that is the question...
From *Hamlet* by William Shakespeare

Shakespeare certainly hit the nail on the head: the most basic question of all is whether or not something exists: an object, a person or a solution that satisfies a given set of constraints. For Shakespeare, human existence was a fundamental question of life; for this book, existence of a feasible solution is a fundamental question of optimization.

Why such interest in feasibility and infeasibility in the context of optimization? Surely it is most important to find the best (i.e. optimum) solution, rather than just any feasible solution? The questions of feasibility and optimality are in fact two sides of the same coin. First, the existence of a feasible solution is a very fundamental question: before you can determine which solution is the best, you must first determine whether or not it is even possible for a feasible solution to exist at all. Second, it is easy to convert an optimization question to a feasibility question (and vice-versa), so the two questions are fundamentally the same. For example, you can pose the feasibility question as to whether or not a solution exists with an objective function value that is at least as good as a certain stated aspiration value. Over a series of iterations this aspiration value can be adjusted until we can definitely answer that no solution exists beyond a certain value. That last feasible solution is the optimum value of the objective function, found by answering a series of feasibility questions.

Looking at this the opposite way, it is common practice to pose feasibility questions as optimization problems. This is the basic idea of any phase 1 technique: create an objective function that measures the degree of violation of the constraints at any given point, and then minimize this function. If a value of zero can be found for this phase 1 optimization problem, then a feasible point exists, otherwise the model may not be feasible.

Third, there are unique and interesting questions associated with feasibility and infeasibility in optimization. For example, given a set of constraints that a solver determines to be infeasible, provide a diagnosis of why this is so. This question has grown in importance in recent years as optimization models have grown larger and more complex in step with the phenomenal increases in inexpensive computing power. One approach to this question is to isolate an *irreducible infeasible subset (IIS)* of the constraints, i.e. a (small) subset of constraints that is itself infeasible, but becomes feasible if one or more constraints is removed. This helps focus the diagnosis and model repair efforts and is especially helpful in very large models. This approach is well summarized by Greenberg's aphorism: "diagnosis = isolation + explanation" (Greenberg 1993). A related diagnostic question is this: given an infeasible model, what is the smallest number of constraints to remove such that the remaining constraints constitute a feasible set? Another is: what is the best way to repair the infeasible system (e.g. what is the smallest set of changes that can be made to the constraint right hand sides such that the set of constraints becomes feasible)?

Many of the algorithms used in answering these diagnostic questions depend on assessing the feasibility of numerous subsets of the original set of constraints. Hence those algorithms operate much more efficiently if the feasibility status of an arbitrary set of constraints can be determined quickly (which is of course a fundamental feasibility question itself). This is not difficult for sets of linear constraints, but it can be extremely difficult and time-consuming to determine feasibility status at all when there are nonlinear constraints or integer restrictions on some or all of the variables. Hence one focus of this book is algorithms for improving the speed with which the first feasible solution can be found (if one exists) for the more difficult cases in optimization.

A fourth major reason for interest in feasibility-related algorithms is the many applications that have been found for them. Some of these applications are surprising: data classification, training of neural networks, radiation treatment planning, analysis of protein folding, automatic test assembly, applications in statistics, etc. Some of these are briefly reviewed in Part III.

Finally, the question of feasibility or infeasibility is a major overlap between the field of optimization and the field of constraint programming. Constraint programming, arising from computer science, has special strength in seeking a yes/no answer to the question of whether a solution exists for a stated set of constraints; this is identical to the feasibility question in optimization. However, because of their different roots and traditions, constraint programming researchers approach the question in a different way and with different techniques. The two fields have begun to merge in recent years, resulting in stronger hybrid techniques. Constraint programming techniques and their links with optimization are addressed at an elementary level.

The emphasis in this volume is on algorithms and computational methods, specifically practical algorithms for solving the feasibility/infeasibility related problems that are the main subject. The book summarizes the main developments over the last twenty years or so, a very active period for the field, spurred by improvements in computing power and an increase in the size and complexity of optimization models. It should prove useful for academics teaching and conducting research in the field and their graduate students, as well as practitioners.

As opposed to a mathematical treatment, we take the involvement of a computer as a given: modern optimization problems are normally of such scale and complexity that they simply cannot be solved without using a computer. The essential element in solving a feasibility or optimization problem via computer is an efficient and effective algorithm. The computer implementation of these algorithms introduces a number of practical issues and complications, such as tolerances. These are also dealt with as they arise.

A Note on Theorems: There is a significant amount of mathematical development underlying the algorithms and computational methods that are the main topic of this book. To keep the focus on algorithms, proofs are generally included where a theorem relates to whether an algorithm functions as intended. However, where theorems relate to mathematical underpinnings, the proof is generally omitted in favour of a simple reference to the original publication containing the proof.

1 Preliminaries

This chapter provides basic definitions and explanations that we need to get started. Many related terms are defined in the online Mathematical Programming Glossary (Holder 2006).

1.1 The Optimization Model

A standard optimization model consists of an objective function, a set of constraint functions, bounds on the variables, and declaration of variable types, as shown in in Eq. 1.1:

$$
\begin{array}{rl}
\textit{objective function:} & \{\min, \max\}\ f(\boldsymbol{x}) \\
 & \textit{subject to:} \\
\textit{(functional) constraints:} & g_i(\boldsymbol{x})\ \{\le, =, \ge\}\ b_i,\ i=1\ldots m \\
\textit{variable bounds:} & \boldsymbol{l} \le \boldsymbol{x} \le \boldsymbol{u} \\
\textit{variable types:} & x_j \text{ is } \{\text{real, integer, binary}\},\ j=1\ldots n
\end{array}
\qquad (1.1)
$$

Vectors and arrays are shown in boldface. The objective of the optimization is to find values for the variables that provide a maximum or minimum value for $f(\boldsymbol{x})$, while respecting all of the restrictions on the values that variables can take, including the constraints, the variable bounds, and the variable types.

There are m functional constraints (often referred to simply as *constraints*). The b_i constant is often referred to as the *right hand side (RHS)* of the constraint while $g_i(\boldsymbol{x})$ is often referred to as the *constraint body* or *left hand side (LHS)*. The constraints may be doubly bounded, e.g. $b_{lower_i} \le g_i(\boldsymbol{x}) \le b_{upper_i}$, but this is easily converted to the form in Eq. 1.1 by using a pair of inequalities. It is common to see the functional constraints expressed in this doubly bounded format, in which case less-than inequalities are constructed by setting $b_{lower_i} = -\infty$, greater-than inequalities by setting $b_{upper_i} = \infty$, and equality constraints by setting $b_{lower_i} = b_{upper_i}$.

There are n variables which may have upper and/or lower bounds. The variable bounds are constants, and may be $-\infty$ and ∞ respectively (i.e. a variable may be completely unbounded, or unbounded in one direction only). A frequently-used variable bound is simple nonnegativity, i.e. $x_i \ge 0$. The lower and upper bounds can be omitted for binary variables since they are implied. Binary variables are a special case of integer variables, and hence most statements about integer variables in this book can be taken to apply equally to binary variables unless a specific distinction is made.

The various classes of optimization models are obtained by suitable choices in Eq. 1.1. For *linear programs* (LP), $f(\mathbf{x})$ and the $g_i(\mathbf{x})$ are all linear in form, and all x_j are real-valued. Because of the linear format, models are often written as matrices in which the functional constraints form the rows and the variables form the columns. Because of this, *rows* is often used as a synonym for functional constraints, and *columns* as a synonym for variables in linear models.

For *nonlinear programs* (NLP), at least one of $f(\mathbf{x})$ or one of the $g_i(\mathbf{x})$ is nonlinear in form, and all x_j are real-valued. *Mixed-integer linear programs* (MIP or MILP) are linear programs in which at least one x_j is integer or binary-valued and a least one variable is real-valued (for simplicity of reference, we will consider mixed-integer programs to include all combinations of at least one integer or binary-valued variable with any number of other real, integer, or binary variables). Suitable definitions can also be created for mixed-integer nonlinear programs, binary nonlinear programs, etc.

This work concentrates on the restrictions placed on the possible variable values by the constraints, the variable bounds, and the variable types. For ease of exposition, we may use "constraints" to mean any of these restrictions, and will make it clear by context when we intend to refer specifically to functional constraints. The objective function is often ignored, but can be important, e.g. when considering how close the first feasible solution is to the optimum value of the objective function.

1.2 Measuring Infeasibility

The most common measure of the infeasibility associated with an individual violated constraint is the difference between $g_i(\mathbf{x})$ and b_i, i.e.:

constraint type	*constraint violation*
$g_i(\mathbf{x}) \geq b_i$	$b_i - g_i(\mathbf{x})$
$g_i(\mathbf{x}) \leq b_i$	$g_i(\mathbf{x}) - b_i$
$g_i(\mathbf{x}) = b_i$	$\lvert g_i(\mathbf{x}) - b_i \rvert$

A similar measure applies to violated variable bounds:

bound type	*bound violation*
$x_j \geq l_j$	$l_j - x_j$
$x_j \leq u_j$	$x_j - u_j$
$l_j = x_j = u_j$	$\lvert {}_j l - x_j \rvert$

In practice, most solvers do not consider an individual constraint or bound to be violated unless the constraint or bound violation, as defined above, exceeds some tolerance ε, frequently on the order of 1×10^{-6}. The constraint or bound violation is defined to be zero for satisfied constraints and bounds.

A *function tolerance test* as described above is easy to implement, but is severely impacted by row scaling issues. Consider the following example:

scaling factor	*constraint*	*constraint violation at* $x = 3.5$
1	$x^2 \leq 9$	$12.25 - 9 = 3.25$
10	$10x^2 \leq 90$	$122.5 - 90 = 32.5$
1×10^{-7}	$(1 \times 10^{-7})x^2 \leq 9 \times 10^{-7}$	$1.225 \times 10^{-6} - 9 \times 10^{-7} = 3.25 \times 10^{-7}$

At a scaling factor of 1, the constraint violation is 3.25 at $x = 3.5$, well above the standard tolerance of 1×10^{-6}, hence the constraint fails the function tolerance test and is considered violated. If the same constraint is multiplied by a scaling factor of 10, then the constraint violation also increases by the scaling factor, giving a constraint violation of 32.5, indicating a severely violated constraint, even though it is simply a scaled version of the original constraint evaluated at the same point. But if a scaling factor of 1×10^{-6} is applied, then the constraint violation is just 3.25×10^{-7}, well *below* the standard tolerance, so the constraint is considered satisfied! This example shows that any constraint can be considered violated, severely violated, or satisfied, depending on its row scaling.

For this and other numerical reasons, most solvers *scale* the model prior to solution by applying multipliers to the rows and columns to try to bring all of the coefficients to about the same scale. This alleviates the problem, but does not prevent the solver from treating constraints differently. Any solver that uses a function tolerance test to assess the degree of infeasibility will work harder to satisfy some constraints than others since some will appear to be more violated due to scaling issues. It may even consider some constraints to be satisfied when they are in fact violated. Issues such as this underlie the phenomenon of one solver considering a solution to an optimization model to be feasible while another considers the identical solution to be infeasible.

Greenberg (2003) points out that relative tolerances are also in common use. For example, if we have a value v compared to some *referent* value V, then v is close enough to V if $|v-V| \leq \varepsilon_r |V|$ where ε_r is the relative tolerance. Absolute tolerances as described earlier can be combined with relative tolerances: v is close enough to V if $|v-V| \leq \varepsilon_r |V| + \varepsilon_a$ where ε_a is the absolute tolerance.

The two most common measures of the infeasibility of a set of constraints in continuous variables are:

- *The sum of the infeasibilities* (SINF): the sum of the constraint violations over all of the constraints and bounds.
- *The number of infeasibilities* (NINF): the number of constraints or bounds whose violations exceed the tolerance ε.

It is possible to have SINF > 0 at the same time as NINF $= 0$ because some of the constraints are violated, but none by more than ε. Note also that NINF is just as affected by scaling issues as SINF because the scaling may affect whether or not the constraint violation exceeds ε, as shown in the example.

The row scaling problem is avoided if infeasibility is measured in the variable space instead of the function space. Infeasibility can be measured as the Euclidean distance between the current point and the closest feasible point, which is the approach taken by projection algorithms, originated by Cimmino (1938) and extensively developed in recent years by Censor et al. (e.g. Censor and Zenios (1997, chapter 5)), among others. In projection algorithms, the *orthogonal projection* of an infeasible point is defined as the closest feasible point (Xiao et al. 2003). As described later, the Euclidean distance from a given point to the closest feasible point on a linear constraint is easily obtained. However it is much more difficult to obtain exactly for nonlinear constraints, though approximations are readily available.

Chinneck (2004) defines the *feasibility vector* for an individual constraint as the vector extending from an infeasible point to its orthogonal projection on the constraint. The length of the feasibility vector is identical to the Euclidean distance between an infeasible point and the closest feasible point on a single violated constraint. This can be extended to define a measure for the total infeasibility of a set of constraints: *the sum of the lengths of the feasibility vectors* (SLVF). As usual for numerical reasons, a tolerance ε may be applied to determine whether or not the constraint is violated. In this case, NINF means the number of constraints whose feasibility vector lengths are longer than ε.

As discussed extensively by Chinneck et al. (Chinneck 2004, Ibrahim and Chinneck 2005), the feasibility vectors (or their approximations in the case of nonlinear constraints) for a set of violated constraints can be combined in numerous ways to create a single *consensus vector* that can be applied to the current in feasible point to move it onto the closest feasible point in a single step. The consensus vector is normally an approximation only, so the process can be repeated in a cycle, which has been proved to terminate under certain conditions (Censor and Zenios 1997). However, the *length of the consensus vector* can also be used as an approximate measure of the total infeasibility of a set of constraints.

As measures of infeasibility, the lengths of the feasibility and consensus vectors and SLVF have the desirable property of being immune to row scaling problems. However they can be affected by column scaling. Fortunately, good column scaling is usually simpler to achieve than row scaling, and has a more intuitive meaning. The variables should be scaled so that a given error (e.g. a 1% error) has about the same impact in every dimension so that the Euclidean distance to feasibility is about as accurate in every dimension. We will return to the use of these variable-space measures of infeasibility in Sec. 5.5.

Thus far we have considered models in continuous variables only. The definition of infeasibility can be different if some or all of the variables are integer or binary valued, as in mixed-integer programming. Now we must consider not only SINF, NINF, SLVF, or consensus vector length, but also how far the integer or binary variables are from integrality. This *integer infeasibility* is defined in various ways, mainly for the purpose of selecting the next node in a branch and bound search tree (more on this in Chap. 3). One common definition is as follows: define the integer infeasibility of an integer variable as the distance to the closest integer value, i.e. where x_j is an integer variable that does not currently have an integer value, its integer infeasibility is the minimum of $(x_j-\lfloor x_j \rfloor, \lceil x_j \rceil - x_j)$, where $\lfloor x_j \rfloor$ is the value of x_j rounded down to the closest integer value, and $\lceil x_j \rceil$ is the value of x_j rounded up to the closest integer value.

The integer infeasibility of the entire model is then taken as the sum of integer infeasibilities over all of the integer variables. Another measure is simply the number of integer variables that do not have integer values at the current point.

Note that tolerances also affect the decision as to whether an integer variable is considered to have an integer value or not. As Greenberg (2003) points out, a relative tolerance is often used: 1,000,000.1 is close enough to 1,000,000 to be rounded to the integer value, but 1.1 is not close enough to 1 for a similar rounding. For integer rounding decisions, it is common to consider v close enough to its integer rounding if $|v - \lfloor v + 0.5 \rfloor \leq \varepsilon_r |v|$.

Other measures of infeasibility are available for specific classes of optimization models. The *duality gap* can be calculated for linear programs; this is the difference between the primal objective function value for a primal feasible solution and the dual objective function value for any dual feasible solution, and can be used as a measure of distance from optimality. When applied during a phase 1 solution this is another measure of the infeasibility of the current point.

Different feasibility-seeking algorithms use different measures of infeasibility. This will be an important theme in Part I of this book.

PART I: SEEKING FEASIBILITY

There are several good reasons for wanting to be able to reach feasibility quickly in a mathematical program. Some solution methods are unable to proceed to optimality without first reaching a feasible solution (most commonly for nonlinear programs). Overall solution speed is increased in some algorithms if a feasible solution is available, e.g. branch and bound solutions of mixed-integer linear programs (MIP) models can be much faster if a feasible incumbent solution is available early to help prune the subsequent tree. For many models, a feasible solution is all that is required (e.g. in scheduling applications). Finally, many methods for analyzing infeasibility require the repeated solution of subsets of the model constraints. Such methods are greatly speeded if the feasibility status of sets of constraints can be decided quickly; reaching feasibility rapidly is very helpful in this effort.

As we will see in Chapters 2–5, there is a wide variety of algorithms for seeking feasibility for all model forms. Most recently, there has been a great deal of progress in algorithms for reaching feasible solutions quickly for nonlinear programs and for mixed-integer linear programs.

A number of researchers (Mitchell et al. 1992, Mammen and Hogg 1997, Conrad et al. 2007) have noticed that the difficulty of determining feasibility status is directly related to how tightly the problem is constrained. They have observed a typical "easy-hard-easy" pattern as the model moves from lightly constrained at one extreme (in which case a feasible solution is easy to find) to tightly constrained at the other extreme (in which case it is easy to determine that the model is infeasible). The hardest feasibility problems, on average, are those in the middle range where the model is neither lightly nor tightly constrained. In the middle range, a great deal of search effort may be required to arrive at a definite determination of the feasibility status of the set of constraints. This pattern holds for many problems, including the classic satisfiability problem, graph colouring, constraint satisfaction programs, the connection subgraph problem, etc.

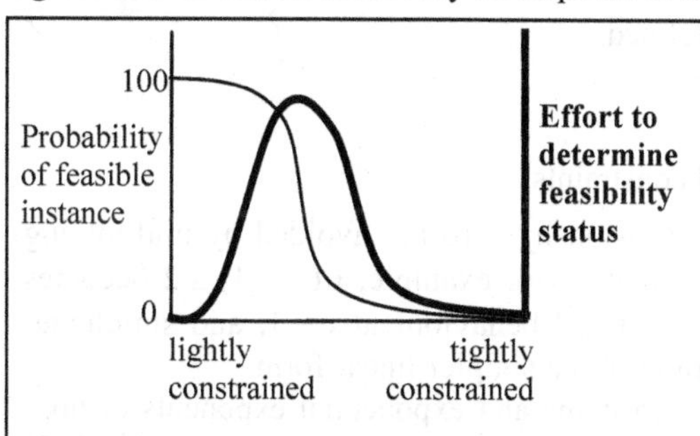

Fig. P1.1. Typical easy-hard-easy pattern for determination of feasibility status

The transition from mostly feasible instances of models to mostly infeasible instances is generally a rather sharp "phase transition". A typical diagram of the phase transition and the average computing effort to solve an instance is shown in Fig. P1.1. The computational effort to determine the feasibility status typically peaks in the region of the phase transition.

Surprisingly, Conrad et al. (2007) also observe in their experiments that proving the infeasibility of infeasible instances can be much harder than proving optimality in the computationally difficult part of the problem space for their particular connection subgraph application. This is likely a side effect of the fact that finding a single feasible solution proves feasibility, but to prove infeasibility you need to investigate all possible feasible regions. In most combinatorial problems this usually necessitates some kind of full expansion of the search tree, e.g. the full expansion of the branch and bound tree for a MIP to show that all leaf nodes are infeasible. In contrast, a single feasible leaf proves that a feasible solution exists and also helps to prune the subsequent tree search for the optimum solution by providing a bound on the optimum solution.

Model Reformulation

It is natural to work directly with the original model as supplied by the modeler. However this may not be the best form of the model for the purposes of seeking either feasibility or optimality. A given constraint can sometimes be algebraically manipulated so that it has the same mathematical properties (in the sense of the combinations of variable values that satisfy it) but has much better characteristics for use in algorithms for seeking a feasible or optimal point. This is true for all model forms, but is especially true for the more difficult NLP and mixed-integer NLP forms. We assume throughout Part I that the model has been put into the most suitable form for feasibility-seeking before the listed algorithms are applied. We next describe some reformulations that render constraints easier to satisfy.

Amarger et al. (1992) developed a system specifically for algebraically reformulating nonlinear models so that they are easier to solve. Their REFORM system preprocesses models formulated in the GAMS algebraic modeling language (Rosenthal 2007) to produce reformulated GAMS code that has better solution properties. The reformulations normally have one of these effects:

- Avoid functions that may be undefined,
- Reduce the degree of nonlinearity,
- Convexify a nonlinear model, or
- Improve scaling of variables and constraints.

Undefined functions caused by division by zero are avoided by multiplying the constraint through by the denominator. For example, $x/(y-1) \leq 2$ becomes $x-2y \leq -2$, which removes the undefined behaviour at $y=1$, and simultaneously converts the nonlinear constraint into a simpler linear form.

Undefined functions caused by logarithms and exponential exponents of non-positive values are avoided by exponential transformations. Amarger et al. (1992) provide this example: $\log(x/(y-z)) \leq d$ is transformed into $x-(y-z)\,e^{d} \leq 0$. Not only is the undefined behaviour avoided, but the transformed constraint is linear when d is fixed. The constraint $y - z \geq 0$ can also be added to the model if needed.

As seen above, nonlinearity can also be removed or reduced by multiplying through by the denominator. Consider the frequently-occurring nonlinear ratio of two functions such as $g_1(\boldsymbol{x})/g_2(\boldsymbol{x}) \leq b$. This is reformulated as $g_1(\boldsymbol{x}) - b \cdot g_2(\boldsymbol{x}) \leq 0$ by this tactic, which is linear if both $g_1(\boldsymbol{x})$ and $g_1(\boldsymbol{x})$ are linear. In addition, approximations that are less nonlinear are known for various commonly used functions, and these can be substituted where appropriate. Finally, some nonlinear terms can be substituted out. For example, if the variable x only ever appears in the term x^3, even if this term appears in several different places, then x^3 can be replaced everywhere by another variable e.g. w. The value of x can be recovered afterwards, when the value of w has been fixed, by solving the expression $x = w^{1/3}$. This idea extends to longer terms as well, especially if they occur frequently.

Convexifying the constraints of a nonlinear model greatly improves the ability of nonlinear solvers to find a feasible point. *Posynomial functions* (these have terms that are products of nonnegative variables with arbitrary exponents and positive coefficients) are easy to convexify by exponential transformations. For example, using the transformations $x = e^X$, $y = e^Y$, and $z = e^Z$, the nonconvex posynomial function $x^{0.6}/y + 1/z \leq b$ is transformed to the convex function $e^{(0.6X-Y)} + e^{-Z} \leq b$. Using exponential functions can make even non-posynomial functions convex, for example $\log(x - a_1) + \log(y - a_2) \leq b$ is convexified to $X + Y \leq b$ using the transformations $(x - a_1) = e^X$ and $(y - a_2) = e^Y$; this forms a convex feasible region if x and y appear only in this function, but the system is nonconvex if the transformation equalities must be included in the model because x and y appear elsewhere. In that case, the system is convex if the transformation equalities can be relaxed to $(x - a_1) \geq e^X$ and $(y - a_2) \geq e^Y$ (examples from Amarger et al. (1992)).

Amarger et al. (1992) also point out the importance of proper model *scaling*. The general idea is to avoid having variable or constraint body values of widely differing scales (e.g. x_1 has a range of 0.00001 to 0.00005 while x_2 has a range of 100,000 to 500,000). Keeping the variable and constraint values in the same general range avoids many numerical problems and improves the ability of solvers to reach both feasibility and optimality. Scaling is normally accomplished by multiplying variables and constraints by constant factors so that their ranges are approximately equalized.

Tightening of the variable bounds is also very helpful, especially for nonlinear models, a topic we will return to in Secs. 5.6.1. and 6.1.1. Nonlinear solvers are much more likely to reach a feasible point if started near the feasible region, and tightening the variable bounds improves the chances of finding a good initial point.

Special rules are also available to reformulate models containing binary or integer variables which result in simpler models for which it is faster to find an integer-feasible point. If there are simple bounds on variables such as $x \leq by$ where x is a continuous variable, y is binary, and b is a constant, then b can be reduced as tighter bounds on x are deduced (Amarger et al. 1992). Reduction rules are available for constraints composed entirely of binary variables, and of the form

$$\sum_{N^+} a_j y_j - \sum_{N^-} a_k y_k \leq b$$

where N^+ and N^- are disjoint sets of variable indices, and all of a_j and a_k are strictly positive constants. Crowder et al. (1983) show that the following deductions can be made:

- If $-\sum_{N^-} a_k > b$ then the constraint cannot be satisfied and the model is infeasible.
- If $\sum_{N^+} a_j \leq b$ then the constraint is always satisfied and can be removed from the model.
- If $a_j > b + \sum_{N^-} a_k$ then y_j can never be equal to 1, and hence can be fixed to 0.
- If $a_{k'} > b + \sum_{N^-} a_k$ where $a_{k'} \in N^-$, then $y_{k'}$ must be equal to 1 and can be fixed to that value.

The bounds on integer variables can also be tightened as a by-product of the tightening of continuous variables. For example if an integer value has an upper bound of 7, but treating it as continuous during bound tightening shows a tightened continuous bound of 6.8, then the integer bound can be reset to 6.

Amarger et al. (1992) give several examples of models in which no feasible solution can be found until the model is reformulated.

2 Seeking Feasibility in Linear Programs

A general linear program has the form {min, max} $\boldsymbol{cx}$, subject to $\boldsymbol{Ax}$ $\{\leq, \geq, =\}$ $\boldsymbol{b}$, $\boldsymbol{l} \leq \boldsymbol{x} \leq \boldsymbol{u}$, where $\boldsymbol{c}$ is a $1 \times n$ row vector, $\boldsymbol{x}$, $\boldsymbol{l}$, $\boldsymbol{u}$, and $\boldsymbol{b}$ are $n \times 1$ column vectors, and $\boldsymbol{A}$ is an $m \times n$ array, all consisting of real numbers. It is simple to find an immediate feasible solution for certain linear programs. For example, the origin (all variables equal zero) is always a basic feasible solution for an LP in a variation of canonical form that consists entirely of $\leq$ inequalities in which every element of $\boldsymbol{b}$ is nonnegative, and all variables are nonnegative. Similarly, network LPs in which the arc flow lower bounds are all zero admit the origin as a feasible solution.

It is more difficult to find a first feasible solution when the general LP is not in this special form, e.g. includes equality or $\geq$ constraints, or has negative entries in $\boldsymbol{b}$. In these cases, the origin is no longer available as a feasible solution, so more advanced methods of seeking feasibility are needed. In the simplex method, the most popular technique for reaching feasibility for general LPs is the two-phase method for reasons of numerical stability. The Big-M method, commonly presented in textbooks, is seldom used in implemented solvers.

More recently, infeasible-path interior point methods have been developed that do not necessarily reach feasibility until they also reach optimality. These techniques are beyond the scope of this book. See Wright (1997).

While reaching feasibility in LPs may seem to be a well-understood problem, there are a variety of heuristics which can speed the process considerably, such as crash starts, warm starts, and crossover from an infeasible solution.

2.1 The Phase 1 Algorithm

Given a basic feasible solution, the simplex algorithm is efficient at moving to a better adjacent basic feasible solution. It simply repeats this operation until recognizing that no further improvement is possible, and returns this final basic feasible solution as the optimum solution. As mentioned above, the difficulty with general LPs is that no basic feasible solution is immediately obvious except in very special cases. The phase 1 method addresses this problem by introducing nonnegative artificial variables into the problem so that a basic feasible solution is immediately available at the origin in the artificial space. A phase 1 objective function is also introduced which reaches its optimum value when the artificial variables are driven to their lowest possible values; if all artifical variables achieve a value of zero, then we are at feasible solution for the original problem Details follow.

Without loss of generality, let us initially assume an LP in which all variables are restricted to be nonnegative, and all of the elements of $\boldsymbol{b}$ have nonnegative values. With these restrictions, the constraints that eliminate the origin as a basic feasible solution are the equality constraints and $\geq$ constraints that have strictly positive entries in $\boldsymbol{b}$. To permit the origin as a feasible point, we introduce a non-negative artificial variable y_i for each such nonstandard constraint i, as follows (i) the inequality $\boldsymbol{a}_i\boldsymbol{x} \geq b_i$ is replaced by $\boldsymbol{a}_i\boldsymbol{x} + y_i \geq b_i$, and (ii) the equality $\boldsymbol{a}_i\boldsymbol{x} = b_i$ is replaced by $\boldsymbol{a}_i\boldsymbol{x} + y_i = b_i$. The phase 1 objective function is to minimize $W = \Sigma y_i$. The origin is a basic feasible solution for this phase 1 LP, hence the simplex method is able to initialize and iterate towards an optimum solution.

If the phase 1 LP terminates at an optimum solution in which $W = 0$, then it has found a point at which the artificial variables can be dropped and all of the original constraints are satisfied, i.e. a feasible point for the original problem. The solution process now initiates phase 2 at the current point by dropping all of the artificial variables and the phase 1 objective function, and re-introducing the original objective function. Ordinary simplex iterations then proceed to the optimum of the original objective function. Note that for efficiency reasons, the original objective function is normally included in the phase 1 matrix and updated as a nonbinding row so that it is in proper form when it comes time to solve the phase 2 problem.

On the other hand, if the phase 1 LP terminates at an optimum solution in which $W > 0$, then we know that the original LP is infeasible. W represents the sum of the violations of the equality and $\geq$ constraints, hence the size of W at the optimum solution is in some sense a measure of the size of the infeasibility. This notion can be generalized if the LP is fully elasticized (see Sec. 6.1.4). Other properties of the phase 1 solution, such as the dual prices of the slack variables, are useful in analyzing the cause of the infeasibility, as explained in later chapters.

There are some minor potential difficulties if the phase 1 solution terminates with $W = 0$ but the solution is degenerate. In this case, an artificial variable may have a value of zero and yet be in the basis. This can happen when the model has redundancies. However this is easy to recognize and handle. Dantzig and Thapa (1997, pp. 81– 82) list three ways to handle this problem, the simplest of which is to simply pivot the artificial variable out of the basis. This is done by choosing a nonzero element in a column for an original variable in the row for which the artificial variable is basic, and performing the pivot.

Note that it is possible to formulate a phase 1 that includes only a single artificial variable, however for implementation reasons this variant is not used in practice. See Nazareth (1987, pp. 147–149) for details.

2.2 The Big-*M* Method

The Big-*M* method requires the introduction of the same artificial variables as in the phase 1 method described above. The difference lies in how the artificial variables are driven out of the basis. "Big-*M*" refers to a large positive multiplier *M*, which is used as a penalty to discourage the inclusion of any artificial variables in the basis. The method works towards feasibility and optimality simultaneously within a single phase by using an appropriate form of the objective function:

- for maximization: $max\ Z = \mathbf{cx} - My$,
- for minimization: $min\ Z = \mathbf{cx} + My$.

As for the two-phase method described above, feasibility is recognized when all of the artificial variables are driven to zero. This may not happen until optimality is also reached.

The practical difficulty with the Big-*M* approach is that the large multiplier introduces numerical difficulties in the solution by dominating the calculations, however if the value of *M* is too small, then the procedure will terminate with an infeasible optimum solution. See Padberg (1999) for guidelines on choosing a suitable value for Big-*M*. Because of the numerical difficulties, the Big-*M* method is seldom used in practice.

2.3 Phase 1 from Any Basis

The phase 1 procedure given in Sec. 2.1 must start at the origin. A procedure that can be invoked from any given starting basis is preferable since it can be invoked when feasibility is lost (e.g. by accumulated rounding errors, or by changes to the model after it has been solved). As shown by Nazareth (1987), such a phase 1 procedure is possible if the upper and lower bounds on the variables are specifically considered (though this method applies equally well to singly-bounded or unbounded variables).

Consider the usual equation format of the LP after any necessary slack and surplus variables have been added: $\mathbf{Ax} = \mathbf{b}$. Partitioning the variables into the set of basic variables $\mathbf{x}_B^0$ and the set of nonbasic variables $\mathbf{x}_N^0$ at a given basis induces a similar partitioning of the $\mathbf{A}$ matrix into $\mathbf{B}^0$, the columns associated with the basic variables, and $\mathbf{N}^0$, the columns associated with the nonbasic variables. The rewritten LP equation is then

$$\mathbf{B}^0\mathbf{x}_B^0 + \mathbf{N}^0\mathbf{x}_N^0 = \mathbf{b}.$$

Now the following relationship holds at any iteration:

$$\mathbf{B}^0\mathbf{x}_B^0 = \mathbf{b} - \mathbf{N}^0\mathbf{x}_N^0.$$

The following phase 1 procedure considers that a variable can be nonbasic at either its upper or lower bound.

Given a basis, the values of the nonbasic variables are known (each nonbasic variable is at one of its bounds), and so all of the constant and variable values on the right hand side of $\boldsymbol{B}^0\boldsymbol{x}_B^0 = \boldsymbol{b} - \boldsymbol{N}^0\boldsymbol{x}_N^0$ are known. Now we can solve for the values of the basic variables:

$$\boldsymbol{x}_B^0 = (\boldsymbol{B}^0)^{-1}(\boldsymbol{b} - \boldsymbol{N}^0\boldsymbol{x}_N^0).$$

Note that it may be numerically convenient to peg some of the nonbasic variables at values between their bounds; these variables are called *superbasic* (see (Nazareth 1987)). After solving for the values of the basic variables, some of them may be outside of their bounds during phase 1, i.e. the solution may be infeasible. The goal of the phase 1 procedure is then to drive all of the basic variables that are currently outside their bounds to within them.

Let us define $\overline{V}$ as the set of basic variables that violate their upper bounds, and $\underline{V}$ as the set of basic variables that violate their lower bounds. Nazareth (1987) shows that if the prices and reduced costs (π) are set to $c_i = 1$ if $i \in \overline{V}$, $c_i = -1$ if $i \in \underline{V}$, and $c = 0$ otherwise, then the sum of the infeasibilities is given correctly and the phase 1 reduced costs reflect the rate of change of the sum of the infeasibilities when a nonbasic variable is introduced into the infeasible basis $\boldsymbol{B}^0$. This means that whenever infeasibility is discovered, the cost vector $\boldsymbol{c}$ is replaced by the vector just described, and the simplex method is able to iterate in the normal manner towards feasibility. The cost component is reset to zero when a variable that is outside its bounds eventually satisfies them.

When variables can violate their bounds, or can be nonbasic at either the upper or the lower bound, there are several conditions to consider when choosing the leaving basic variable during simplex iterations (see Greenberg (1978)):

- A variable may be basic, outside its bounds and moving away from them, and hence will never be chosen as the leaving basic variable.
- A variable may be basic, outside its bounds and moving towards them, in which case it may pass through the violated bound and become nonbasic at the opposite bound.
- A variable may be basic and within its bounds, in which case it may become nonbasic at the first bound it meets.

These conditions are checked when determining the leaving basic variable, and the basic variable that most restricts the change in the value of the entering basic variable is chosen as the leaving basic variable, as usual. Note that an entering basic variable may be decreasing in value. Any variables that satisfy their bounds are kept inside their bounds by this procedure, while variables that violate their bounds are gradually made to satisfy them. In other words, the *number of infeasibilities* (NINF) is gradually reduced, eventually to zero if the LP is feasible.

While this procedure is effective, the fact that it keeps a variable within its bounds once it satisfies them can be overly restrictive. In some cases it is preferable to allow an entering basic variable to increase beyond the point at which the first currently-feasible basic variable encounters a bound because the overall sum of infeasibilities is still decreasing. When choosing the leaving basic variable, there are up to two thresholds associated with every basic variable:

- No thresholds if the basic variable is currently outside its bounds and moving away from them.

- One threshold if the basic variable is currently within its bounds. Beyond this threshold, the variable contributes to the sum of the infeasibilities.
- Two threshholds if the basic variable is currently outside its bounds and moving towards them. The first threshhold moves the basic variable into its feasible range, but is not blocking; beyond this threshold the variable no longer contributes to the sum of the infeasibilities. The second threshold is at the second bound and beyond this point the variable again contributes to the sum of the infeasibilities.

A more advanced procedure for choosing the leaving basic variable first sorts all of the thresholds in order from smallest to largest. It then looks at the rate of change of the sum of the infeasibilities in the zone between each threshold. The threshold dividing the last zone that shows a rate of decrease in the sum of the infeasibilities from the first zone that shows a rate of increase in the sum of the infeasibilities identifies the leaving basic variable. This emphasizes the decrease in the sum of the infeasibilities at the possible expense of increasing NINF. It is also possible to combine the two goals by examining the thresholds to reduce the sum of the infeasibilities as much as possible while not increasing NINF. This is done by choosing the threshold that is latest in the sorted list that does not increase NINF. Note that while you may pass through a threshold that causes a currently feasible basic variable to violate its bounds, a later threshold may cause a variable that currently violates its bounds to satisfy them, hence there is no net impact on NINF.

Nazareth (1987) describes the practical details of an efficient implementation of this scheme, including ways to immediately eliminate variables from consideration as the leaving basic variable, and ways to combine the calculations into a single pass through the candidate variables.

2.4 Crash Start Heuristics

A *crash start* in the context of linear programming is a procedure for generating a high quality initial basis. It may not be feasible, but it should be as close to feasibility as possible and have other helpful characteristics such as providing a nearly triangular matrix (which speeds the calculations). An LP with m independent rows and n original variables is normally converted to a form having $n + m$ variables, where one slack variable is added for each row. The main operation in crashing the initial basis is selecting m of the variables to be in the initial basis.

Once the basis is selected, the current values of the basic variables can be calculated. Then an appropriate phase 1 cost structure is assigned, as described in Sec. 2.3 and the phase 1 procedure iterates to feasibility.

The FortMP software (Ellison et al. 1999) describes a fairly standard crash procedure. The *unit basis* consisting of the slack variables is first set up, and then nonbasic original variables are gradually exchanged for basic slack variables. A basic slack variable is a candidate for an exchange with a nonbasic original variable if the pivot element at the intersection of the row for the basic slack variable and the column for the nonbasic original variable is nonzero. To avoid the work involved in updating the matrix to check this condition, various heuristics are applied, using the fact that there has been no update to the pivot element if the variable columns selected in previous exchanges have nothing but zeroes on the current pivot row. If the rows and columns in the revised basis are ordered in the same order as their selection into the basis, this leads to a triangular basis.

The row selected for an exchange should have as few nonzero elements as possible in columns that are candidates for exchange into the basis, on the principle described above. A variable is then selected for exchange into the basis, and all other candidate nonbasic variables that have a nonzero pivot element in the current row are marked as unsuitable for exchange into the basis later (because, if selected, a matrix update would be required).

There are many ties for the selection of the row corresponding to the basic variable and the column corresponding to the nonbasic variable to be exchanged when the selection is based on sparsity as described. According to Ellison et al. (1999), the way in which ties are broken has a big impact on the feasibility of the final basis. Ties for the basic variable row are broken in favour of equality constraints (so that artificial variables are removed from the basis), and after that according to the degree of restriction, from most to least (i.e. basic variables that have a smaller range are exchanged first). Rows having free variables are never selected. Ties for the nonbasic variable column are broken by preferring to exchange variables that have the largest range, with first consideration being given to free variables (those without bounds). Fixed columns are never selected for exchange into the basis. The crash procedure can also be adjusted, primarily by changing the tie-breaking rules, to reduce the amount of degeneracy in the crashed basis.

If the phase 1 procedure uses artificial variables, then the crashing procedure can be designed to reduce the number of artificial variables in the basis. Only rows corresponding to basic artificial variables can be selected. The nonbasic variable is chosen so that the pivot element is of reasonable size; this helps avoid basis singularity. In this same vein, most solvers include a parameter that allows the user to select a minimum size for any pivot, usually set as a minimum fraction of the largest element in the column.

2.5 Crossover from an Infeasible Basis

Crossover normally refers to the process of moving from a feasible point provided by an interior point LP algorithm to a nearby feasible basis (the basic solution is desirable because it gives access to sensitivity analysis, etc.). However, if an advanced infeasible basis can be provided, e.g. by a crash procedure, then it is sometimes possible to crossover from that basis to a nearby feasible basis. This opens the possibility of using heuristic methods to generate an initial solution that is reasonably close to feasibility and then crossing over to a nearby feasible basis. The FortMP software (Ellison et al. 1999) includes techniques for providing a close-to-feasible initial point and for the subsequent crossover.

At a basis provided by a crash start, the solution is likely to include a certain number of superbasic variables (nonbasic variables that are not equal to one of their bounds, but instead lie between their bounds). So-called *purify* or *push* algorithms are then used to move superbasic variables to either a basic or nonbasic status, i.e. to arrive at a feasible basis. In FortMP (Ellison et al. 1999) there are separate push algorithms to remove primal superbasic variables and to remove dual superbasic variables. Both function in essentially the same way. The main idea is to examine the effect on the basic variables when the value of a superbasic variable is adjusted (in a manner similar to examining the effect of an entering basic variable on the existing basic variables). If the superbasic variable reaches one of its bounds before any basic variable does, then the superbasic is simply switched to nonbasic status. If a basic variable reaches one its bounds before the superbasic does, then a basis change is made, in which the basic variable is made nonbasic and the superbasic is made basic.

The version of the crash heuristics that tries to eliminate artificial variables is preferred for use with the push heuristics since it helps reduce the amount of work during the push phase. In addition, during the push phase, any original variables that are at their bounds after the crash are temporarily fixed at those values.

An approximate solution that is even closer to feasibility can be supplied by improving the output of the crash step before purifying. FortMP uses a *successive overrelaxation* (SOR) algorithm (Press et al. 1992), an iterative technique for solving systems of linear equations (see Sec. 2.8), for this purpose. The overall procedure has three steps: (i) apply the crash heuristic to create an approximately lower triangular basis, (ii) apply the successive overrelaxation algorithm to improve the point provided by the crash heuristic, and (iii) apply the push algorithms to cross over to a feasible basis. With luck the SOR procedure produces a feasible solution directly, which eases the crossover to a basic solution. If it does not produce a feasible solution, then the push algorithms may yet do so, though this is not guaranteed.

2.6 Advanced Starts: Hot and Warm Starts

If the LP solution process is stopped for any reason, the current basis and associated information may be stored. If the solution process is restarted later, this stored information provides a *hot start* which allows the solver to begin where it left off without repeating the set of iterations, including the phase 1 feasibility-seeking iterations, which originally generated the stored basis.

It frequently happens that minor changes are made to the LP model before it is restarted. This may happen because the conditions being modeled have changed, but it is an essential part of two important procedures. In solving mixed-integer programs via branch and bound, numerous LPs are solved in a tree-structured search for a solution that is both LP-feasible and integer-feasible. Each LP is identical to a previous LP except that a bound on one variable has been adjusted so that the previous LP solution is rendered infeasible. In LP infeasibility analysis (see Sec. 6.2), several algorithms require the solution of sequences of LPs that differ by the addition or removal of one or several of the constraints or bounds. In cases such as these where the next LP to be solved is substantially similar (but not identical) to a previous LP, then a *warm start* that makes use of the previous solution and basis may be effective. This usually means that you can arrive at a new feasible (and optimal) solution in only a few iterations.

In warm-starting, if the changes made to the model have not rendered the warm-start point infeasible, then the primal simplex iterations just pick up where they left off and continue iterating to optimality. However, if the changes to the model have made the warm-start point primal-infeasible (normally by a change to a constraint or bound, or by the addition of one), then the warm-start point will still be dual feasible. The solver then switches to the dual simplex method and will quickly reach primal feasibility at the dual optimum point, normally in a small number of iterations.

Warm-starting an interior point method is considerably more difficult, but progress is being made. See Yildirim and Wright (2002) and John and Yildirim (2006) for details.

2.7 Seeking Feasibility and Optimality Simultaneously

An option often provided in simplex-based LP solvers is the ability to seek feasibility and optimality simultaneously. This is what happens when using the big-*M* feasibility-seeking algorithm, of course, but there are better ways to combine the two that avoid the numerical difficulties associated with big-*M*.

The simplest approach is to use a *composite objective* that weights the objective function and a measure of infeasibility, normally the sum of the infeasibilities. The MINOS software (Murtagh and Saunders 1987) uses a composite objective of the form *minimize* $\sigma w(\mathbf{cx})$ + (sum of infeasibilities), where $\sigma = 1$ for a minimization objective function and $\sigma = -1$ for a maximization objective function and w is a user-specified weight. If the LP solver reaches an optimum solution for that objective function while the original model remains infeasible, then w is reduced by a factor of 10, and up to five such reductions are allowed before the algorithm gives up.

Infeasible-path interior point algorithms for linear programming have been the subject of a great deal of research in the past decade. Also known as *primal-dual* interior point methods, these algorithms maintain an interior point that satisfies all of the inequality constraints, but that do not necessarily satisfy all of the equality constraints at any point before the optimum is reached. Details are beyond our scope here, but see e.g. Andersen et al. (1996) or Wright (1997).

2.8 Projection Methods

There is a rich and extensive literature on *projection methods* for finding feasible points for sets of constraints that form a convex set, of which sets of linear constraints are an important special case. The properties of these methods are well-studied, including guarantees of convergence for sets of convex inequalities. An excellent reference on this class of methods is Censor and Zenios (1997). Projection methods, under the name of *constraint consensus* methods, are also used as a heuristic technique for reaching near-feasible points in general sets of nonlinear constraints for which the convexity properties are not known (see Sec. 5.5); convergence cannot be guaranteed under these conditions, but the algorithms are remarkably effective.

All methods in this category employ some form of a projection for each violated constraint, most commonly a projection in the gradient or anti-gradient direction. The main idea is to use the gradient of the violated constraint at the current infeasible point, easily given by $\boldsymbol{a}_i$, the ith row of the constraint matrix $\boldsymbol{A}$ in the set of linear constraints $\boldsymbol{Ax}\{\leq,\geq,=\}\boldsymbol{b}$, to calculate the closest point that satisfies the constraint. This closest feasible point is called the *orthogonal projection* of the violated point, and is obtained by moving in the gradient or anti-gradient direction, as appropriate, to the limiting value of the violated constraint (see Sec. 1.2). The vector showing how to move from the current infeasible point to the orthogonal projection point onto an individual violated constraint is sometimes called the *feasibility vector* (Chinneck 2004), and denoted by $\boldsymbol{fv}_i$ for the ith constraint c_i. As has been shown by Xiao et al. (2003) and others, $\boldsymbol{fv}_i = v_i d_i \nabla c_i(\boldsymbol{x}) / \|\nabla c_i(\boldsymbol{x})\|^2$ where:

- $\nabla c_i(\boldsymbol{x})$ is the gradient of the constraint, and $\|\nabla c_i(\boldsymbol{x})\|$ is its length.
- v_i is the *constraint violation* $|c_i(\boldsymbol{x}) - b_i|$, or zero for satisfied constraints,
- d_i is +1 if it is necessary to increase $c(\boldsymbol{x})$ to satisfy the constraint, and −1 if it is necessary to decrease $c_i(\boldsymbol{x})$ to satisfy the constraint.

The squared term in the denominator seems unexpected, but is easily explained. $d_i \nabla c_i(\boldsymbol{x}) / \|\nabla c_i(\boldsymbol{x})\|$ is a unit vector in the gradient or anti-gradient direction, as necessary to reach feasibility. $V_i / \|\nabla c_i(\boldsymbol{x})\|$ is the number of units to move in the appropriate gradient or anti-gradient direction to reach feasibility; the product is $v_i d_i \nabla c_i(\boldsymbol{x}) / \|\nabla c_i(\boldsymbol{x})\|^2$. The length of the feasibility vector for the ith constraint is denoted by $\|\boldsymbol{fv}_i\|$.

The feasibility vectors for the violated constraints are used in different ways in the numerous varieties of projection algorithms (Censor and Zenios 1997; Censor, Elfving and Herman 2001). In all variants, the feasibility vectors must be combined in some way to arrive at an update vector; this final vector is sometimes called the *consensus vector* (Chinneck 2004). Some main algorithm variants are:

- *Sequential* projection algorithms update the current point by finding and applying the feasibility vector for one violated constraint at each iteration. The process continues until feasibility is achieved. The simplest version is cyclic (see below), but other variants are possible, see control sequences below.

- *Simultaneous* projection algorithms calculate the feasibility vector for every violated constraint and then combine them using some form of weighting to determine a final update consensus vector. This process if repeated until feasibility is achieved.
  - In the usual simultaneous projection algorithm, the complete set of feasibility vectors for the violated constraints is combined in a weighted average. *Component averaging* (Censor, Gordon and Gordon 2001) on the other hand, realizes that not all of the constraints contain all of the variables. The final movement vector is therefore computed component-wise, and only the constraints which contain a particular variable are considered when the movement in that dimension is calculated.
- *Control sequences* may be used to adjust which constraints are assessed at each iteration. In a *cyclic* control sequence, a sequential algorithm assesses the constraints in a round-robin fashion. The control sequence may also be *almost cyclic* (constraints or sets of constraints appear in every cycle, but not necessarily in the same order) or *repetitive* (Censor and Zenios 1997). Control sequences may be applied to individual constraints or to sets of constraints.
  - The *most violated constraint control* determines which constraint is currently most violated and uses that constraint in a sequential update algorithm. A similar idea applies in the case of the *remotest set control* which determines a set of constraints that is most violated and uses those constraints in a simultaneous projection algorithm.
  - *Voting heuristics* may be used to determine which subsets of constraints to combine in a simultaneous algorithm (Ibrahim and Chinneck 2005). For example, if the feasibility vectors of more constraints have positive values for some component x_j than negative values, then increase the x_j component by the average value of only the positive x_j components in the feasibility vectors. Several variants of voting methods are described in Sec. 5.5.
- *Relaxation parameters* may be used adjust the length of the consensus vector, either lengthening or shortening it.
- *Oblique projections* may be used instead of orthogonal projections.

A simple example showing several steps in a cyclic orthogonal projectionl projection algorithm for three equality constraints is shown in Fig. 2.1. Fig. 2.2 shows the consensus vector resulting from the component-wise combination of the three orthogonal feasibility vectors.

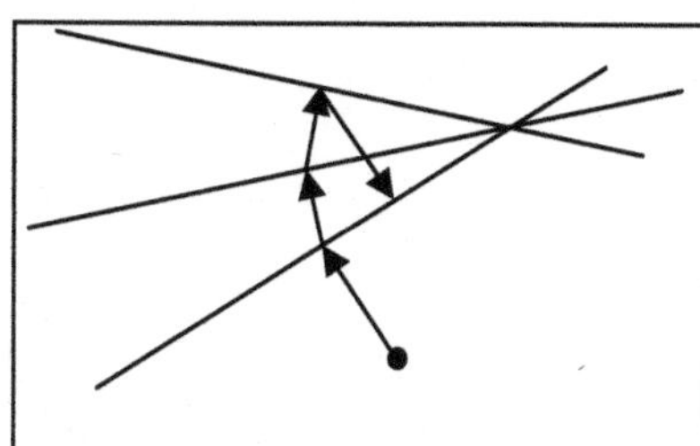

Fig 2.1. Several steps in a cyclic orthogonal projection method

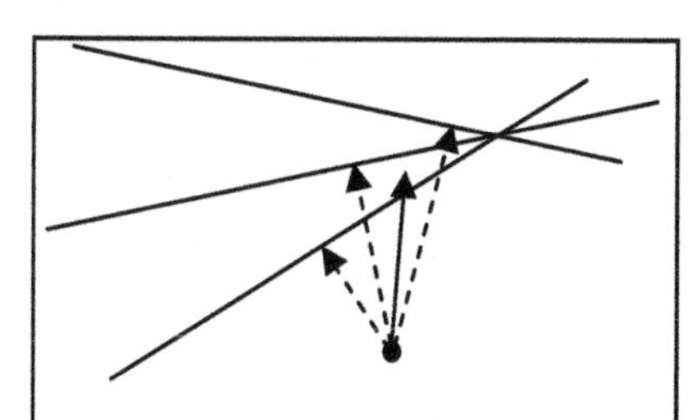

Fig. 2.2. The consensus vector (solid) results from the component-wise averaging of the three feasibility vectors (dashed)

Versions of these algorithms have been introduced by many authors. One of the earlier methods for linear equalities is by Kaczmarz (1937), a cyclic orthogonal projection method. Cimmino (1938) first suggested a fully simultaneous method for linear constraints. Another influential early development was the *relaxation method* for linear inequalities due to Agmon (1954), Motzkin and Schoenberg (1954), which consisted of a cyclic orthogonal projection method with relaxation. See Censor and Zenios (1997) for complete coverage of all related methods.

While projection methods could potentially be used in a feasibility-seeking phase 1 procedure for general linear programs, they have not been adopted for this purpose in commercial LP solvers (though a successive overrelaxation procedure is optionally used as part of a phase 1 procedure in at least one solver: see Sec. 2.5). Instead they have been applied in special-purpose feasibility seeking applications in radiation therapy planning, image reconstruction, etc. many of which are convex nonlinear problems if not linear.

A variant of projection methods known as *randomized thermal relaxation algorithms* is used in the context of finding a maximum cardinality feasible subset for an infeasible set of linear constraints (see Sec. 7.6 for details on the algorithm). Experiments with feasible models comprised of large numbers of linear inequalities show that the method is capable of reaching feasibility or near-feasibility very quickly (Amaldi et al. 2005).

3 Seeking Feasibility in Mixed-Integer Linear Programs

Mixed-integer linear programs (MIPs or MILPs) are much harder to solve than linear programs. The requirement that some variables take on integer or binary values means that simple linear programming cannot be used directly since it yields fractional values for the integer variables. The initial temptation is to relax the integrality restrictions, solve as an LP and simply round the solutions for the integer variables to the closest integer values. This frequently causes constraint violations or yields non-optimum solutions, and hence is ineffective in general (though there are a few simple special cases such as assignment problems for which LP is guaranteed to yield integral solutions).

In general, MIPs are solved by a solution space subdivision strategy, normally via a *branch and bound* or *branch and cut* algorithm. Branch and bound has a long history, dating to the 1960s (Land and Doig 1960) and has been extensively developed since then (e.g. Johnson et al. (2000)). The general steps of the method, summarized in Alg. 3.1, are fairly standard, but there are numerous variations in the details. Branch and bound generates a tree structure. At each node in the tree an LP-relaxation of the MIP which ignores the integrality restrictions is solved. If the LP relaxation solution does not provide integer values for all of the integer variables, the solution space is subdivided and the process continues.

Two of the most important aspects of branch and bound are the selection of the next node for expansion (Step 5), and selection of the branching variable (Step 2). Both can have a significant impact on the speed of the solution. After solving the LP relaxation associated with the chosen node in Step 6, the list of *candidate variables* for branching is known: it consists of the integer variables that have fractional values at the optimum solution of the LP relaxation. In Step 2, one of the candidate variables is chosen for branching, thereby creating two new child nodes. Each child node is created by adding a new variable bound to the model in the parent node. For example, if some variable x_j is chosen for branching, then it must be an integer variable that has a fractional value f in the LP relaxation solution of the parent LP, i.e. $k_L \leq f \leq k_U$, where k_L is the first integer below f and k_U is the first integer above f. One child node is created by adding the variable bound $x_j \leq k_L$ to the model in the parent node, and the other child node is created by adding the variable bound $x_j \geq k_U$ to the model in the parent node.

A common node selection scheme for solving MIPs via branch and bound is depth-first, in which one of the two just-created child nodes is always selected for expansion next (or failing that, the most recently created node). This has the advantage of providing an immediate advanced start based on the LP relaxation solution for the parent node, thereby increasing the overall speed of solution. There are several common ways to choose between the two child nodes: (i) *branch down*, in which the child node with the added bound $x_i \leq k_L$ is chosen next, (ii) *branch up*, in which the child node with the added bound $x_i \geq k_U$ is chosen next, and (iii) other schemes, e.g. based on whether f is closer to k_L or k_U.

INPUT: Mixed-integer linear program.
0. Incumbent solution = ϕ. List of unexplored nodes = ϕ.
1. Root node is the original model. Solve the LP relaxation of the root node.
   1.1 IF LP relaxation is infeasible THEN exit with "infeasible" outcome.
   1.2 IF LP relaxation is integer-feasible THEN exit with relaxation solution as optimum.
2. Choose a candidate variable in the current node for branching.
3. Create two child nodes from the current node by branching on the selected variable and add these new nodes to the list of unexplored nodes.
4. IF list of unexplored nodes is empty THEN:
   4.1 IF incumbent = ϕ THEN exit with "infeasible" outcome.
   4.2 Optimum is incumbent solution: exit with "optimal" outcome.
5. Choose a node from the list of unexplored nodes for expansion.
6. Solve the LP relaxation for the chosen node.
   6.1 IF LP relaxation is infeasible THEN discard the node and go to Step 4.
   6.2 IF LP relaxation is feasible and integer-feasible THEN:
      6.2.1 IF LP relaxation objective function value is better than incumbent objective function value THEN replace incumbent with this solution.
      6.2.2 Go to Step 4.
   6.3 Go to Step 2.

OUTPUT: MIP status (optimal or infeasible) and solution.

Alg. 3.1. General steps in the branch and bound method for solving MIPs

MIP solutions via branch and bound have several important characteristics. First, it is possible that the solution process will not terminate, as illustrated in Chap. 6 in Fig. 6.5 (this outcome is omitted from Alg. 3.1 for simplicity). Second, if the MIP is infeasible, this is proven only by a full expansion of the tree in which the LP-relaxation at every leaf node is infeasible. Finally, the branch and bound tree can vary widely depending on the choice of the node selection strategy, the branching variable selection strategy, and the branching direction. One of the themes of this chapter is setting these policies in ways that promote reaching feasibility quickly.

Branch and cut is an addition to branch and bound in which new functional inequality constraints are added to the model. These new constraints have the effect of eliminating part of the feasible region for the LP relaxation, including the current LP-relaxation solution, without eliminating any of possible integer solutions. See e.g. Rardin (1998) for details.

Reaching MIP feasibility quickly is important for several reasons. In some cases, a feasible solution is the only goal. When optimality is the goal, very difficult models may terminate before reaching an optimum, so finding a feasible solution quickly increases the likelihood that the solver will at least be able to report a usable solution. Finding a feasible incumbent solution quickly permits early pruning and hence the development of a smaller search tree. Feasible solutions are also needed so that local search heuristics such as *relaxation-induced neighbourhood search* (RINS) (Danna et al. 2005) can be used. Finally, some methods for analyzing infeasibility in MIPs require the repeated solution of variations of the original MIP in which only the feasibility status of the variant MIP is required (see Sec. 6.3); finding a feasible solution quickly terminates the assessment, thereby speeding the analysis.

This chapter reviews the state of the art in algorithms for seeking feasibility quickly in MIPs and binary programs. A useful comparison of the performance of a number of the methods described here is provided by Berthold (2006).

3.1 Pivot-and-Complement and Pivot-and-Shift Heuristics

The pivot-and-complement procedure (Balas and Martin 1980) can be applied to a *binary integer linear program* (BIP or BILP) in which all integer variables are of the binary type. It has an initial phase that tries to find a "good" binary-feasible point for an inequality-constrained BIP. The complete algorithm works towards optimality, but we will restrict our attention to the feasibility-seeking initial phase. The heuristic relies on the fact that a pure binary program has an equivalent LP with the additional requirement that all slack variables be basic, other than those in the upper bounding constraints. The algorithm first solves the LP relaxation of the binary problem, and then pivots to move all of the relevant slack variables into the basis. Some details follow.

A binary program $\max\{\boldsymbol{cx}|\boldsymbol{Ax}\leq\boldsymbol{b},\ x_j \text{ binary}, j\in N\}$, where $\boldsymbol{A}$ is $m\times n$, $\boldsymbol{b}$ is $m\times 1$, and $\boldsymbol{c}$ is $1\times n$ is equivalent to $\max\{\boldsymbol{cx}|\boldsymbol{Ax}+\boldsymbol{y}=\boldsymbol{b},\ \boldsymbol{0}\leq\boldsymbol{x}\leq\boldsymbol{1},\ \boldsymbol{y}\geq\boldsymbol{0},\ y_i \text{ basic for } i=1\ldots m\}$. The two formulations are equivalent because binary variables can be nonbasic at either the lower bound (0) or upper bound (1). Forcing the slack variables to be basic forces all of the binary variables to be nonbasic and hence either 0 or 1. It is assumed that all elements of $\boldsymbol{c}$ are integers.

The simplex tableau at any point is represented as

$$x_i = a_{i0} + \sum_{j\in J} a_{ij}(-x_j),\ i\in\{0\}\cup I$$

where I and J are index sets for the basic and nonbasic variables and 0 is the index of the objective function row.

Five types of operations help achieve binary feasibility:

1. *Type 1 pivots* maintain primal feasibility of the LP relaxation while exchanging a nonbasic slack for a basic binary variable. The pivot occurs in the

nonbasic slack column q and a row p for a basic binary variable such that

$$a_{p0}/a_{pq} = \min\left\{ \min_{i \in I, a_{iq} > 0} a_{i0}/a_{iq}, \min_{i \in I \cap N, a_{iq} < 0} (1 - a_{i0})/|a_{iq}| \right\}$$

2. *Type 2 pivots* maintain primal feasibility of the LP relaxation and do not affect the number of basic binary variables. A slack is exchanged for a slack or a structural variable is exchanged for a structural variable while reducing the sum of the integer infeasibilities, defined as $\sum_{i \in I \cap N} \min\{a_{i0}, 1 - a_{i0}\}$, by a positive Δ.
3. *Type 3 pivots* exchange a nonbasic slack for a basic binary variable while sacrificing primal feasibility. The slack variable must enter the basis with a positive value.
4. *Complements* involve flipping the values of a set of 1 or 2 binary variables. During the feasibility-seeking initial phase, variables are complemented to reduce a measure of infeasibility defined as $\sum_{i \in I} \max\{0, -a_{i0}\}$. A set S of nonbasic binary variables of size 1 or 2 can be considered for complementing if $\sum_{i \in I} \max\{0, -a_{i0}\} - \sum_{i \in I} \max\left\{0, -a_{i0} + \sum_{j \in S} a_{ij}\right\} \geq \Delta > 0$.
5. *Rounding and truncating solutions.*

The initial search phase which tries to achieve a first binary-feasible solution is summarized in Alg. 3.2. Balas and Martin (1980) report very good results when the feasibility-seeking initial phase is paired with a standard branch and bound method.

Pivot-and-shift (Balas and Martin 1986, Balas et al. 2004) is a later extension of pivot-and-complement that can handle general MIPs. The initial integer-feasibility seeking search phase is a staged rounding procedure. It runs through a cycle of rounding and pivot and shift procedures such as pivoting out basic integer variables, reducing the number of basic integer variables, improving the objective without increasing integer infeasibility, and reducing integer infeasibility. Small neighbourhood searches are also used.

Let $\boldsymbol{x}$ be the current solution at any point in the procedure. Initially $\boldsymbol{x}$ is the optimum solution of the initial LP relaxation. As before, I and J are sets of basic and nonbasic variables respectively. I_1 and J_1 are the sets of basic and nonbasic integer variables. The integer infeasibility at $\boldsymbol{x}$ is defined as

$$ZI = \sum_{i \in I_1} \min\{x_i - \lfloor x_i \rfloor, \lceil x_i \rceil - x_i\}.$$

There are 3 types of pivots analogous to those for pivot-and-complement:

- *Type 1 pivots* reduce $|I_1|$ while leaving the primal solution feasible. A nonbasic continuous variable is exchanged with a basic integer variable.
- *Type 2 pivots* improve the objective function and remain primal feasible while leaving $|I_1|$ unchanged. A nonbasic continuous variable is exchanged with a

basic continuous variable, or a nonbasic integer variable is exchanged with a basic integer variable.

- *Type 3 pivots* reduce ZI while leaving $|I_1|$ unchanged and maintaining primal feasibility. Continuous variables are exchanged with continuous variables or integer variables are exchanged with integer variables.

All three pivots happen in a column chosen by the specific pivot rule, and the row selected by the minimum ratio rule.

INPUT: inequality-constrained BIP.
1. Solve the LP relaxation for the binary model.
   IF solution is binary THEN exit successfully.
2. IF a type 1 pivot exists THEN perform the type 1 pivot that yields the largest objective function value and go to Step 4.
3. IF a type 2 pivot exists THEN perform the first such pivot,
   ELSE go to Step 5.
4. IF current solution is binary THEN exit successfully, ELSE go to Step 2.
5. Try rounding or truncating current basic solution to see if that yields a feasible binary solution. IF yes THEN exit successfully.
6. Perform a type 3 pivot that minimizes the value of the infeasibility measure.
7. Search for a single nonbasic binary variable whose complementing reduces the value of the infeasibility measure.
   7.1 IF none exists THEN go to Step 9.
   7.2 Complement the variable yielding the largest improvement in the measure of infeasibility.
8. IF the current solution is infeasible THEN go to Step 7.
   8.1 Check whether rounding or truncating current solution yields a feasible binary solution. IF yes THEN exit successfully, ELSE go to Step 2.
9. IF there is a pair of nonbasic variables whose complementing reduces the current value of the infeasibility measure THEN complement the first such pair and go to Step 8.
10. Exit with failure message.

OUTPUT: Binary-feasible point or failure message.

Alg. 3.2. The pivot-and-complement heuristic for binary programs (Balas and Martin 1980)

A rounding procedure is carried out at certain times to see whether a nearby integer-feasible solution exists. A small neighbourhood search is also conducted under certain conditions. This is normally done by imposing a linear constraint to restrict the search to a local neighbourhood of a feasible solution and running a MIP solver for this restricted problem (see Fischetti and Lodi (2003)). However since there is no integer-feasible solution available, Balas et al. (2004) define a neighbourhood around a close-to-integer-feasible solution, specifically

$$S = \{i \in I_1 : \min\{x_i - \lfloor x_i \rfloor, \lceil x_i \rceil - x_i\} \leq \alpha$$

for a small α such as 0.1. x_i^* denotes the value obtained from x_i by rounding. The neighbourhood restriction is composed of the pair of constraints

$$\sum_{j\in S} x_j^* - 1 \le \sum_{j\in S} x_j \le \sum_{j\in S} x_j^* + 1 .$$

The logic of the integer-feasibility seeking initial search phase is summarized in Alg. 3.3. In practice, the feasibility-seeking stage is run for a limited amount of time. If no solution is found, or if the integer solution is found by rounding and its value is 40% or more worse than the unrounded solution, then the heuristic is abandoned in favour of the feasibility-seeking routines in the commercial MIP solver Xpress (Dash Optimization 2006).

INPUT: MIP model.
0. Solve the LP relaxation of the original MIP.
1. IF rounding is successful, THEN exit successfully.
2. Continue making type 1 pivots as long as they are available.
3. If rounding is successful, THEN exit successfully.
4. Continue making type 3 pivots as long as they are available.
5. IF there was at least one successful type 3 pivot THEN:
   5.1 Continue making type 2 pivots as long as they are available.
   5.2 Go to Step 3.
6. Conduct a small neighbourhood search.
   IF successful THEN exit successfully.
7. Exit unsuccessfully.

OUTPUT: Integer-feasible point or failure message.

Alg. 3.3. The pivot-and-shift integer-feasibility seeking search phase (Balas et al. 2004)

Empirical tests reported by Balas et al. (2004) show that combining the pivot-and-shift heuristic with Xpress is quite effective in reducing the time to optimality for general MIP models.

3.2 The OCTANE Heuristic

Balas et al. (2001) developed the OCTAhedral Neighbourhood Enumeration (OCTANE) heuristic for generating feasible solutions for pure binary programs (all variables are binary) within a branch-and-cut framework. The heuristic uses an n-dimensional octagon circumscribing the n-dimensional cube to associate facets with binary solutions: each facet of the octagon is associated with exactly one fully binary solution. Given a fractional solution, usually from the current LP-relaxation solution, directions for improvement from this point (i.e. closer to feasibility in our case) are proposed. Movement in this direction crosses the extended facets of the octagon, and based on the binary solutions associated with these facets, heuristic solutions are proposed. The central idea is to explore the binary solutions that are in the neighbourhood of the current fractional point. The authors report good empirical results.

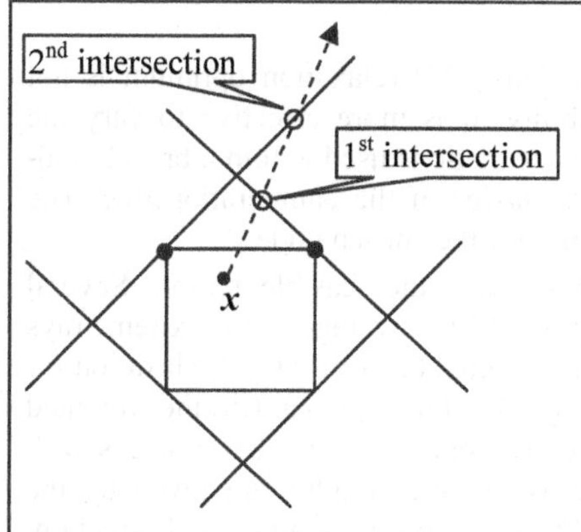

Fig. 3.1. Two-dimensional octagon around unit cube

The OCTANE heuristic applies to binary programs of the form $\min\{\boldsymbol{cx}|\boldsymbol{Ax} \geq \boldsymbol{b},\ x_i \text{ binary},\ i = 1...n\}$. Given a fractional solution $\boldsymbol{x}$ and a direction from $\boldsymbol{x}$, the heuristic works by finding the first k intersections with the extended facets defining the octagon around the unit cube. The binary points associated with the intersected facets provide a list of possible solutions that can be checked for binary feasibility. A simple 2-dimensional example is sketched in Fig. 3.1. The associated binary solutions that are generated are shown by the black dots.

The unit cube K in the theoretical development of the algorithm is actually centred at the origin, with the binary solutions offset by ½:

$$K = \left\{ \boldsymbol{x} \in \Re^n : -\frac{1}{2} \leq \boldsymbol{x} \leq \frac{1}{2} \right\}.$$

The regular octagon K^* circumscribing this n-dimensional cube is given by

$$K^* = \left\{ \boldsymbol{x} \in \Re^n : \boldsymbol{\delta x} \leq \frac{1}{2}n, \forall \boldsymbol{\delta} \in \{\pm 1\}^n \right\}.$$

There is a one-to-one correspondence between the binary points at $K + ½\cdot\mathbf{1}$ and the facets $\boldsymbol{\delta}$ of $K^* + ½\cdot\mathbf{1}$, given by $\boldsymbol{x} = ½\boldsymbol{\delta} + ½\cdot\mathbf{1}$.

The main steps in the heuristic are summarized in Alg. 3.4. Sophisticated procedures are used to reduce the effort in finding the first k intersected facets, see Balas et al. (2001) for details. Efficiencies are also introduced by avoiding the enumeration of facets that lead to infeasible binary solutions.

INPUTS: a fractional point $\boldsymbol{x}$.

0. $\underline{\boldsymbol{x}} = \boldsymbol{x} - ½\cdot\mathbf{1}$.
1. Choose a direction vector $\boldsymbol{a}$ and consider the half-line $\boldsymbol{r} = \underline{\boldsymbol{x}} + \lambda\boldsymbol{a}$, $\lambda \geq 0$.
2. Find the first k facets of K^* intersected by $\boldsymbol{r}$ and their corresponding associated binary points.
3. FOR each of the k binary points found in Step 2:
   3.1 IF feasible THEN exit (success).
4. Exit (failure).

OUTPUTS: a binary-feasible point or a failure message.

Alg. 3.4. Main steps in the OCTANE heuristic (Balas et al. 2001)

Some important implementation details:

- Using directions that begin only at the initial LP relaxation optimum is not particularly effective in reaching feasibility. It is more effective to vary the initial point as well, hence the OCTANE heuristic is used within a branch-and-cut framework by running it at different nodes of the enumeration tree. The starting point is the LP relaxation optimum for the chosen node.
- The best direction $\boldsymbol{a}$ is one that points inside the feasible region. Several different methods of constructing $\boldsymbol{a}$ are used: (i) the average of the extreme rays (normalized) of the cone C defined by the optimal basis of the LP relaxation for the node, (ii) the inward normal to the objective function, and (iii) the weighted average of the extreme rays of C that correspond to the nonbasic slacks with positive reduced cost at the current $\boldsymbol{x}$, where the weights are given by the inverse of the reduced costs. Berthold (2006) proposes an additional direction, the average normal ray, that performs well in the tests he carries out.
- A heuristic is used to determine the number k of intersections to investigate. If there is an inequality that is violated at all of the first 10 binary points returned by the heuristic, then the enumeration is abandoned. Otherwise at most 100 intersections are enumerated.
- The enumeration takes place in the fractional variable space. In other words, the binary variables at $\boldsymbol{x}$ are fixed and only the fractional variables are reset to binary values.
- The OCTANE heuristic is not run at every node of the branch-and-cut tree. It is run for every node in the first five levels of the tree, and thereafter at every eighth node.

The empirical results by Balas et al. (2001) show that OCTANE is competitive with pivot-and-shift, at least in terms of the number of branch-and-cut nodes. CPU time may be less, however but this could not be directly compared for implementation-related reasons.

3.3 The Feasibility Pump

Fischetti et al. (2005) developed the Feasibility Pump heuristic as a way of finding a feasible solution for a MIP problem without branch and bound. The method alternates between LP-relaxations (which satisfy the linear constraints) and "nearby" integer-feasible roundings of the LP-relaxation solutions (which satisfy the integrality restrictions). The back-and-forth action gradually "pumps" the intermediate solutions towards a final integer-feasible outcome. The authors report very good results on binary MIP problems.

The feasibility pump assumes a MIP of the form $\min\{\boldsymbol{c}^{\mathrm{T}}\boldsymbol{x} \mid \boldsymbol{A}\boldsymbol{x} \geq \boldsymbol{b},\ x_j \text{ integer } \forall j \in I\}$ where $\boldsymbol{A}$ is an $m \times n$ matrix. $\mathrm{P} = \{\boldsymbol{x} \mid \boldsymbol{A}\boldsymbol{x} \geq \boldsymbol{b}\}$ is the polyhedron associated with the LP relaxation of this MIP. A point $\boldsymbol{x}$ is integer-feasible if x_j is integer for all $j \in I$. A rounding $\tilde{\boldsymbol{x}}$ of a point $\boldsymbol{x}$ is given by setting $\tilde{x}_j = [x_j]$ if $j \in I$, otherwise

$\tilde{x}_j = x_j$, where [·] represents rounding to the nearest integer value. Note that a point is integer-feasible if $\boldsymbol{x} = \tilde{\boldsymbol{x}}$.

The L1 norm is used to measure the distance between a given point $\boldsymbol{x}$ on the polytope P and an integer point (not necessarily integer-feasible for the original MIP):

$$\Delta(\boldsymbol{x},\tilde{\boldsymbol{x}}) = \sum_{j \in I} | x_j - \tilde{x}_j |.$$

If the MIP includes bounds on the integer variables $l_j \le x_j \le u_j$ for all $j \in I$, then the definition of the L1 norm is modified as follows:

$$\Delta(\boldsymbol{x},\tilde{\boldsymbol{x}}) = \sum_{j \in I : \tilde{x}_j = l_j} (x_j - l_j) + \sum_{j \in I : \tilde{x}_j = u_j} (u_j - x_j) + \sum_{j \in I : l_j < \tilde{x}_j < u_j} (x_j^+ + x_j^-)$$

and additional constraints are added to the MIP model:

$$x_j = \tilde{x}_j + x_j^+ - x_j^-,\ x_j^+ \ge 0,\ x_j^- \ge 0,\ \forall j \in I : l_j < \tilde{x}_j < u_j .$$

With this formulation, given a rounded point $\tilde{\boldsymbol{x}}$, the closest point $\boldsymbol{x}^*$ on the polytope P can be found by solving the LP

$$\min\{\Delta(\boldsymbol{x},\tilde{\boldsymbol{x}}) : \boldsymbol{Ax} \ge \boldsymbol{b}\}.$$

The feasibility pump heuristic then alternates between points $\boldsymbol{x}^*$ that are LP-feasible (but not integer-feasible) and rounded points $\tilde{\boldsymbol{x}}$ that are integer-feasible (but not LP-feasible) in the hope that the two trajectories of points will converge at a point that is both LP-feasible and integer-feasible. The steps in the basic feasibility pump are summarized in Alg. 3.5.

INPUT: MIP model.
0. $\boldsymbol{x}^* \leftarrow$ solution of the LP-relaxation of the original MIP model.
1. $\tilde{\boldsymbol{x}} \leftarrow [\boldsymbol{x}^*]$.
2. IF $\tilde{\boldsymbol{x}} = \boldsymbol{x}^*$ THEN exit with $\boldsymbol{x}^*$ as a feasible solution for the MIP.
3. $\boldsymbol{x}^* \leftarrow \arg\min\{\Delta(\boldsymbol{x},\tilde{\boldsymbol{x}}) : \boldsymbol{Ax} \ge \boldsymbol{b}\}$.
4. Go to Step 1.

OUTPUT: an integer-feasible solution to the MIP, $\boldsymbol{x}^*$.

Alg. 3.5. Simplified feasibility pump algorithm

Note that limits on the process such as a maximum number of iterations or a time limit have been omitted from Alg. 3.5 for simplicity. The algorithm can fail to converge, so such safeguards are necessary. Stalling can happen when $\tilde{\boldsymbol{x}}$ and $\boldsymbol{x}^*$ do not change between iterations, so further safeguards are added to the algorithm as described below.

When all integer variables are binary, Fischetti et al. assume that the functional constraints include the bounds $0 \le x_j \le 1$ for all $j \in I$. This obviates the need for the x_j^+ and x_j^- variables and reduces the distance evaluation to

$$\Delta(x,\tilde{x}) = \sum_{j\in I:\tilde{x}_j=0} x_j + \sum_{j\in I:\tilde{x}_j=1}(1-x_j).$$

To avoid stalling in this binary case, a certain number of the binary variables are flipped, as described in Alg. 3.6.

The inputs in Alg. 3.6 include a time limit *TimeLimit*, a maximum number of iterations *MaxItns*, and a parameter to control the number of binary flips in case of stalling *T*. In Step 1.4, if stalling is detected when the new rounding is identical to the old rounding, then in Step 1.5 a random number of variables with the largest integer infeasibilities are flipped. Not shown in Alg. 3.6 is a further anti-cycling mechanism: if there is cycling in the most recent *R* iterations, then a random perturbation is applied as follows. For each $j\in I$ generate a uniform random value $\rho_j\in[-0.3, 0.7]$ and flip $\tilde{x}_j$ if $|x_j^* - \tilde{x}_j| + \max\{\rho_j, 0\} > 0.5$.

Fischetti et al. (2005) report very promising results on a variety of binary MIP models. The feasibility pump compares favourably to the Cplex 8.1 root node heuristics, reaching feasibility more often and with better optimality gaps.

INPUTS: binary MIP, *TimeLimit*, *MaxItns*, *T*.
0. $Itn \leftarrow 0$.
   $x^* \leftarrow$ solution of the LP-relaxation of the original binary MIP model.
   IF x^* is integer THEN exit with x^* as a binary-feasible solution.
   $\tilde{x} \leftarrow [x^*]$.
1. WHILE time < *TimeLimit* and *Itn* < *MaxItns* DO:
   1.1 $Itn \leftarrow Itn + 1$.
   1.2 $x^* \leftarrow \arg\min\{\Delta(x,\tilde{x}) : Ax \geq b\}$.
   1.3 IF x^* is integer THEN exit with x^* as a binary-feasible solution.
   1.4 IF $\exists j \in I : [x_j^*] \neq \tilde{x}_j$ THEN $\tilde{x} \leftarrow [x^*]$.
   1.5 ELSE flip *rand* $(0.5T, 1.5T)$ elements $\tilde{x}_j$ with highest $|x_j^* - \tilde{x}_j|$.
2. Exit with failure message.
OUTPUT: a binary-feasible solution x^* or a failure message.

Alg. 3.6. The feasibility pump for binary MIPs (Fischetti et al. 2005)

Bertacco et al. (2005) extend these ideas to better handle general mixed-integer models. In addition, they use the information provided by the feasibility pump to drive an enumeration process. Their extended version of the feasibility pump algorithm is given in Alg. 3.7. Note that the system $\tilde{A}x \geq \tilde{b}$ represents the original system of inequalities augmented with the additional inequalities required to handle bounds on the integer variables.

The score calculated in Step 1.3.2.1 of Alg. 3.7 measures the likelihood that $\tilde{x}_j$ will move from its current value to $\tilde{x}_j + 1$ if $x_j^* > \tilde{x}_j$ or to $\tilde{x}_j - 1$ if $x_j^* < \tilde{x}_j$. If cycling is detected in Step 1.3.1 then this score is calculated for each integer variable, and in Step 1.3.2.2 a random number of the variables with the largest scores are moved up or down accordingly. The cycling check in Step 1.3.2.3 initiates a restart in either of two cases: (i) $\tilde{x}$ is unchanged from the previous iteration, or (ii) $\Delta(x^*,\tilde{x})$ has decreased by less than 10% over the last *KK* iterations, where *KK* is a prespecified parameter.

The rounding function is also altered as a further anti-cycling measure. Normal rounding is defined by $[\tilde{x}_j] = \lfloor x_j + \tau \rfloor$ where $\tau = 0.5$. A random τ is used instead, based on ω, a uniformly distributed random variable in (0, 1), and defined as $\tau(\omega) = 2\omega(1-\omega)$ if $\omega \leq 0.5$ or $\tau(\omega) = 1 - 2\omega(1-\omega)$ if $\omega > 0.5$. This gives a value of τ that is most likely near 0.5, but also has small probabilities of being elsewhere in the range of zero to one.

INPUTS: general MIP, *MaxItns*, *T*.

0. *Itn*←0.
   $x^* \leftarrow$ solution of the LP-relaxation of the original binary MIP model.
   IF x^* is integer THEN exit with x^* as an integer-feasible solution.
   $\tilde{x} \leftarrow [x^*]$.
1. WHILE $\Delta(x,\tilde{x}) > 0$ and *Itn* < *MaxItns* DO:
   1.1 *Itn*←*Itn* + 1.
   1.2 $x^* \leftarrow \arg\min\{\Delta(x,\tilde{x}) : \tilde{A}x \geq \tilde{b}\}$.
   1.3 IF $\Delta(x,\tilde{x}) > 0$ THEN:
      1.3.1 IF $\exists j \in I : [x_j^*] \neq \tilde{x}_j$ THEN
         1.3.1.1 $\tilde{x} \leftarrow [x^*]$.
      1.3.2 ELSE
         1.3.2.1 FOR each $j \in I$ define score $\sigma_j \leftarrow | x_j^* - \tilde{x}_j |$
         1.3.2.2 Move *rand* (0.5*T*,1.5*T*) components x_j with largest σ_j.
         1.3.2.3 IF cycling detected THEN perform random restart.
   1.4 ELSE exit with x^* as an integer-feasible solution.
2. Exit with failure message.

OUTPUT: an integer-feasible solution x^* or a failure message.

Alg. 3.7. The feasibility pump for general MIPs (Bertacco et al. 2005)

Bertacco et al. (2005) execute the feasibility pump in two stages. They deal first only with the binary variables while ignoring the general integers. When this stage is complete they then deal with all integer variables (including the binary variables) simultaneously. Alg. 3.6 is used during the binary stage with minor changes to the restart operation. The binary stage is exited when either (i) a binary-feasible solution has been found, or (ii) $\Delta(x^*,\tilde{x})$ has not changed in the last *KK* iterations, where *KK* = 70 in their empirical tests. The point giving the smallest $\Delta(x^*,\tilde{x})$ during the binary stage is used as the initial point in the second stage.

A third stage applies if no integer-feasible solution has been obtained by the end of the second stage. This is an enumeration around $\tilde{x}=[x^B]$ where x^B is the best solution available at end of Stage 2, i.e. the x^* associated with the smallest $\Delta(x^*,\tilde{x})$. This is carried out using a general-purpose MIP solver applied to the original MIP, but with the objective function replaced by $\min\Delta(x,x_B)$.

The empirical results reported by Bertacco et al. (2005) for this version of the feasibility pump are comparable to those for the general purpose MIP solvers Cplex 9.1 (Ilog 2006) and Xpress Optimizer 16.01.05 (Dash Optimization 2006). Stage 1 which focuses only on the binary variables is surprisingly effective even though all tested models include at least one general integer variable. Not only does stage 1 increase the overall speed, but it frequently finds an integer-feasible solution for the entire model, including the general integer variables. Alg. 3.6 can be applied to general MIPs if the integer variables are converted to sums of binary variables. Experiments comparing Alg. 3.7 to Alg. 3.6 confirm that Alg. 3.7 is much faster for general MIPs.

Achterberg and Berthold (2005) extend Alg. 3.7 so that it produces feasible solutions that are closer to the optimum. This is accomplished by taking the objective function into account during the course of the algorithm. The main idea is to gradually reduce the influence of the original objective function and gradually increase the influence of the $\Delta(x^*,\tilde{x})$ measure as the algorithm proceeds. See Achterberg and Berthold (2005) for details.

3.3.1 The Feasibility Pump for Mixed-Integer Nonlinear Programs

Bonami et al. (2006) adapt the feasibility pump heuristic for use in finding feasible solutions for inequality-constrained mixed-integer convex nonlinear programs (MINLP). Similar in approach to the feasibility pump for MIPs, the nonlinear version alternates between solutions that satisfy the constraints in the continuous relaxation of the MINLP and points that satisfy the integer restrictions in a linear approximation of the NLP.

The integer variables are denoted by the set x and the real-valued continuous variables are denoted by the set y. The starting point is a feasible solution for the continuous relaxation of the MINLP. A linear approximation to the NLP constraints is constructed at this initial point, and a complete MIP is solved to find a point $(\hat{x}^i,\hat{y}^i)$ that satisfies the linear approximation as well as the integer restrictions (though it will not satisfy all of the original nonlinear inequalities in general). Finally another NLP is solved to find a point $(\bar{x}^i,\bar{y}^i)$ that satisfies the continuous relaxation of the MINLP at step i and that is closest to the MIP point. The process iterates between solving a MIP based on an updated linear approximation at the current point and solving an NLP to find the closest point to the MIP solution that satisfies all of the constraints in the continuous relaxation. The process is summarized in Alg. 3.8. Note that it is not necessary to solve the continuous relaxation in Step 0 to optimality if only a feasible solution is needed.

The linear approximation to the NLP that is used in Step 1 uses a technique by Duran and Grossmann (1986). It is an outer approximation based on the linearization of the nonlinear constraints around the points produced by the solutions of the continuous relaxations. This is a simple truncated Taylor's series expansion around each continuous relaxation solution point, i.e.

$$\boldsymbol{g}(\overline{\boldsymbol{x}}^k, \overline{\boldsymbol{y}}^k) + \boldsymbol{J}_g(\overline{\boldsymbol{x}}^k, \overline{\boldsymbol{y}}^k)\left(\begin{pmatrix} \boldsymbol{x} \\ \boldsymbol{y} \end{pmatrix} - \begin{pmatrix} \overline{\boldsymbol{x}}^k \\ \overline{\boldsymbol{y}}^k \end{pmatrix}\right) \leq \boldsymbol{b}, \forall k = 0,...,i-1$$

where $\boldsymbol{g}(\boldsymbol{x},\boldsymbol{y}) \leq \boldsymbol{b}$ is the set of inequality constraints and $\boldsymbol{J}_g(\boldsymbol{x},\boldsymbol{y})$ is the Jacobian matrix for the constraints. Note that the set of linear constraints includes all of the linearizations around the continuous relaxation solutions from Step 0 through Step i-1. It is a valid assumption that the constraint linearizations from previous steps continue to apply if all of the constraint inequalities are convex. In this case, the continually growing set of linear approximations of the constraints simply makes a better and better outer approximation of the original nonlinear constraints.

INPUT: MINLP model.

0. Solve the continuous relaxation of the MINLP using an NLP solver to obtain $(\overline{x}^0, \overline{y}^0)$. $i \leftarrow 1$.
1. Solve the MIP with objective function $\min \left\| x - \overline{x}^{(i-1)} \right\|_1$ and constraints based on the linear approximation of the NLP at the points $(\overline{x}^{(i-1)}, \overline{y}^{(i-1)})...(\overline{x}^0, \overline{y}^0)$ to obtain $(\hat{x}^i, \hat{y}^i)$.
   IF the MIP is infeasible THEN exit with failure message.
2. IF $(\hat{x}^i, \hat{y}^i)$ satisfies all of the original constraints THEN exit with $(\hat{x}^i, \hat{y}^i)$ as a feasible point for the MINLP.
3. Solve the continuous relaxation of the MINLP solver to minimize $\left\| x - \hat{x}^i \right\|_2$, obtaining the point $(\overline{x}^i, \overline{y}^i)$.
4. IF $(\overline{x}^i, \overline{y}^i)$ satisfies all of the integrality restrictions THEN exit with $(\overline{x}^i, \overline{y}^i)$ as a feasible point for the MINLP.
5. $i \leftarrow i+1$. Go to Step 1.

OUTPUT: an integer-feasible point for the MINLP or a failure message.

Alg. 3.8. The feasibility pump for convex mixed-integer nonlinear programs

Other valid linear inequalities can be added to the MIP approximation when it is known that all of the constraints are convex. At each iteration we have $(\hat{x}^i, \hat{y}^i)$ which is outside the convex feasible region formed by the convex constraints, and the associated closest point $(\overline{x}^i, \overline{y}^i)$ which satisfies all of the original constraints. Thus the constraint $(\overline{x}^k - \hat{x}^k)^T (x - \overline{x}^k) \geq 0$ is a valid inequality: it represents the hyperplane orthogonal to $\bar{x}^k - \hat{x}^k$ through $\bar{x}^k$. Bonami et al. propose a strengthened version of the MINLP feasibility pump that adds this new cut to the linearization at each iteration.

It is also possible that the set of nonlinear constraints forms a convex set even though some or all of the individual constraints are not everywhere convex. Bonami et al. show that the linear approximations to the constraints continue to be valid for any constraint $g_j(x,y)$ that is nonconvex, provided that it is at its limiting value, i.e. $g_j(x,y) = b_j$. They use this fact to alter the feasibility pump algorithm slightly: they use the constraint linearizations derived only from constraints that are everywhere convex, or that hold with equality at the linearization around $(\bar{x},\bar{y})$. The additional inequalities added by the strengthened version of the feasibility pump are always valid for every pair of points $(\hat{x}^i,\hat{y}^i)$ and $(\bar{x}^i,\bar{y}^i)$: the cut construction works for any exterior point and its closest feasible point on the boundary of the convex feasible region.

When the region defined by the constraints is nonconvex, then the feasibility pump can fail to find an integer-feasible solution even if one exists. This is because the linear approximations to the constraints may construct an infeasible LP or MIP. However, if a particular constraint qualification holds, then Bonami et al. prove that neither the basic nor the enhanced feasibility pump can cycle. The constraint qualification concerns those inequalities that hold with equality at the current point $(\bar{x}^i,\bar{y}^i)$. If the gradient vectors for those constraints are linearly independent at $(\bar{x}^i,\bar{y}^i)$ then the constraint qualification holds. Note that this is different than for the original feasibility pump for MIPs, which can cycle, but is easily explained by the fact that a complete MIP is solved in Step 1 of Alg. 3.8. The essential feature of the feasibility pump for linear MIPs is that it entirely avoids solving the complete MIP.

The computational results reported by Bonami et al. for a selection of convex MINLPs are excellent. In most cases the feasibility pump (basic or enhanced) finds a feasible solution within a second.

The authors also propose an optimizing version of the feasibility pump which iteratively adds constraints based on the objective function that require the next solution to be better than the current one. The subproblems that are solved are also slightly different in that they include the original objective function as well; see Bonami et al. (2006) for details.

The convexity of the individual constraints and the convexity of their intersection may be difficult to assess analytically. Techniques for empirical evaluation of the convexity of both constraints and regions are available however, see Sec. 5.2.

3.4 Branching Variable Selection by Active Constraints Methods

Patel and Chinneck (2006) develop a new approach to selecting the branching variable that shows significant improvement over existing state of the art methods in finding the first integer-feasible solution in a MIP quickly. Changing the policy for branching variable selection can have a dramatic effect on the speed to first feasible solution. For example, for the MIPLIB2003 (Achterberg et al. 2006) *momentum1* model, Cplex 9.0 with all default heuristics turned on times out after 28,800 seconds, while one of the active constraints methods reaches a feasible node in just 67 nodes and 74.61 seconds.

Most branching variable selection methods choose the candidate variable that maximizes the degradation of the objective function value at the optimal solution of the child node LP relaxation (Benichou et al. 1971, Dakin 1965, Eckstein 1994, Gauthier and Ribiere 1977, Linderoth and Savelsbergh 1999). This gives a tighter bound on the unsolved nodes. As pointed out by Linderoth and Savelsbergh (1999), most branching variable selection methods either estimate degradation in the objective function value of the LP relaxation or provide bounds on the degradation. Many estimation methods are based on pseudo-costs introduced by Benichou et al. (1971). None of these methods focus on finding an integer-feasible solution quickly.

Strong branching (attributed to Bixby by Linderoth and Savelsbergh (1999)) performs a number of dual simplex pivots to get a better lower bound on the degradation in the objective function value at the LP relaxation optimal solution of the child nodes, prior to selecting a child node for expansion. Branching variable selection can also be based on *Special Ordered Sets* (Beale and Tomlin 1970).

In contrast to objective-oriented methods, the *active-constraints* methods recognize that the solution point in an LP relaxation is determined by the constraints that are active at the optimum. To move the optima of the child nodes as much as possible, choose the candidate variable that has the most impact on the active constraints in the parent node LP-relaxation optimum solution, instead of choosing the variable that has the most impact on the objective function. The general idea is that the child node relaxation optima should be far apart, so that they are as dissimilar as possible in the hopes that one of the child nodes will never be expanded.

The active constraint methods are related to the concept of *surrogate constraints* due to Glover (1968, 2003). In the most basic form, a surrogate constraint is any linear combination of a set of linear constraints. When the constraints are all inequalities, their linear combination yields a single linear knapsack inequality. This gives a heuristic method for estimating the impact of a variable on the objective function by calculating the ratio between the objective function coefficient and the constraint coefficient for each variable in the resulting knapsack constraint (the "bang for the buck"). Various weightings of the individual constraints can be used in constructing the linear combination. Numerous sophisticated methods for selecting the weightings and applying the heuristic have been developed.

The set of "active constraints" includes all equality constraints and all inequalities that hold with equality at the current point (Greenberg 1996b). This means that all tight inequalities are included among the active constraints, both those associated with nonbasic variables, and those that are tight due to degeneracy. The point in question is the optimum point for the current LP relaxation.

The active constraints methods estimate the impact that an individual candidate variable has on the active constraints by looking at two components: (i) how much influence the variable has within a particular active constraint, and (ii) how much a particular active constraint can be influenced by a single variable.

Measures of the influence of a variable within an active constraint include (i) simple presence of a candidate variable in an active constraint, (ii) magnitude of the coefficient of a candidate variable in an active constraint, and (iii) normalizations of (ii) e.g. by the sum of the magnitudes of all of the coefficients in the active constraint (or the sums of the magnitudes of the coefficients of just the integer variables, or of just the candidate variables).

Measures of how much an active constraint can be influenced include (i) equal valuation for each active constraint, (ii) inverse of the sum of the magnitudes of all of the coefficients in the active constraint (or the sums of the magnitudes of the coefficients of just the integer variables, or of just the candidate variables), or (iii) inverse of the number of variables in the active constraint (or the number of integer variables or the number of candidate variables).

A weight w_{ij} is assigned to candidate variable j in active constraint i, based on some combination of the measures mentioned above. The variable having the highest total weight over all of the active constraints is chosen as the branching variable. Variations on the basic schemes include biasing the weights using the dual costs of the active constraints, looking at the single highest w_{ij} instead of the total weight, and a voting scheme. Ties are broken by selecting the variable with maximum infeasibility (defined as minimum distance from integrality); if still tied, the variable with the lowest solver-determined index is chosen.

Patel and Chinneck (2006) developed and tested 20 methods using various combinations of the measures listed above, but reported on a smaller subset of the best-performing methods, described below. Several methods not presented here have comparably good results and some omitted methods have inferior overall results, but perform spectacularly well on individual models.

The following MIP example is used to illustrate the different schemes:

$$
\begin{array}{ll}
\text{maximize} & z = 3y_1 - 4x_1 + y_2 - 2y_3 \\
\text{subject to:} & \text{P: } 8y_1 + y_2 - y_3 \leq 9 \\
 & \text{Q: } -x_1 + 2y_2 + y_3 \leq 5 \\
 & \text{R: } 3y_1 + 4x_1 + 2y_2 \leq 10 \\
 & x_1, y_1, y_2, y_3 \geq 0 \\
 & x_1 \text{ real; } y_1, y_2, y_3 \text{ integer}
\end{array}
$$

The LP relaxation optimal solution at the root node of the branch and bound tree is $z(y_1, x_1, y_2, y_3) = z(0.8125, 0, 2.5, 0) = 4.9375$. The candidate branching variables are y_1 and y_2. P and Q are the active constraints at the LP relaxation optimum and their dual costs are 0.375 and 0.3125 respectively.

Method A uses a simple count of the number of active constraints in which a candidate variable occurs. For candidate variable j in active constraint i, $w_{ij} = 1$ if the candidate variable appears in the active constraint, and $w_{ij} = 0$ if the candidate variable does not appear in the constraint. The total weight is a simple count of the number of active constraints that the candidate variable appears in. In the example, the weights of the candidate variables are found as follows:

Active constraint *i*	$w_{i(y1)}$	$w_{i(y2)}$
P	1	1
Q	0	1
Total:	*1*	*2*

y_2 has the highest total weight and is selected as the branching variable.

Method B recognizes that constraints are relatively easier or more difficult to influence via a single variable. This effect is estimated by noting the sum of the magnitudes of the coefficients of all of the variables in the active constraint. The weight associated with a particular active constraint, instead of being 1 as in Scheme A, is taken as $1/\Sigma_j|a_{ij}|$, where the coefficient of variable j in constraint i is a_{ij}. Active constraints with many coefficients of large magnitude thus have lower weights since they are likely less influenced by a single variable. $w_{ij} = 0$ if candidate variable j does not appear in active constraint i. In the example, the weights of the candidate variables are found as follows:

Active constraint *i*	$\Sigma_j\|a_{ij}\|$	$w_{i(y1)}$	$w_{i(y2)}$
P	10	0.1	0.1
Q	4	0	0.25
Total:		*0.1*	*0.35*

y_2 has the highest total weight and so is selected as the branching variable.

Method L adjusts the relative weight of each active constraint according to the number of integer variables in the constraint. The idea is that constraints that have many variables are less influenced by changes in a single variable because the other variables may be able to compensate. The weight associated with a particular active constraint is taken as $1/N^I_i$ where N^I_i is the number of integer variables in constraint i. $w_{ij} = 1/N^I_i$ if candidate variable j appears in constraint i and $w_{ij} = 0$ if candidate variable j does not appear in active constraint i. In the example, method L yields the following weights:

Active constraint *i*	N^I_i	$w_{i(y1)}$	$w_{i(y2)}$
P	3	0.333	0.333
Q	2	0	0.5
Total:		*0.333*	*0.833*

y_2 has the highest total weight and is selected as the branching variable.

Method M is identical to method L except that N^I_i is replaced by N^F_i, the number of fractional or candidate variables in constraint i. In the example, method M yields the following weights:

Active constraint *i*	N^F_i	$w_{i(y1)}$	$w_{i(y2)}$
P	2	0.5	0.5
Q	1	0	1
Total:		*0.5*	*1.5*

y_2 has the highest total weight and is selected as the branching variable.

Method O considers both the size of the coefficient associated with a candidate variable in an active constraint and the number of variables. The idea is that larger coefficients indicate a greater impact on the active constraint while more variables indicate a smaller impact. The weight associated with candidate variable j in active constraint i is $w_{ij} = |a_{ij}| / N^I_i$ where N^I_i is the number of integer variables in constraint i, and $w_{ij} = 0$ if candidate variable j does not appear in active constraint i. In the example, method O yields the following weights:

Active constraint *i*	N^I_i	$w_{i(y1)}$	$w_{i(y2)}$
P	3	2.667	0.333
Q	2	0	1.000
Total:		*2.667*	*1.333*

y_1 has the highest total weight and is selected as the branching variable.

Method P is identical to method O except that it considers only the candidate variables in the active constraint. The weight associated with candidate variable j in active constraint i is $w_{ij} = |a_{ij}| / N^F_i$ and $w_{ij} = 0$ if candidate variable j does not appear in active constraint i. In the example, method P yields the following weights:

Active constraint *i*	N^F_i	$w_{i(y1)}$	$w_{i(y2)}$
P	2	4.000	0.500
Q	1	0	2.000
Total:		*4.000*	*2.500*

y_1 has the highest total weight and is selected as the branching variable.

Method H looks for the maximum impact of a candidate variable on a *single* active constraint when using a particular method. The variable having the largest weight in an individual active constraint is selected as the branching variable. When applied to scheme M for example, the resulting scheme is designated H_M. In the example results for method P above, the highest individual weight of 4.000 belongs to variable y_1 in active constraint P, hence y_1 is chosen as the branching variable by method H_P. Patel and Chinneck (2006) concentrate on methods H_M and H_O in their experiments.

The empirical results for the active constraint methods are very good. Versions of the algorithms were built into a framework that calls Cplex 9.0 as the MIP solver except when a branching variable must be selected. In a first experiment, all of the Cplex internal heuristics are turned off to approximate a basic branch and bound arrangement. Fig. 3.2 provides a performance profile using the total number of simplex iterations as the metric for comparison. As shown in the figure, the active constraint methods are in general much faster than Cplex 9.0 in reaching feasibility, though some solutions were not completed within the imposed time limits, mostly for implementation reasons. It is also possible that the active constraints method is not a good match for the model in some cases.

See Patel and Chinneck (2006) for performance profiles using the number of branch and bound nodes as the performance metric. The active constraints methods use far fewer branch and bound nodes than Cplex 9.0.

In a second experiment, all of the many Cplex 9.0 internal heuristics are turned on. These heuristics sometimes have unpredictable effects; its own internal heuristics cause worse results for Cplex itself in about half of the models not solved by the root node heuristics. This underscores the need for expert advice in choosing combinations of heuristics and matching them carefully to the model at hand. Active constraints methods B, L, and P give good results in the small study conducted by Patel and Chinneck, and are the best candidates for further integration with the internal Cplex heuristics in order to yield more consistent results. Fig 3.3 provides performance profiles based on the number of simplex iterations. Method P is interesting in that it is very quick on those models that it completes successfully, but it has a higher rate of premature termination. This again emphasizes the need to carefully choose the variable selection scheme based on the characteristics of the model.

The active constraints methods do not have a negative impact on the quality of the first feasible solution returned (as measured by the optimality gap) compared to Cplex 9.0. In fact they return a higher quality first feasible solution more often than not, sometimes consistently so (method P returns a lower optimality gap than Cplex 9.0 for 78% of the compared models in the first experiment).

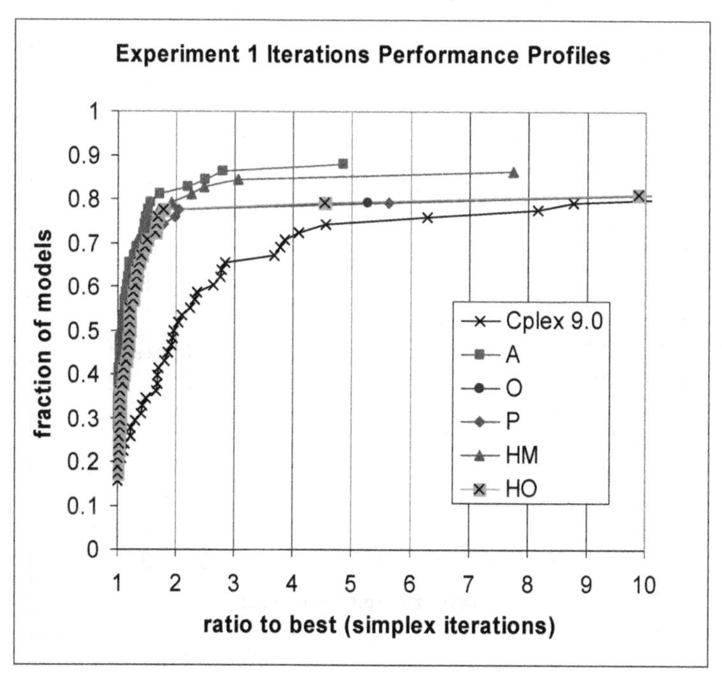

Fig. 3.2. Simplex iterations performance profiles with Cplex internal heuristics off (Patel and Chinneck 2006)

Research on active constraints methods is ongoing. A main goal is to determine how to match a model to the best active constraints method for its solution. Work is also ongoing to determine the best active constraints method to apply at a particular node in the branch and bound tree, based on the characteristics at the node.

3.5 Conflict Analysis

A common technique in constraint programming is *constraint learning* or *nogood learning* in which the cause of infeasibility at a node in the search tree is used to construct additional constraints that are added to the model to steer the subsequent development of the tree away from generating the same infeasibility again (see Chap. 4). This improves the efficiency of the search for a feasible solution.

Achterberg (2007) adapts these ideas for MIP by creating *conflict constraints* that are added to the MIP. Conflict constraints can be created for sets of bound changes that conflict with the original bounds. It is important for efficiency reasons to involve as few bounds as possible. Finding a minimum-cardinality Irreducible Infeasible System (IIS) (see Part II) would be ideal, but Achterberg opts to keep the time requirements low by simply analyzing the available dual solution

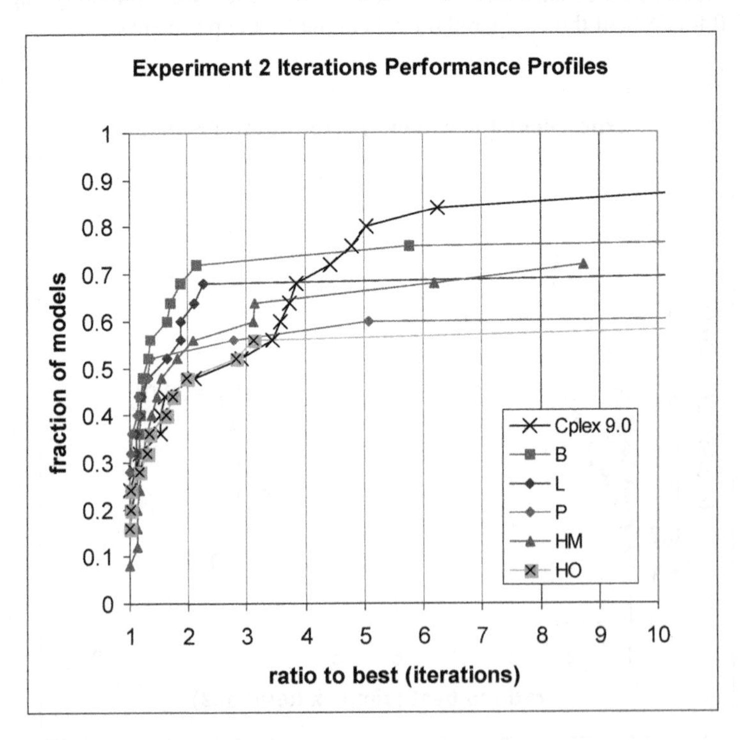

Fig. 3.3. Simplex iterations performance profiles with all Cplex 9.0 heuristics turned on (Patel and Chinneck 2006)

information. He applies a kind of deletion filter (see Sec. 6.1.2) to the dual in an effort to find a small infeasible subsystem. There is some difficulty in dealing with non-binary variables in the conflict set. After some experimentation, Achterberg concludes that most improvement is gained by adding only conflict constraints that include only binary variables. For binary variables, the resulting conflict constraint expresses the fact that at least one of the binary variables must have a different value. Adding this constraint prevents the re-occurrence of this infeasible combination of binary variables in another branch of the search tree.

For feasible MIP instances, the solution time using Achterberg's method is generally increased due to the added work for finding and constructing the conflict constraints, and for assessing them thereafter. However for infeasible MIP instances, the solution time is generally reduced.

Similar ideas are proposed by Davey et al. (2002) in the context of intelligent backtracking for binary linear programs. This is discussed in detail in Sec. 11.7. See also related work by Sandholm and Shields (2006).

3.6 Market Split Problems

Williams (1978) introduced *market split* integer programming problems that address the problem of splitting the sales between two divisions of a large company such that a particular market split fraction is achieved. The company sells m products to n vendors. Vendors can be assigned to either division with the goal of achieving a specified market split for every product. Given a market split fraction b $(0 \leq b \leq 1)$ for one division, the problem is to find $\boldsymbol{x}$ such that $\boldsymbol{a}_i\boldsymbol{x} = b\sum_j a_{ij}$ for $i = 1...m$, where $x_j \in \{0,1\}$ for $j = 1...n$ and $\boldsymbol{a}_i$ is the ith row of the $\boldsymbol{A}$ matrix defining the constraints. Cornuéjols and Dawande (1998) showed that instances of the market split problem constructed in a particular way are exceptionally difficult to solve. The Cornuéjols-Dawande instances are constructed for a given m by setting $n = 10(m\text{-}1)$. The a_{ij} coefficients are randomly chosen from the interval [0,99] and the right-hand side coefficients are set as

$$d_i = \left\lfloor \tfrac{1}{2}\sum_{j=1}^{n} a_{ij} \right\rfloor, i = 1...m.$$

Aardal et al. (2000) present a generalized version of the market split feasibility problem: is there a solution $\boldsymbol{x}$ that satisfies $\boldsymbol{Ax} = \boldsymbol{d}$, $\boldsymbol{l} \leq \boldsymbol{x} \leq \boldsymbol{u}$, where $\boldsymbol{x},\boldsymbol{d},\boldsymbol{l},\boldsymbol{u} \in Z^n$, $\boldsymbol{A} \in Z^{m\times n}$, and $m \leq n$? Further, the greatest common divisor of the elements of any row $\boldsymbol{a}_i$ of $\boldsymbol{A}$ is 1, and $\boldsymbol{A}$ has full row rank. They use a *basis reduction* approach to solve this problem. Briefly, they look for an initial basis such that a solution vector $\boldsymbol{x}_d$ for the basis satisfies $\boldsymbol{Ax}_d = \boldsymbol{d}$, but does not necessarily satisfy the variable bounds, along with $n\text{–}m$ linearly independent vectors $\boldsymbol{x}_0$ for which $\boldsymbol{Ax}_0 = \boldsymbol{0}$. The basis reduction techniques are used to find these in polynomial time. In some cases the solution $\boldsymbol{x}_d$ directly satisfies all of the variable bounds, but otherwise it is known that $\boldsymbol{A}(\boldsymbol{x}_d + \lambda\boldsymbol{x}_0) = \boldsymbol{d}$ for an integer multiplier λ and a vector $\boldsymbol{x}_0$ for which

$\boldsymbol{Ax_0} = \mathbf{0}$. Branching then occurs on integer linear combinations of $\boldsymbol{x}_0$ vectors that satisfy $\boldsymbol{Ax_0} = \mathbf{0}$. See Aardal et al. (2000) for details.

Empirical testing of the basis reduction algorithm by Aardal et al. (2000) shows that LP-based branch and bound solvers are able to solve market share problems much faster using the reformulation described above, which branches on the λ variables instead of the original $\boldsymbol{x}$ variables. This approach is also able to solve larger versions of the problem.

4 A Brief Tour of Constraint Programming

A main motivation in *constraint programming*, especially in the *constraint satisfaction problem*, is to find a feasible solution to a stated set of constraints, or to prove that no such solution exists, hence it has a great deal of overlap with the subject of this book. Indeed, a main thrust of research on the constraint satisfaction problem is achieving a feasible solution as quickly as possible. As will be seen, a number of the techniques used in constraint programming are related to methods well known in optimization, yet others are novel. In recent years, constraint programming and mathematical programming (i.e. optimization) have begun to cross-fertilize, developing more capable hybrid methods along the way. See Chinneck (2002a) for example. For an excellent up-to-date summary of how optimization and constraint programming have merged, see the book *Integrated Methods for Optimization* by John Hooker (2007). Lustig and Puget (2001) also provide a concise explanation of the relationship between mathematical programming and constraint programming.

This chapter presents a very brief summary overview of constraint programming based on material by Bartak (1999), Kumar (1992), Dechter and Rossi (2002), Russell and Norvig (2002) and Miguel (2001), among others. References to the original publications on the techniques described herein can be found in those sources. This topic deserves an in-depth treatment in light of the subject of this book, possibly in a companion volume of about the same size, but that is a project for another time and an author more versed in the subject matter. The purpose of this brief tour is simply to make the reader aware of the rich body of algorithms relevant to issues of feasibility and infeasibility that is available in the constraint programming literature. The reader is urged to investigate further.

In the most general sense, constraint programming is language allowing the declaration of a set of constraints defined over a set of variables, and the associated computational systems that seek to find a feasible solution or to prove that none exists. More recently (e.g. the OPL language (Van Hentenryck 1999)), constraint programming systems also tackle optimization. The earliest general systems were for *logic programming,* consisting of Boolean literals that can take on true/false values and a set of logical statements (i.e. constraints) involving these literals and their negations. The goal in logic programming is to find a set of truth values for the literals that proves the goal assertion is true, or to show that no such assignment of truth values exists. Confusingly, logic programming is sometimes referred to as LP, which optimizers would of course take to mean "linear programming". *Constraint Logic Programming* (CLP) extends basic logic programming to include continuous variables and much more general types of constraints, including linear and nonlinear mathematical constraints as well as other forms. Certain special constraints are particularly useful for capturing common restrictions in a graceful, declarative manner, such as the *alldiff* constraint that specifies that a set of variables must all have different values.

The fundamental *constraint satisfaction problem* is defined over a set of variables, each having a *domain* of possible values. A variable domain is normally a discrete set, e.g. binary, integer, or simply members of a set such as various cities, or a set of letters from the alphabet. A distinction is sometimes made between constraint satisfaction and *constraint solving* in which a variable domain can also be continuous. It would be unusual to find a constraint satisfaction or constraint solving problem that included no discrete variables: this would amount to a standard linear or nonlinear programming problem in the optimization literature.

The set of constraints restricts the values that the variables can assume in the usual way. Constraints are sometimes represented in a graph format, e.g. with a square representing the constraint being linked by arcs to circles which represent the variables. Reasoning may be performed on this graph. Constraints may be categorized as *unary* (a possible value is simply excluded from the domain of a single variable), *binary* (a constraint relates two variables), or *higher-order* (a constraint relates several variables). *Preference constraints* indicate that certain solutions are preferred over others; these would normally be dealt with via an objective function in an optimization formulation.

The major techniques for solving a constraint satisfaction problem are backtracking search, propagation, and local search, as well as numerous special-purpose heuristics. Most of these techniques have a direct counterpart in common use in optimization.

Backtracking search builds up a solution gradually by setting the values of variables one at a time. This results in a solution tree that is reminiscent of a branch and bound solution tree: one variable is considered at each node of the solution tree, and child nodes are created for each value assigned to that variable. A complete solution, feasible or infeasible, is reached only at a leaf node of the tree. Backtracking search implies a depth-first exploration of the search tree that backtracks when an infeasible leaf is reached (normally signaled when none of the values in the domain of at least one variable are legal, given the assignments made thus far). As in branch and bound optimization, the efficiency of the backtracking search is greatly affected by the ordering of the variables and their values, and by the backtracking method (i.e. node selection in branch and bound terminology). Propagation of variable values, discussed later, also improves efficiency by reducing the domains of other variables as a consequence of fixing a particular variable value at a node.

There are several useful heuristics for improving efficiency when selecting the branching variable. The *minimum remaining values* heuristic (sometimes called the *fail-first* heuristic) chooses the variable that has the fewest remaining legal values as the branching variable. This helps prune the search tree at higher levels, avoiding fruitless exploration of hopeless nodes. The *degree* heuristic selects the variable that appears in the largest number of constraints on other variables whose values have not yet been assigned.

When the branching variable is chosen, the next decision is which of its possible values to branch on first. The *least constraining value* heuristic can be used to make this choice. This heuristic chooses the variable value that eliminates the fewest values for the other unassigned variables that appear in constraints that the two variables share. The idea is to retain the greatest chance for a feasible solution at a descendant node.

Propagation is the use of the constraints to eliminate possible values from the domains of the variables by logical implication, and is similar to the techniques used in presolving for optimization (see Sec. 6.1.1). Propagation is typically applied just after a variable value is fixed, and also, as in optimization, before the backtracking search even begins. In *forward checking* each variable that appears in a constraint with a variable x whose domain has just been reduced (either by fixing as part of the backtracking search, or as a logical consequence of propagation) is checked to see whether its domain has been reduced due to the domain reduction in x. This process may result in a cascade of variable domain reductions, and can shorten the backtrack search considerably in some cases.

If all of the constraints are binary (i.e. relate just two variables), then it is easy to construct a graph in which each node represents a variable and each arc represents a constraint that relates the two variables that it connects. Given this graph, constraint implications can be propagated via *arc consistency*. This is a directional concept. Given a binary constraint connecting some variable x and some variable y, the x to y arc is consistent if, for every value in the domain of x there exists some consistent value in the domain of y. Arc consistency is checked in both directions. If not consistent, an arc can often be made consistent by removing a value from the domain of the variable at the tail of the arc, or the process may show that consistency is not possible, in which case infeasibility of this node in the search tree is proved. *MAC* (*maintaining arc consistency*) algorithms check the arc consistency after each variable assignment, and reduce variable domains as necessary to achieve consistency for all arcs.

As for the usual propagation processes, checking and maintaining arc consistency can entail a long cascade of variable domain reductions. There are a variety of arc consistency algorithms known as AC-1 through AC-7 that vary in the details concerning which arcs are rechecked during the cascade of domain reductions, and hence in the extent of CPU and memory required.

k-consistency is a generalization of arc consistency. A problem is k-consistent if for any set of k-1 variables a consistent value can be assigned to a related kth variable. 2-consistency is the same as arc consistency. If the values assigned to any pair of related variables always allow a consistent value for a related third variable, then the problem is 3-consistent. The higher the value of k the greater the effort in checking consistency, so in practice there is some trade-off between consistency checking and simply exploring the search tree. A constraint graph is *strongly k-consistent* if it is also j-consistent for all $j < k$. *Node-consistent* is the same as strong 1-consistency, arc-consistent is the same as strong 2-consistency, and *path-consistent* is the same as strong 3-consistency.

There are several algorithms for backtracking when the search tree arrives at an inconsistent leaf node. The simplest approach is to backtrack to the preceding variable and choose a different value from its domain. In contrast, *intelligent backtracking* focuses on backtracking to the causes of the failure, the so-called *conflict set*. A conflict set is defined in various ways, but always includes the idea that it is a subset of the variables and constraints in the problem that contribute to the inconsistency at the leaf node in the search tree. Intelligent backtracking realizes that it is much more efficient to backtrack to the variables involved in the conflict set rather than simply the last variable instantiated. When conducting a tree search, it is obvious that the last variable assigned is part of the conflict set. A conflict set can be constructed in various ways, for example, the conflict set for a given variable, normally the one that just caused the node failure, could consist of all of the previously-assigned variables that share a constraint with it.

Conflict-directed backtracking takes this idea one step further. It first identifies a *minimal conflict set*, and then backtracks on this step. A minimal conflict set is a set of constraints that is infeasible, but becomes feasible if any single constraint is removed. This is identical to the concept of an Irreducible Infeasible Subset (IIS) of constraints that is the main focus of Chap. 6. We will return to methods of finding minimal conflict sets in constraint satisfaction problems in Sec. 6.5.

Constraint learning (sometimes called *nogood learning*) techniques use the information in the conflict sets to add constraints to the problem so that the tree search will not repeat the mistake after backtracking.

The structure of the constraint satisfaction problem, as represented by its variable and constraint graph, provides information that helps direct the solution process. For example, there may be disconnected components that can be solved as individual problems. If the graph has a tree shape, then the solution process is greatly simplified since backtracking is not required. One class of methods seeks to reduce more complex problem structures to constraint trees by eliminating or collapsing nodes. This can happen at any point during the tree search phase: it may be simple to reduce the remaining constraint graph after a particular variable is assigned a value, for example. The overall graph can also be subdivided into tree-structured parts which are solved independently and then combined.

In backtracking tree search, variables are assigned values one at a time. In contrast, *local search* methods work with complete solutions in which all variables are assigned values. Of course the complete solution is inconsistent because the process halts when the first feasible solution is found. The idea is to adjust the values of some of the variables until a feasible solution is reached. The *min-conflicts* heuristic is the most common: given a variable, choose the new value that results in the fewest conflicts with other variables. This process continues until feasibility is reached. In *hill-climbing* the variable value is chosen so that the number of violated constraints is reduced.

As will be familiar to optimizers, local search methods may be randomized to avoid becoming trapped at a local minimum. *Stochastic local search* methods such as *random walks* choose a random value from the domain of a variable and also randomly apply the min-conflicts heuristic. Well-known local search methods from the optimization literature such as *tabu search* (e.g. Glover (1990)) or *simulated annealing* (Kirkpatrick et al. 1983) may also be applied.

4.1 Branching in the Satisfiability Problem

The *satisfiability* (SAT) problem is a particular form of constraint satisfaction problem. It is defined over Boolean literals or their negations and consists of the conjunction of clauses in which each clause is a disjunction of literals or their negations. The goal is to find an assignment of true/false values to the literals which satisfies the constraint, or to prove that no such assignment exists. For example is there a true/false setting for each literal that makes the logic statement $(A \vee B) \wedge (\neg A \vee C \vee D) \wedge (\neg B \vee \neg D) \wedge (\neg C)$ true? In the *maximum satisfiability* (MAXSAT) problem, the goal is find a true/false assignment for each literal such that the maximum number of clauses is satisfied. Many difficult problems can be transformed to SAT or MAXSAT, hence solution methods for these problems have been studied intensively. A good survey of solution approaches is available in Nadel (2002).

As a special type of constraint satisfaction problem, solutions for SAT and MAXSAT make use of backtracking tree search, notably the Davis-Putnam-Logemann-Loveland algorithm (Davis and Putnam 1960, Davis et al. 1962). This operates by choosing a Boolean literal, assigning a truth value to it, and simplifying the formula by propagating the newly-chosen value. If the formula is now true, then the search exits with success. Otherwise the branch assigning the opposite value of the chosen literal can be followed. If the formula is not yet true or false, then another literal can be chosen in a recursive manner. Simplifications include checking for literals which appear everywhere unnegated or everywhere negated, in which case the appropriate value can be assigned (set to true if unnegated everywhere or set to false if negated everywhere). Another simplification identifies clauses that have just one literal, in which case the true/false value that the literal must take is known (in the example above, C is the only literal in the fourth clause, and hence must be false to make the clause true).

A number of interesting rules for the selection of the branching variable have been developed to improve the speed in finding a feasible solution. Examples of branching literal selection rules include the following (Lagoudakis and Littman 2001):

- MAXO (maximum occurences) selects the literal that occurs the most often in the satisfiability formula. The idea is that the literal has a widespread effect.
- MOMS (maximum occurrences of minimum size) selects the literal that appears the most often in all clauses of minimum size (i.e. all clauses that have the smallest number of literals). The idea is that the literal has a widespread effect on the most tightly constrained clauses.
- MAMS combines MAXO and MOMS by adding their scores for each literal and selects the literal having the highest total.
- *Jeroslaw-Wang* calculates the weight for each literal l as $\sum_{j:l \in C_j} 2^{-n_j}$ where n_j is the number of literals in clause C_j. This gives small clauses more weight; the literal with the largest weight is selected.

- UP (unit propagation) makes a test assignment for each unassigned literal and counts the number of unit propagations that are triggered. The literal that triggers the most unit propagations is selected. This is a computationally expensive method.
- In GUP (greedy unit propagation), if a test assignment causes feasibility or infeasibility then it is selected, otherwise the rule is the same as for UP.
- SUP (selective unit propagation) tries to reduce the number of literals that are tested via the UP rule. It does this by first running all four of MAXO, MOMS, MAMS and Jeroslaw-Wang to produce a set of up to four candidate literals. The final selection among the candidate literals is made by the UP rule.

These branching rules are particularly interesting since they show promise for use in branch and bound for solving MIPs. This is the subject of ongoing research by the author.

5 Seeking Feasibility in Nonlinear Programs

We are concerned here with seeking feasibility in models that include at least one nonlinear constraint; the form of the objective function is irrelevant. For ease of reference we will refer to *nonlinear programs* (NLPs) with this concept in mind.

Finding a feasible point quickly is important because many optimization algorithms require one before they can even initialize (e.g. the Generalized Reduced Gradient algorithm (Abadie and Carpentier 1969, Lasdon and Waren 1978, Drud 1994), feasible sequential quadratic programming (Lawrence and Tits 2001), or methods of feasible directions (Lasdon 1970)), and so reaching feasibility is an important goal in itself. Additionally, a feasible solution is sometimes all that is required by the modeller, and using an algorithm that treats optimization and feasibility simultaneously may be computationally wasteful.

Finding a feasible point in an NLP can be notoriously difficult. There may be multiple disconnected feasible regions for example, possibly at extreme distances from each other. Many feasibility-seeking algorithms rely on optimizing a phase 1 objective that minimizes a penalty function, reaching zero at a feasible point. This is just as tricky as solving any NLP, and is subject to the same difficulties, such as the possibility of multiple local optima, including some which trap the phase 1 solution process, but which are not actually feasible points. It may be difficult to solve for the intersection of nonlinear constraints, or difficult to get correct derivatives.

Unless the constraints have specific properties (e.g. form a convex set), there is no guarantee that a particular algorithm will be able to find a feasible point when started from an arbitrary initial point. This means that it is very difficult to conclude that a given model is infeasible: it may simply mean that you have not started your solver in the right place. The best approach is to use knowledge about the problem such as a previous solution to a similar problem or logical reasoning to provide the solver with a "good" initial point. As the joke goes, the best way to solve an NLP is to start at the optimum. If that approach fails, the only recourse may be to start the solver in many different places, i.e. a multi-start or scatter search approach, hoping that it will be able to reach feasibility from one of those initial points.

There are a few relatively simple cases however. A feasible point is easily found if each constraint is everywhere convex or everywhere concave and the collection of constraints forms a convex set, for example. However the general problem of finding a first feasible point quickly in any given LP is very difficult.

Many NLP solvers use a *penalty function* approach to guide the search towards the feasible region. Penalty functions evaluate the constraint violations at the current point and calculate a penalty based on the degree of violation; the sum of the penalties forms the objective function in this unconstrained problem (see Sec. 5.1). The minimum of the penalty function in a feasible model is zero; a point having this value of the penalty function satisfies all constraints. However, the success of this, or any other approach, depends heavily on the characteristics of the nonlinear functions, such as their convexity or concavity. These characteristics can be estimated empirically via methods described in Sec. 5.2. If the feasible region is likely to be convex and full-dimensional, then simple bootstrapping methods can be used to find an initial feasible point (Sec. 5.3).

In the more general case, modern NLP local solvers are reasonably effective at reaching feasibility if given an initial point that is sufficiently close to a feasible region. A variety of methods now exist for choosing a good starting point, including single point heuristics (Sec. 5.4), and methods for improving the initial point prior to passing it to the NLP solver (Sec. 5.5). When single-start methods fail, multiple starting points must be tried. The first step in a multistart method (Sec. 5.7) is determining a good region in which to launch trial starting points (Sec. 5.6). More sophisticated bootstrapping methods are also available for nonlinear models with special structure (Sec. 5.8). Finally, relatively slow global optimization methods that fully explore the variable space can be used to find a feasible point as a last resort (Sec. 5.9).

5.1 Penalty Methods

A very common approach to finding a feasible solution for a set of nonlinear constraints is to minimize an unconstrained function that assigns a nonnegative penalty for each constraint violation at a given point. The idea is to work towards the minimum of the penalty function in the hopes of reaching a point at which the total penalty function value is zero, i.e. a feasible point. In general, penalty functions have the form $p_i(\boldsymbol{x}) = 0$ if $\boldsymbol{x}$ satisfies constraint i, and $p_i(\boldsymbol{x}) > 0$ if $\boldsymbol{x}$ violates constraint i (see e.g. section 14.5 of (Rardin 1998)). It is also important that the penalty function be monotonic, i.e. if the violation (see Sec. 1.2) of some constraint i at point $\boldsymbol{x}_1$ is greater than the violation at point $\boldsymbol{x}_2$, then $p_i(\boldsymbol{x}_1) > p_i(\boldsymbol{x}_2)$.

Common choices for the penalty function are the sum or sum of squares of the constraint violations:

- Sum of constraint violations penalty function: $p_i(\boldsymbol{x}) = \max\{0, b_i - g_i(\boldsymbol{x})\}$ for $g_i(\boldsymbol{x}) \geq b_i$, $p_i(\boldsymbol{x}) = \max\{0, g_i(\boldsymbol{x}) - b_i\}$ for $g_i(\boldsymbol{x}) \leq b_i$, and $p_i(\boldsymbol{x}) = |g_i(\boldsymbol{x}) - b_i|$ for $g_i(\boldsymbol{x}) = b_i$.
- Sum of squared constraint violations penalty function: $p_i(\boldsymbol{x}) = (\max\{0, b_i - g_i(\boldsymbol{x})\})^2$ for $g_i(\boldsymbol{x}) \geq b_i$, $p_i(\boldsymbol{x}) = (\max\{0, g_i(\boldsymbol{x}) - b_i\})^2$ for $g_i(\boldsymbol{x}) \leq b_i$, and $p_i(\boldsymbol{x}) = |g_i(\boldsymbol{x}) - b_i|^2$ for $g_i(\boldsymbol{x}) = b_i$.

Elwakeil and Arora (1995) carry out an empirical evaluation of several penalty methods. Given a set of constraints defined as $S = \{x \mid g_i(x)=0, i=1\ldots p; g_i(x) \leq 0, i=p+1\ldots m\}$ they evaluate five different penalty functions:

- For inequalities only, $\phi_1 = \sum_{i \in N} g_i(x) + r \sum_{j \in K} -1/g_j(x)$, where N is the set of indices of all active or violated constraints, K is the set of indices of all of the strictly inactive constraints, and $r > 0$ is a penalty parameter that is decreased throughout the procedure.
- $\phi_2 = 0.5 \sum_{i=1}^{p} [g_i(x)]^2 + 0.5 \sum_{i=p+1}^{m} [g_i(x) + |g_i(x)|]^2$, which has a value of zero for any individual satisfied constraint.
- $\phi_3 = \frac{1}{r} \sum_{i=p+1}^{m} e^{rg_i(x)-1}$. This is an exponential penalty function for inequality-constrained models.
- $\phi_4 = \frac{1}{r} \sum_{i=p+1}^{m} v_i e^{rg_i(x)-1}$ is identical to ϕ_3 except that different weights can be assigned to the different inequalities. These are also updated at each iteration.
- $\phi_5 = \frac{1}{r} \ln \sum_{i=p+1}^{m} e^{rg_i(x)-1}$. This is a logarithmic-exponential variation on ϕ_3.

Elwakeil and Arora (1995) report that none of these penalty functions dominates on the small and well-behaved models they studied.

A penalty function minimization can be carried out as a phase one procedure whose sole purpose is to find a feasible point. Alternatively, a penalty function term with an appropriate sign can be added to the objective function so that the solution works towards feasibility and optimality simultaneously. In this case, the combined expression has this form:

$$\text{max or min } F(x) = f(x) \pm \mu \sum_i p_i(x)$$

where $f(x)$ is the original objective function. The positive μ parameter is chosen so that an optimal solution to the combined function $F(x)$ yields a feasible and optimal solution to the original constrained problem. Penalty functions that have this property are called *exact*. In general the sum of squared violations penalty functions are not exact, but the simple sum of violations penalty functions are.

Solution algorithms for unconstrained nonlinear functions (e.g the penalty function by itself or combined with the original $f(x)$) are beyond the scope of this book, but will be found in any standard textbook on nonlinear programming. A sequential approach is often used in which a sequence of unconstrained problems is solved with the value of μ increased after each iteration.

5.2 Determining the Characteristics of an NLP

Many algorithms for solving NLPs, including finding an initial feasible point, depend on the model having specific characteristics such as including only quadratic nonlinear constraints or the feasible region consisting of a convex set defined by the constraints. While algebraic properties such as quadratic constraints are easy to check, shape properties such as constraint convexity and concavity and the convexity of the resulting feasible region (if it exists) are much harder to determine analytically. For a brief review of convexity and concavity properties of functions, see Greenberg (2003a) or Greenberg and Pierskalla (1971).

As pointed out by Pardalos (1994), "there is no known computable procedure to decide convexity", let alone the other shape possibilities. It is easy to check the shape of functions of one or two variables by visual inspection of a plot, but more sophisticated methods are required for functions of higher dimension.

An airtight conclusion about the convexity/concavity of a nonlinear function can sometimes be obtained by analytic evaluation of the function statement. This is the approach taken by the Dr. AMPL tool (Fourer and Orban 2007), which operates on a model written in the AMPL mathematical programming language (Fourer et al. 2003). AMPL represents the constraints and the objective function internally as a directed acyclic graph in which the leaf nodes are either constants or variables, and the internal nodes are operators (such as +, –, /, exponentiation, etc.). Given the graph and a set of rules that govern the convexity of combinations of terms, traversing the graph from leaves to root can prove that a given function is convex. Examples of convexity proving rules include (Fourer and Orban 2007):

- f and g convex implies that $f+g$ is convex,
- fg is convex when both have the same monotonicity and f and g are nonnegative and convex (or f and g are nonpositive and concave),
- e^f is convex if f is convex,
- $\cosh(f)$ is convex if f is linear or f is convex and nonnegative (or f is concave and nonpositive),
- $\sqrt{f}$ is nonconvex in general but $\sqrt{e^x}$ is convex.

Similar rules are included for all of the operators in the AMPL language. If a complete traversal of the graph for a function encounters only rules that preserve convexity, then the function is proven to be convex.

If the convexity-proving rules do not apply, then a different tack is taken to try to disprove convexity by showing negative curvature in the Hessian matrix of the function. This involves solving at least one, if not several, nonlinear optimization problems of the form

$$\min_d \nabla f(\boldsymbol{x})^T \boldsymbol{d} + \tfrac{1}{2}\boldsymbol{d}^T \nabla^2 f(\boldsymbol{x})\boldsymbol{d}$$

using a trust-region method. If negative curvature is found, then convexity of the function is disproved, but if negative curvature is not found, then there is no definite conclusion at all. However in studies of the shape properties of the objective functions for a number of NLPs, this approach returned a definite outcome in most cases. If the indeterminate case when convexity can be neither proved nor disproved, other methods must be applied, such as the sampling methods described next.

It is possible to gain a good idea of the shape, and the extent of the shape (e.g. highly concave or just barely concave) via empirical sampling approaches. Early work on sampling approaches concentrated mostly on the discovery of redundancy. Boneh's PREDUCE system (Boneh 1983) is mainly for the identification of redundant constraints, but it also discovers several general characteristics of NLPs such as boundedness, convexity, and the dimensionality of the feasible region. Information on the size of the facets of the feasible region and the bounds on the variables is also provided. Chinneck (2001, 2002) developed extensive sampling techniques and associated software (MProbe) for estimating the shape properties of nonlinear functions and assembling this information into conclusions about the shape of any possible feasible region.

A brief review of the techniques used in MProbe follows. The first step is to bracket a region of interest (normally the feasible region, if one exists) to create a *sampling enclosure* in which samples are taken. The first and simplest way to do this is to use the upper and lower bounds on the variables in the NLP, creating a sampling enclosure in the shape of a box. Uniform sampling is simple to conduct inside a box-shaped enclosure. However the results of the sampling analysis are more accurate the more closely the sampling enclosure surrounds the region of interest, so for this reason Chinneck develops additional forms of sampling enclosures as described later.

Given a sampling enclosure such as a box, the next step is to estimate the shape properties of individual constraints (convexity, concavity, etc.) via sampling. Convexity and concavity of functions are defined as follows. Construct a line segment by connecting any two points in the variable space. Estimate the value of the function at any point on the line segment by interpolating the function values at the two end points. A function is convex if the interpolated values at all points on every such line segment are greater than or equal to the actual function value at the same point. A function is concave if the interpolated values at all points on every such line segment are less than or equal to the actual function value at the same point.

These basic definitions provide a way of estimating function shape via sampling. The function's variable space is sampled in a region of interest by randomly scattering line segments, and then comparing the interpolated value at various points on the line segment with the actual value of the function at the same point. Defining the difference as (interpolated value) – (actual function value), convexity will show as a positive difference, concavity as a negative difference, and linearity as zero difference. Random line segments are constructed by connecting two uniformly distributed random points in the sampling enclosure. Difference calculations are made at a specified number of points arranged at fixed intervals along the line segment, as illustrated in Fig. 5.1.

The difference information collected over a large number of random samples is presented as a histogram, which provides useful information on the extent and range of convexity and concavity. Different thresholds are used to help identify

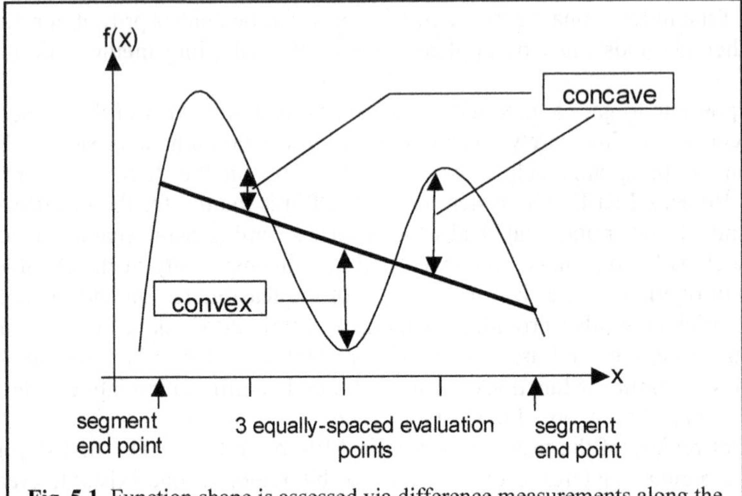

Fig. 5.1. Function shape is assessed via difference measurements along the line segment (Chinneck 2002)

candidates for approximation by more convenient functions. As is necessary in numerical calculations, a small tolerance $\varepsilon_=$ is needed in any assessment of the equality of two floating-point numbers. A difference histogram with all entries within $\pm\varepsilon_=$ indicates a completely linear function. A somewhat larger tolerance ε_{almost} helps to identify functions that are "almost" concave or "almost" convex. Functions whose difference histograms include values in these tolerance regions may be candidates for approximation.

For the purpose of feasibility-seeking we are interested in only the following shape outcomes, which can be distinguished by the MProbe sampling approach:

- *linear*: all differences are within $\pm\varepsilon_=$.
- *convex*: all differences are above $-\varepsilon_=$ and at least one is above $\varepsilon_=$.
- *concave*: all differences are below $\varepsilon_=$, and at least one is below $-\varepsilon_{almost}$.
- *convex and concave*: at least one difference is above ε_{almost}, and at least one difference is below $-\varepsilon_{almost}$.

Other combinations of the tolerances provide outcomes such as convex almost linear, almost convex, concave almost linear, almost concave, convex and concave almost linear. These are useful in determining which constraints are candidates for approximation via a simpler shape and reinsertion in the original model. Strict definitions can be obtained by setting ε_{almost} equal to $\varepsilon_=$. See Chinneck (2002) for details on how the same samples for testing function shape also provide information on function value ranges, multidimensional "slope", etc.

Once the shapes of the individual constraints are estimated via the sampling procedure, conclusions can be drawn about how well a simple solver algorithm such as steepest descent will be able to find a feasible point, and about the shape of the feasible region, if it exists, i.e. whether it forms a convex set or not. A convex set of points is defined as one in which the straight line connecting any two points in the set is contained entirely within the set.

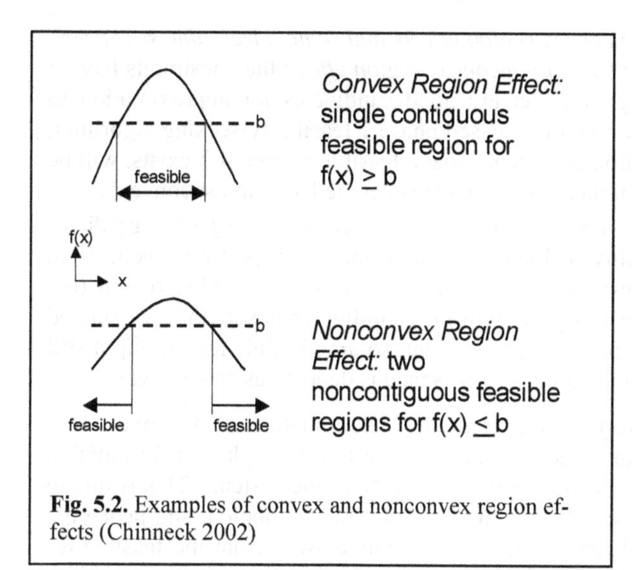

Fig. 5.2. Examples of convex and nonconvex region effects (Chinneck 2002)

The *constraint region effect* is the effect that each individual constraint has on the possibility of a convex feasible region, and is deduced from the empirical function shape and the constraint type (≤,≥,=), as follows:

- *Convex region effect*: contributes to a convex constrained region. This is given by (i) any linear constraint, (ii) convex inequalities of ≤ type, (iii) concave inequalities of ≥ type.
- *Almost convex region effect*: given by (i) almost linear equality constraints (ii) almost convex inequalities of ≤ type, (iii) almost concave inequalities of ≥ type.
- *Nonconvex region effect*: given by all constraints whose empirical shape is not "convex" or "almost convex".

A convex region effect is reported when the function shape combines with the constraint type (≤, ≥, =) in a way that results in a single contiguous convex feasible region for the variables in the constraint (see Fig. 5.2 for some one-dimensional concave function examples). The individual constraint region effects are now combined to assess the shape of the overall feasible region, if it exists.

The *constrained region* refers to the interaction of the constraints within the sampling enclosure. This is a broader concept than the feasible region, which, if it exists, is a subset of the constrained region. To assess the convexity of the feasible region itself requires that only feasible points be sampled; however it is virtually impossible to randomly generate feasible points, especially when there are equality constraints. Instead, we can draw various conclusions by combining the independently evaluated region effects of the individual constraints.

The conclusions are based upon how a standard gradient-based phase 1 feasibility-seeking algorithm is likely to perform if started at an arbitrary point in the sampling enclosure. In some cases, conclusions are also drawn about the shape of the feasible region, if one exists. Assuming that the independent constraint region effects are evaluated correctly, the three primary conclusions that can be drawn are:

- *If all constraints have convex region effects*: a gradient-based phase 1 feasibility seeking algorithm will accurately determine the feasibility of the constraint set. Further, a feasible region, if one exists, will be a convex set. The constrained region shape is denoted as "convex".

- *If some constraints have convex region effects and some have "almost convex" region effects, and none have a nonconvex region effect*: the constraints having an "almost convex" region effect are good candidates for approximation to improve the behavior of a gradient-based phase 1 feasibility-seeking algorithm. Appropriate approximation also means that a feasible region, if it exists, will be a convex set. The constrained region shape is denoted as "almost convex".
- *If there is at least one constraint having a nonconvex region effect*: a gradient-based phase 1 feasibility-seeking algorithm may not perform well. No conclusions can be drawn about the shape of a possible feasible region (the nonconvexity may occur in a portion of the sampling enclosure that is rendered infeasible by the action of another constraint, so any feasible region might still be a convex set). The constrained region shape is denoted as "nonconvex".

The most significant fact arising from the determination of a convex constrained region shape is that a steepest descent algorithm for a phase 1 formulation is guaranteed to accurately determine the feasibility of the system. This is due to the fact that for each individual constraint the constraint violation, and hence the phase 1 measure, grows steadily higher as you move away from the feasible region for that constraint. Hence, combining the constraints, there is no possibility of a local minimum in the phase 1 measure. If the set of constraints is feasible, then the local minima are also the global minima where the phase 1 measure reaches zero. If the set of constraints is infeasible, any local minimum will have a nonzero phase 1 measure. Hence finding any local minimum will accurately determine the feasibility status of the set of constraints.

An entirely different approach to identifying the shape properties of a nonlinear model is to build it in such a way that favourable shape characteristics are guaranteed. This is the main idea of *disciplined convex optimization* (Grant 2004, Grant et al. 2006) which provides a library of function "atoms" and a set of rules to allow a modeler to build up a model that will have a convex feasible region.

5.2.1 Convex Sampling Enclosures

For the most accurate function shape estimates, the sampling enclosure should tightly bound the region of interest. Boxes defined by bounds on the variables are easy to construct and to sample uniformly, but they may not bound the region of interest tightly. A better alternative is to construct a general convex enclosure by choosing appropriate constraints from the model. There are two disadvantages: increased complexity of the sampling procedures, and finding an initial point within the sampling enclosure.

Random sampling procedures of the type described previously require a full-dimensional convex sampling enclosure. Finding such an enclosure is straightforward if box enclosure sampling is already available. First sample in the box enclosure and identify all inequalities that have a convex region effect in the box. These constraints, along with the variable bounds forming the box, then form the convex sampling enclosure. One pitfall in determining a general convex enclosure in this manner is the possibility that an implied equality may eliminate the full-dimensionality of the resulting enclosure.

Sampling inside this new convex enclosure may in fact show that other constraints that previously showed nonconvex region effects when sampled in the original box actually have convex region effects when sampled in the smaller convex enclosure. They can then be added to the list of enclosing constraints. A large cardinality convex sampling enclosure is built up in this manner.

Equality constraints (even if linear) are excluded from the convex sampling enclosure due to the virtual impossibility of satisfying them during sampling. Constraints that have effectiveness of 1.0 (see Sec. 6.1.7) are also not permitted as part of the sampling enclosure since they are impossible to satisfy. The hit-and-run methods described below depend on the ability to generate an initial point that is feasible relative to the enclosure constraints. The convexity of the enclosure is also essential to ensure uniform sampling via hit-and-run methods.

In the current MProbe implementation a first feasible point inside the convex enclosure is identified via the bootstrapping method described in Sec. 5.3.

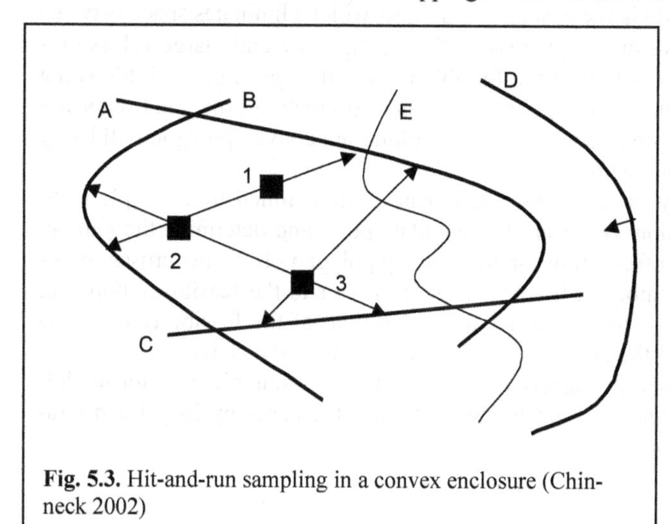

Fig. 5.3. Hit-and-run sampling in a convex enclosure (Chinneck 2002)

5.2.2 Hit-and-Run Methods

Hit-and-run methods (Berbee et al. 1987) allow sampling of the interior and perimeter of a general full-dimension convex enclosure. Starting at an arbitrary Point 0 (x_0) that is feasible relative to the enclosure, a *spanning line segment* is created by generating a random ray rooted at x_0. Point 1 (x_1), the point at which this ray meets the first enclosure constraint, is noted, as is Point 2 (x_2) the point at which the oppositely directed ray meets the first enclosure constraint. x_1 and x_2, the two *hit points,* define a spanning line segment.

There are various options for choosing a new x_0 for generating the next spanning line segment. It can be chosen at a random point on the last spanning line segment, or at a fixed point on the last spanning line segment (e.g. the center). In a *stand-and-hit* algorithm, x_0 is a single fixed point. The ray direction from x_0 is constructed by choosing a random point on the unit hypersphere surrounding x_0, though other choices are also possible.

An illustration of hit-and-run sampling is given in Fig. 5.3. Inequality constraints A-D constitute a convex sampling enclosure. Constraint E is not part of the convex sampling enclosure, so the usual data can be collected about it (constraint shape, effectiveness, etc.). The dark squares indicate the various $\boldsymbol{x}_0$ points used to generate hitting rays; the numbers indicate their order of use.

5.2.3 Approximating Nonconvex Feasible Regions

Banerjee and Ierapetritou (2005) address the problem of approximating the feasible region in the face of constraints that render it nonconvex. The specific application in this case is the operation of chemical processes, and a main goal is determining how much flexibility the process has for deviation from an initial setpoint, i.e. how much flexibility the operating point has for movement within the feasible region. The general approach is to approximate the convex hull of the feasible region.

Banerjee and Ierapetritou first sample the feasible region using a genetic algorithm to help guide the sample point placement. An approximation to the feasible region is then obtained via the α-*shape* technique, which eliminates space between feasible sample points using a sphere of radius α. At sufficiently large values of α this amounts to a convex hull of the feasible region, though in general this is not achieved. The feasible sample points are joined to create a polygonal outer approximation of the feasible region, i.e. the resulting nonconvex polygon will likely contain some infeasible regions.

Now the feasibility of candidate points can be determined by a simple technique: generate a random ray from the candidate point and determine the number of crossings it has with faces of the approximating polygon. If the ray crosses polygonal faces an odd number of times, then it is interior to the feasibility polygon, else it is exterior to it. This simplifies the assessment of the feasibility of points when the evaluation of the original constraint functions is expensive.

The method proposed by Banerjee and Ierapetritou is suitable only for models of low dimension having few constraints. See also the paper by Goyal and Ierapetritou (2003).

5.3 Bootstrapping in a Convex Constrained Region

As defined in Sec. 5.2, a convex constrained region consists entirely of constraints which have a convex region effect. Finding an initial feasible point $\boldsymbol{x}_0$ in a convex constrained region consisting entirely of inequalities is important for two reasons. First, real problems may have a convex constrained region, so reaching feasibility quickly is important for this class. Second, an initial feasible point is needed to initiate hit-and-run sampling for assessing the shape and other characteristics of constraints in the resulting convex sampling enclosure.

Given an initial $\boldsymbol{x}_0$ that is feasible relative to the enclosure constraints; the feasibility of subsequent $\boldsymbol{x}_0$'s is maintained thereafter by the hit-and-run method and the convexity of the enclosure. Alg. 5.1 uses this property in an efficient *boot-*

strapping method that adjusts an initial point to satisfy a monotonically increasing number of the constraints until a feasible point is reached. Random points are generated and tested against the constraints; when a constraint is satisfied all subsequent random points also satisfy that constraint because the hit-and-run method is used to generate the later random points. Feasibility relative to all of the enclosure constraints is built up gradually.

The model is deemed infeasible if no generated point satisfies all of the enclosure constraints simultaneously. A reasonably large number of points should be sampled before this conclusion is reached. Note that the procedure becomes more and more accurate in its sampling as constraints are moved from the *NotSat* set to the *Sat* set.

The shape of the sampling enclosure can make it difficult to find an initial feasible point. Long and thin sampling enclosures are especially difficult because the $\boldsymbol{x}_0$ launch point tends to stay in one region of the sampling enclosure and does not tend to move along the length of the sampling enclosure. This is because the probability of a random ray being oriented along the length of the thin enclosure is very small. Hence only one part of the enclosure is sampled, and if the feasible region is not in that part, then an initial feasible $\boldsymbol{x}_0$ cannot be found.

This difficulty occurs in practice. Many models have tight bounds on some variables along with variables that are unbounded or have very large bounds. The initial sampling box is then extremely long and thin, as are the subsequent sampling enclosures built up during the operation of Alg. 5.1. It is possible, however, to take advantage of the fact that these very common long and thin enclosures are axis-aligned.

The solution is to bias the ray-generation probabilities so that there is a much higher probability of generating a ray that points along the length of the long and thin enclosure. In the case of axis-aligned enclosures this is easy to do by multiplying the search direction vector produced by a random hypersphere around $\boldsymbol{x}_0$ by the lengths of the variable ranges. This converts the hypersphere to an axis-aligned hyperellipse that has a much larger probability of generating rays that are oriented along the long axes of the enclosure. Another option is to generate search directions by simply choosing two points in the variable box (even during Step 2 of Alg. 5.1). The search direction is then set as the difference of the two points. This also has a much higher probability of generating rays that are oriented along the long axes of the enclosure. The greater efficiency of both methods as compared to random hypersphere directions has been shown experimentally (Chinneck 2002). The MProbe software described in Section 5.2 uses the random hyperellipse directions method.

There is no such remedy for long thin enclosures that are not axis-aligned. If an enclosing box that aligns with the long axes of the enclosure can be found, then similar random hyperellipse or box-direction methods can be applied. However a technique developed by Chinneck (2002) to approximate the *prime analytic centre* (PAC) can be used to at least drive the $\boldsymbol{x}_0$ hit-and-run launch point away from satisfied constraints and towards relatively unexplored areas of the enclosure. This technique uses the prime analytic centre objective function, also called the logarithmic barrier function:

INPUTS: *NotSat:* the set of inequality constraints having convex region effects.

Step 1 (initialization):
Sat = the set of variable lower and upper bounds.
Do the following a specified number of times:
Generate a random point $\boldsymbol{x}_0$ satisfying *Sat* using box sampling.
IF any constraints in *NotSat* are satisfied at $\boldsymbol{x}_0$ THEN:
Move the satisfied constraints from *NotSat* to *Sat*; go to Step 2.
Issue an infeasibility message and exit.

Step 2 (satisfy general inequalities):
Do the following a specified number of times:
IF *NotSat* = ϕ THEN exit (success).
Generate a random line segment satisfying *Sat* from $\boldsymbol{x}_0$ using the hit-and-run method.
Select a random point on the line segment, label this $\boldsymbol{x}_0$.
IF any constraints in *NotSat* are satisfied at $\boldsymbol{x}_0$ THEN:
Move the satisfied constraints from *NotSat* to *Sat*.
Issue an infeasibility message and exit.

OUTPUTS: a point satisfying all of the constraints having convex region effects or a failure message.

Alg. 5.1. Bootstrapping procedure to achieve initial feasibility of a convex constrained region (Chinneck 2002)

$$\sum_{i=1}^{m} \ln(b_i - B_i x)$$

where $\boldsymbol{Bx} < \boldsymbol{b}$ (Caron et al. 2002) to evaluate $\boldsymbol{x}_0$ points. Candidate $\boldsymbol{x}_0$ points with higher values of the PAC objective function over all necessary constraints are closer to the PAC. At the same time, higher values of this barrier function correspond to points that are farther away from the limiting values of the inequalities. At an intermediate stage of the bootstrapping process, some subset of the constraints are satisfied, and these can be used in the PAC objective function. A candidate $\boldsymbol{x}_0$ point can then be assessed by evaluating the PAC objective function using only the satisfied subset of the constraints. An $\boldsymbol{x}_0$ having a higher value of the logarithmic barrier function indicates a point that is farther away from any satisfied constraints. See Chinneck (2002) for details on how necessary constraints are identified for use in moving towards the PAC.

This suggests a method of moving away from satisfied constraints and towards unexplored areas of the sampling enclosure. As $\boldsymbol{x}_0$'s are generated, the highest value of the PAC objective function associated with any $\boldsymbol{x}_0$ is recorded. If a proposed $\boldsymbol{x}_0$ has a greater value of the PAC objective function than the existing best value, then the new $\boldsymbol{x}_0$ is accepted and the best value is updated. Otherwise, the new x_0 is rejected, and the old x_0 is again used to generate the next hitting rays. The PAC objective function barrier-like repelling effect on the placement of subsequent x_0's moves the sampling towards unexplored areas of the enclosure. This method tends to keep the x_0 from becoming stuck in "corners" of the sampling enclosure.

This constitutes a method that is somewhere between hit-and-run and stand-and-hit. x_0 usually moves fairly frequently at the beginning, but gradually settles into longer and longer stays at particular points. Satisfaction of an additional constraint may cause some movement. Performance is improved when the new x_0 candidates are generated by taking the midpoint of the last spanning line segment, rather than a random point on the last spanning line segment.

5.4 Initial-Point Placement Heuristics

In practice, the crux of nonlinear programming is the initial-point placement. As the joke goes, the best way to solve a nonlinear programming problem is to start at the optimum. The joke is equally true for the feasibility problem: the best way to reach feasibility in an NLP is to start at a feasible point. A better result is usually obtained if information about the nature of the problem, external reasoning, or previous solutions of similar problems is available to guide the placement of the initial point. Many nonlinear solvers are able to find a feasible point if given an initial point that is close to feasibility, but may fail if the initial point is far from feasibility. If no external information is available to guide the initial-point placement, various heuristics can be used.

Some solvers provide a nonlinear *crash* heuristic to set the initial point, see e.g. the procedure used in MINOS (Murtagh and Saunders 1987), but details differ between implementations and are often confidential. A widely-applied heuristic (referred to here as the *standard heuristic*) is as follows:

- if the variable is doubly bounded: set at midpoint,
- if the variable is singly bounded: set on the bound,
- if the variable is unbounded in both directions: set at zero.

There are two main problems with the standard heuristic for initial-point placement. First, it sets many variables to zero, which can cause numerical errors, e.g. for a constraint that includes a term such as $1/x$. Second, since many variables are given similar bounds by the modeller (e.g. unbounded or singly bounded), many of the variables are also given the same initial values. This can also cause numerical errors, e.g. in constraints that include terms like $1/(x_1 - x_2)$.

For these reasons, Ibrahim and Chinneck (2005) developed a simple modification to the standard heuristic that superimposes a random perturbation Δ on the initial values proposed above. The *randomized standard heuristic* operates as follows:

- if the variable is doubly bounded: set at midpoint + Δ,
- if the variable has a single lower bound: set at bound + Δ,
- if the variable has a single upper bound: set at bound – Δ,
- if the variable is unbounded in both directions: set at zero + Δ,

Δ is a uniformly distributed random number between 0 and 1 (or suitably smaller if the bounds on the variable define a smaller range). Note that a positive perturbation is applied when the variable is unbounded in both directions. This avoids numerical problems caused by some functions (e.g. square root) when the variable really should have been specified as nonnegative, or even as positive (e.g. the derivative of the square root blows up at zero). The randomized heuristic avoids many of the numerical problems associated with the original heuristic while retaining its main features.

Ibrahim and Chinneck (2005) carried out a study of a number of initial-point placement heuristics:

- Random placement of initial points within the variable bounds.
- The origin (all variables set at 0.0).
- The standard heuristic.
- The randomized standard heuristic.

The points provided by these heuristics are compared with the initial points supplied with the models by determining the frequency with which a variety of nonlinear solvers are able to reach feasibility when launched from these points. Tests are carried out over a large number of models from the CUTE test set (Bongartz et al. 1995). Initial points are provided for most of these models; unspecified variables are set to zero.

In these experiments, the origin and the standard heuristic most frequently provide a point that is immediately feasible, while random points and points provided by the randomized standard heuristic are the least likely to be immediately feasible. This simply reflects the fact that the origin is a feasible point for many models. The much more important question is: which of the initial point heuristics most frequently permits a nonlinear solver to reach feasibility? The experimental results show that the randomized standard heuristic is much more effective than the competing alternatives in providing an initial point from which a variety of nonlinear solvers are able to reach feasibility, closely approaching the success frequency given when the modeler-supplied initial points in the CUTE set are used.

The ordering of the heuristics, from least to most effective in terms of providing initial points that allow solvers to reach feasibility, is the same as given in the list above. The randomized version is significantly more effective than the standard heuristic. The success rates for reaching feasibility from the provided initial points vary, depending on the solvers:

- Random placement: solvers find feasible points for 36.4% to 71.4% of the models
- The origin: solvers find feasible points for 59.3% to 74.9% of the models.
- The standard heuristic: solvers find feasible points for 60.2% to 74.9% of the models.
- The randomized standard heuristic: solvers find feasible points for 79.2% to 90.0% of the models.

The randomized standard heuristic is clearly the preferred initial-point placement heuristic.

5.5 Constraint Consensus Methods for Approximate Feasibility

The initial point supplied to a nonlinear solver may originate from knowledge of the model, from a previous solution to a similar model, or may be generated by an initial point heuristic such as described in Sec. 5.4. It may even be generated randomly by a naïve modeler. It is certainly possible to pass this initial point directly to the solver, but better results can be obtained if the initial point is instead passed to an inexpensive point improvement algorithm first. The point output by the point improvement algorithm is then finally passed to the full-scale, accurate, computationally expensive solver. The overall process can be more effective and much faster if the point improvement algorithm is reasonably accurate, and if inexpensive point improvement computations can be substituted for expensive full-scale solver iterations.

There are relatively complex feasibility-seeking procedures to be used as part of a solver phase-one procedure (e.g. Elwakeil and Arora (1995)). However there are few inexpensive methods for improving a given initial point, not necessarily all the way to feasibility. Chen and Kostreva (1999) describe a feasible directions method that is limited to solving nonlinear inequalities, for use prior to optimization via the method of feasible directions. Gertz et al. (2004) describe an approach that computes an affine scaling step by solving a system of linear equations related to a Newton iteration. Their algorithm is specifically for interior point methods in that it also provides initial values of other multipliers and parameters used by such methods.

The Constraint Consensus methods (Chinneck 2004, Ibrahim and Chinneck 2005) are point improvement algorithms that are effective at moving from a point that is very far away from feasibility to a point that is very near to feasibility. They are also very cheap, consisting almost entirely of function and gradient evaluations without line searches, GRG iterations, LP approximations, matrix inversions etc. As such they are ideal point improvement algorithms and are in fact the only algorithms in this class.

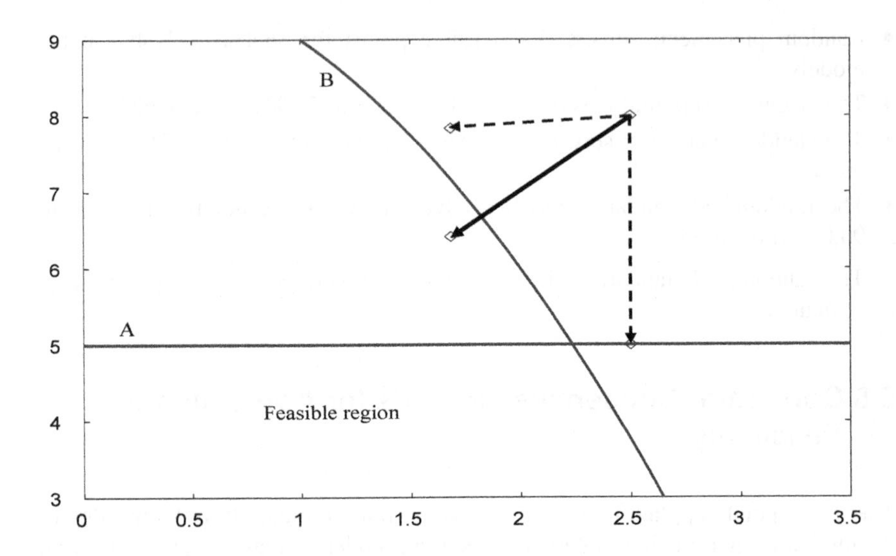

Fig. 5.4. Example iteration of the Constraint Consensus method (Chinneck 2004)

The Constraint Consensus algorithms are variations of *projection algorithms* (Sec. 2.8) that rely on the Euclidean distance to feasibility to gauge the extent of infeasibility (see Sec. 1.2). The *feasibility vector* for an individual constraint is defined as the vector extending from an infeasible point to its orthogonal projection (closest feasible point) on the constraint (Chinneck 2004). As described in Sec. 2.8, both the direction and the distance of movement necessary to achieve feasibility for an individual constraint are captured by the feasibility vector. Adding the feasibility vector to an infeasible point yields the closest point that satisfies the constraint, i.e. the orthogonal projection. The length of the feasibility vector is called the *feasibility distance*. The *gradient-projection* feasibility vector described in Sec. 2.8 is exact for linear constraints, but just an estimate for nonlinear constraints, and is naturally affected by the curvature of the constraint at the estimation point. However it can be used quite effectively in a heuristic method for reaching feasibility in NLPs, as shown below.

The individual feasibility vectors for all of the violated constraints are combined to arrive at the *consensus vector* that is actually used to make the updating move from the current point. This is done in a component-averaging manner: only the violated constraints that include a particular variable in $c(\mathbf{x})$ are able to "vote" on the movement in that dimension. In the original basic version of the algorithm the movement in each dimension is obtained by averaging the relevant component of each eligible feasibility vector; the resulting consensus vector specifies both the direction and distance of movement. The current point is updated by applying the consensus vector. The process iterates until the stopping conditions are met.

Fig. 5.4 provides an example of the update step in the simplest component-averaging simultaneous constraint consensus algorithm. The two feasibility vectors are shown as dashed arrows; the consensus vector is the solid arrow. Note that the feasibility vector is exact for linear constraint A, but just an estimate for nonlinear constraint B. Both feasibility vectors contribute to the consensus vector vertical component, but only the feasibility vector for constraint B contributes to the consensus vector in the horizontal component. Note that feasibility will be achieved in the next iteration of the method: only linear constraint A is violated after the update, and hence the Constraint Consensus method will make an exact move to satisfy it at the next iteration, thereby reaching the feasible region.

The algorithm terminates successfully if the length of every feasibility vector is less than the *feasibility distance tolerance* α, and unsuccessfully if either (*i*) the first condition is not met and the length of the consensus vector is less than the *movement tolerance* β or (*ii*) a preset number of iterations μ is exceeded. When successful, the final point is within an estimated Euclidean distance α of satisfying every constraint, where α might be quite large (e.g. 100) depending on the purpose at hand (e.g. finding the order of magnitude of a suitable starting point for the nonlinear solver). The movement tolerance β is used to detect situations in which the algorithm gets stuck or is proceeding very slowly.

The basic Constraint Consensus method is shown in Alg. 5.2. NINF is the number of violated constraints ("Number of INFeasibilities") at the current point, s_j is the sum of the feasibility vector components in the jth dimension, n_j is the number of violated constraints that involve variable j, and $\boldsymbol{t}$ is the consensus vector. For

Inputs:

- a set of I constraints $c_1 \ldots c_I$, in J variables $x_1 \ldots x_J$
- an initial point $\boldsymbol{x}$,
- a feasibility distance tolerance α,
- a movement tolerance β,
- a maximum number of iterations μ.

1. Repeat μ times:
   1.1. NINF = 0; for all $j : n_j = 0,\ s_j = 0$.
   1.2. For every constraint c_i:
      1.2.1. If c_i is violated then:
         1.2.1.1. Calculate feasibility vector $\boldsymbol{fv}_i$ and the feasibility distance $||\boldsymbol{fv}_i||$
         1.2.1.2. If $||\boldsymbol{fv}_i|| > \alpha$ then:
            - NINF = NINF + 1.
            - For every variable x_j in c_i: $n_j \leftarrow n_j + 1$; $s_j \leftarrow s_j + fv_{ij}$
   1.3. If NINF = 0, then exit successfully.
   1.4. For every variable x_j:
      1.4.1. If $n_j \neq 0$ then $t_j = s_j/n_j$, else $t_j = 0$.
   1.5. If $||\boldsymbol{t}|| \leq \beta$ then exit unsuccessfully.
   1.6. $\boldsymbol{x} \leftarrow \boldsymbol{x} + \boldsymbol{t}$.
   1.7. If any x_j exceeds its bounds, reset onto the nearest bound.
2. Exit unsuccessfully.

Alg. 5.2: The basic Constraint Consensus algorithm (Chinneck 2004) simplicity, details of how the basic algorithm tolerates numerical errors are not shown. Briefly, the algorithm ignores constraints that experience a numerical error at the current point (e.g. divide by zero) and carries on, hoping that the problem will not recur at the newly updated point. If the algorithm returns a final point at which at least one constraint experiences a numerical error, then the termination is deemed unsuccessful. See Chinneck (2003) for details.

The basic constraint consensus method treats all of the eligible feasibility vectors equally. However there may be value in emphasizing the effect of the longest or shortest feasibility vector. It may also be valuable to consider the *number* of constraints voting for a movement in the positive versus negative direction in a particular component. This is the basis of the algorithm variations developed by Ibrahim and Chinneck (2005) and described below.

Inputs:

- a set of I constraints $c_1 \ldots c_I$, in J variables $x_1 \ldots x_J$
- an initial point $\boldsymbol{x}$
- a feasibility distance tolerance α
- a movement tolerance β
- a maximum number of iterations μ
- *mode* (*near*, *far*)

1. Repeat μ times:
   1.1. NINF = 0; $k = 0$; for all j in $\boldsymbol{x}$: $n_j = 0$, $s_j = 0$, $z_j = 0$.
   1.2. If $mode = near$ then $fd = \infty$, else $fd = 0$.
   1.3. For each constraint c_i:
      1.3.1. If c_i is violated then:
         1.3.1.1. Calculate feasibility vector $\boldsymbol{fv}_i$ and feasibility distance $\|\boldsymbol{fv}_i\|$.
         1.3.1.2. If $\|\boldsymbol{fv}_i\| > \alpha$ then:
            1.3.1.2.1. NINF = NINF + 1
            1.3.1.2.2. For each variable j in c_i:
            - $s_j = s_j + fv_{ij}$; $n_j = n_j + 1$
            - If (($mode = near$) and ($\|\boldsymbol{fv}_i\| < fd$)) or (($mode = far$) and ($\|\boldsymbol{fv}_i\| > fd$)) then:
              - $k = i$
              - $z \leftarrow \boldsymbol{fv}_i$
   1.4. If NINF = 0, exit successfully with final point $\boldsymbol{x}$.
   1.5. For each variable x_j:
      1.5.1. If x_j appears in c_k then $t_j = z_j$.
      1.5.2. Else if $n_j \neq 0$ then $t_j = s_j / n_j$, else $t_j = 0$.
   1.6. If $\|\boldsymbol{t}\| < \beta$, then exit unsuccessfully.
   1.7. $\boldsymbol{x} \leftarrow \boldsymbol{x} + \boldsymbol{t}$
   1.8. If any x_j exceeds its bounds, reset onto nearest bound.
2. Exit unsuccessfully.

Alg. 5.3. Feasibility-distance based Constraint Consensus (FDnear, FDfar) (Ibrahim and Chinneck 2005)

The feasibility-distance based variations use the length of the feasibility vector associated with each violated constraint to determine the consensus vector. In the "near" mode used in the *FDnear* algorithm, the consensus vector is set equal to the shortest feasibility vector on the assumption that it is better to move to satisfy the smallest violation first because this keeps the point in a region where the gradients are good approximations of the functions. In the "far" mode used in the *FDfar* algorithm, the opposite assumption is made and the consensus vector is set equal to the longest feasibility vector because this is likely to provide the most rapid movement towards feasibility. In both cases, dimensions that do not appear in the selected shortest or longest feasibility vector are set by averaging as in the basic constraint consensus scheme. Details are shown in Alg. 5.3, where *fd* is the maximum or minimum feasibility distance, and *z* is the shortest or longest feasibility vector. The FDfar approach is related to the "remotest set control" class of projection algorithms (Censor and Zenios 1997, p. 80).

Inputs:

- a set of I constraints $c_1 \ldots c_I$, and J variables $x_1 \ldots x_J$
- an initial point $\boldsymbol{x}$
- a feasibility distance tolerance α
- a movement tolerance β
- maximum number of iterations μ

1. Repeat μ times:
    - 1.1. NINF = 0; for all j: $s^+_j = 0$, $s^-_j = 0$, $n^+_j = 0$, $n^-_j = 0$.
    - 1.2. For each constraint c_i:
        - 1.2.1. If c_i is violated then:
            - 1.2.1.1. Calculate feasibility vector $\boldsymbol{fv}_i$ and feasibility distance $\|\boldsymbol{fv}_i\|$.
            - 1.2.1.2. If $\|\boldsymbol{fv}_i\| > \alpha$ then
                - 1.2.1.2.1. NINF = NINF + 1
                - 1.2.1.2.2. For each variable j in c_i:
                    - If $fv_{ij} > 0$ then $s^+_j = s^+_j + fv_{ij}$ and $n^+_j \leftarrow n^+_j + 1$
                    - If $fv_{ij} < 0$ then $s^-_j = s^-_j + fv_{ij}$ and $n^-_j \leftarrow n^-_j + 1$
    - 1.3. If NINF = 0 then return successfully with final point $\boldsymbol{x}$.
    - 1.4. For each variable x_j:
        - 1.4.1. If $n^+_j = n^-_j$ and $(n^+_j + n^-_j) > 0$ then $t_j = (s^+_j + s^-_j) / (n^+_j + n^-_j)$
        - 1.4.2. Elseif $n^+_j > n^-_j$ then $t_j = s^+_j / n^+_j$
        - 1.4.3. Else $t_j = s^-_j / n^-_j$
    - 1.5. If $\|\boldsymbol{t}\| < \beta$, then exit unsuccessfully.
    - 1.6. $\boldsymbol{x} \leftarrow \boldsymbol{x} + \boldsymbol{t}$
    - 1.7. If any x_j exceeds its bounds, reset onto nearest bound.
2. Exit unsuccessfully.

Alg. 5.4 Average direction-based (DBavg) Constraint Consensus (Ibrahim and Chinneck 2005)

The direction-based algorithms conduct a "vote" on whether to move in the positive or the negative direction for each dimension prior to deciding how far to move. In some variants, the direction vote is the simple count of how many violated constraints would prefer an increase in a dimension versus how many would prefer a decrease in that dimension. In other variants, the vote is settled by the size of the largest proposed movement in the positive versus negative direction: whichever direction has the largest proposed movement wins the vote. Once this vote settles the question of whether to increase or to decrease in the dimension, there are several ways to decide how far to move.

The *DBavg* method decides the direction of movement in a dimension by a simple count of the number of votes for positive or negative movement, and the magnitude of the movement is decided by averaging the projections in the winning

Inputs:

- a set of I constraints $c_1 \ldots c_I$, and J variables $x_1 \ldots x_J$
- an initial point $\boldsymbol{x}$
- a feasibility distance tolerance α
- a movement tolerance β
- maximum number of iterations μ

1. Repeat μ times:
   - 1.1. NINF = 0; for all j: $s^+_j = 0$, $s^-_j = 0$, $n^+_j = 0$, $n^-_j = 0$.
   - 1.2. For each constraint c_i:
     - 1.2.1. If c_i is violated then:
       - 1.2.1.1. Calculate feasibility vector $\boldsymbol{fv}_i$ and feasibility distance $\|\boldsymbol{fv}_i\|$.
       - 1.2.1.2. If $\|\boldsymbol{fv}_i\| > \alpha$ then
         - 1.2.1.2.1. NINF = NINF + 1
         - 1.2.1.2.2. For each variable j in c_i:
           - If $fv_{ij} > 0$ then
             - o $n^+_j \leftarrow n^+_j + 1$
             - o If $fv_{ij} > s^+_j$ then $s^+_j \leftarrow fv_{ij}$
           - Else if $fv_{ij} < 0$
             - o $n^-_j \leftarrow n^-_j + 1$
             - o If $fv_{ij} < s^-_j$ then $s^-_j \leftarrow fv_{ij}$
   - 1.3. If NINF = 0 then return successfully with final point $\boldsymbol{x}$.
   - 1.4. For each variable x_j:
     - 1.4.1. If $n^+_j = n^-_j$ then $t_j = (s^+_j + s^-_j) / 2$
     - 1.4.2. Else if $n^+_j > n^-_j$ then $t_j = s^+_j$
     - 1.4.3. Else $t_j = s^-_j$
   - 1.5. If $\|\boldsymbol{t}\| < \beta$, then exit unsuccessfully.
   - 1.6. $\boldsymbol{x} \leftarrow \boldsymbol{x} + \boldsymbol{t}$
   - 1.7. If any x_j exceeds its bounds, reset onto nearest bound.
2. Exit unsuccessfully.

Alg. 5.5. Maximum direction-based (DBmax) Constraint Consensus (Ibrahim and Chinneck 2005) direction, as shown in Alg. 5.4. s^+_j and s^-_j are the sums of the feasibility vector components in the positive and negative directions for variable j and n^+_j and n^-_j are the number of violated constraints that vote for movement in the positive and negative directions for variable j.

In the *DBmax* variant, the direction vote is decided by the size of the largest proposed movement in each of the positive and negative directions: the largest proposed movement determines both the direction and the size of the component in the consensus vector. See Alg. 5.5. This is again related to the "remotest set control" projection algorithm but is applied in a component-wise manner.

In the bound-type direction-based variant *DBbnd*, the direction vote is settled by a simple count of the number of votes for an increase or a decrease in the component. The size of the movement in each component depends on the types of constraints that include that variable. Movements in the selected direction suggested by equality constraints are totaled; for inequalities only the largest movement in the selected direction is added because the largest movement will satisfy all of the inequalities. The resulting total is then reduced to an average. See Alg. 5.6. $n^{=+}_j$ and $n^{=-}_j$ represent the number of votes for the positive and negative directions for the jth variable recorded by violated equality constraints and max^+_j and max^-_j represent the largest positive and negative feasibility vector components for the jth variable in violated inequality constraints.

While all of the constraint consensus methods deal well with constraint scaling, they are vulnerable to discrepancies in variable scaling. The effect of uneven variable scaling could be more pronounced in the DBmax and DBbnd variants, so care should be taken with variable scaling prior to application of these heuristics.

Ibrahim and Chinneck (2005) carry out a very large empirical study of the Constraint Consensus variants. The DBmax variant exits successfully the most often at all values of α (with FDfar a close second), while FDnear is consistently the worst. It's clear that focusing on the larger violations gives the best performance.

The real test is whether the points returned by a Constraint Consensus algorithm, whether the algorithm concludes successfully or not, provides a useful starting point for a full-scale nonlinear solver. Here the results of empirical tests by Ibrahim and Chinneck (2005) are unequivocal. Applying a Constraint Consensus method prior to launching a nonlinear solver has two effects: (i) it significantly increases the probability that a nonlinear solver will reach a feasible point, and (ii) it dramatically reduces the number of nonlinear solver iterations required to reach feasibility. In fact, combining the randomized standard initial point heuristic with the FDfar Constraint Consensus variant gives success rates that are close to and in some cases better than the success rates obtained when very good modeler-supplied points are used to launch the solver. Obtaining an improved point by applying a Constraint Consensus algorithm is always a good thing to do prior to launching a nonlinear solver.

Inputs:

- a set of I constraints $c_1 \ldots c_I$, and J variables $x_1 \ldots x_J$
- an initial point $\boldsymbol{x}$
- a feasibility distance tolerance α
- a movement tolerance β
- maximum number of iterations μ

1. Repeat μ times:
   1.1. NINF = 0; for all j: $s^+_j=0$, $s^-_j=0$, $n^+_j=0$, $n^-_j=0$, $n^{=+}_j=0$, $n^{=-}_j=0$, $max^+_j=0$, $max^-_j=0$.
   1.2. For each constraint c_i:
      1.2.1. If c_i is violated then:
         1.2.1.1. Calculate feasibility vector $\boldsymbol{fv}_i$ and feasibility distance $\|\boldsymbol{fv}_i\|$.
         1.2.1.2. If $\|\boldsymbol{fv}_i\| > \alpha$ then
            1.2.1.2.1. NINF = NINF + 1
            1.2.1.2.2. For each variable j in c_i:
               - If $fv_{ij} > 0$ then
                  - $n^+_j \leftarrow n^+_j + 1$
                  - If c_j is an equality constraint then $s^+_j \leftarrow s^+_j + fv_{ij}$ and $n^{=+}_j \leftarrow n^{=+}_j + 1$
                  - Else if $fv_{ij} > max^+_j$ then $max^+_j \leftarrow fv_{ij}$
               - If $fv_{ij} < 0$ then
                  - $n^-_j \leftarrow n^-_j + 1$
                  - If c_j is an equality constraint then $s^-_j \leftarrow s^-_j + fv_{ij}$ and $n^{=-}_j \leftarrow n^{=-}_j + 1$
                  - Else if $fv_{ij} < max^-_j$ then $max^-_j \leftarrow fv_{ij}$
   1.3. If NINF = 0 then return successfully with final point $\boldsymbol{x}$.
   1.4. For each variable x_j:
      1.4.1. If $max^+_j \neq 0$ then
         1.4.1.1. $s^+_j \leftarrow s^+_j + max^+_j$
         1.4.1.2. $n^{=+}_j \leftarrow n^{=+}_j + 1$
      1.4.2. If $max^-_j \neq 0$ then
         1.4.2.1. $s^-_j \leftarrow s^-_j + max^-_j$
         1.4.2.2. $n^{=-}_j \leftarrow n^{=-}_j + 1$
      1.4.3. If $n^+_j = n^-_j$ then
         1.4.3.1. $t_j = (s^+_j + s^-_j)/(n^{=+}_j + n^{=-}_j)$
      1.4.4. Else if $n^+_j > n^-_j$ then $t_j = s^+_j / n^{=+}_j$
      1.4.5. Else $t_j = s^-_j / n^{=-}_j$
   1.5. If $\|\boldsymbol{t}\| < \beta$, then exit unsuccessfully.
   1.6. $\boldsymbol{x} \leftarrow \boldsymbol{x} + \boldsymbol{t}$
   1.7. If any x_j exceeds its bounds, reset onto nearest bound.
2. Exit unsuccessfully.

Alg. 5.6 Direction-based and bound-based (DBbnd) Constraint Consensus (Ibrahim and Chinneck 2005)

One caveat is in order. Most of the Constraint Consensus variants tend to produce output points that are on or very close to the limiting values of inequality constraints. This can be a problem for solvers that use barrier methods. A small adjustment of the output point to move it away from the limiting values of the constraints should be applied prior to submitting the point to a barrier-method solver.

5.6 Finding a Good Sampling Box for Multistart

If single-start methods are unsuccessful in providing the solver with an initial point from which it can reach feasibility, then multistart methods must be tried. Multistart methods typically provide the solver with random starting points within the hyperbox defined by the variable bounds. The efficiency of the multistart method can be greatly increased if it is given a sampling box that is highly likely to include a feasible point, i.e. a smaller and more focused hyperbox within the variable bounds. This section addresses the issue of finding a good sampling box that is likely to include a feasible point.

The first step in finding a good sampling box is to apply logical presolving methods to tighten the variable bounds prior to solving the model (see Sec 6.1.1). Following that, there are two other ways to provide a more focused sampling box; these are the subject of this section. First, random sampling methods can be used to heuristically tighten the variable bounds (Sec. 5.6.1), then a heuristically-effective sampling box within those bounds can be used initially (Sec. 5.6.2)

5.6.1 Tightening the Variable Bounds

The variable bounds provided by the modeler may define a box that is far larger than the feasible region(s) for the NLP. The first step in defining a good sampling box for multistart methods is to tighten the bounds as much as possible. Sampling can also be used for this purpose. Chinneck (2002) describes several techniques of this type, which have been implemented in the MProbe software. Consider nonlinear inequality constraints first. At random points inside the initial sampling enclosure, the feasibility of the nonlinear inequality is evaluated. The values of the constraint variables are noted at any points that satisfy the inequality. Over numerous feasible sample points, the minimum and maximum value pair is recorded for each variable. On exit, each minimum/maximum pair provides an overtightening of the bounds on the variable. It is an overtightening because the probability of attaining the true maximum and true minimum of each variable in this way is small. The probability of serious error is greatest when the variable bounds are unduly large or unbounded. The process is illustrated in Fig. 5.5a.

Each constraint determines a range for the variables that appear in it. The intersection set of the ranges for each variable, as determined by the constraints that it appears in, forms the tightened set of bounds on that variable. Since the variable

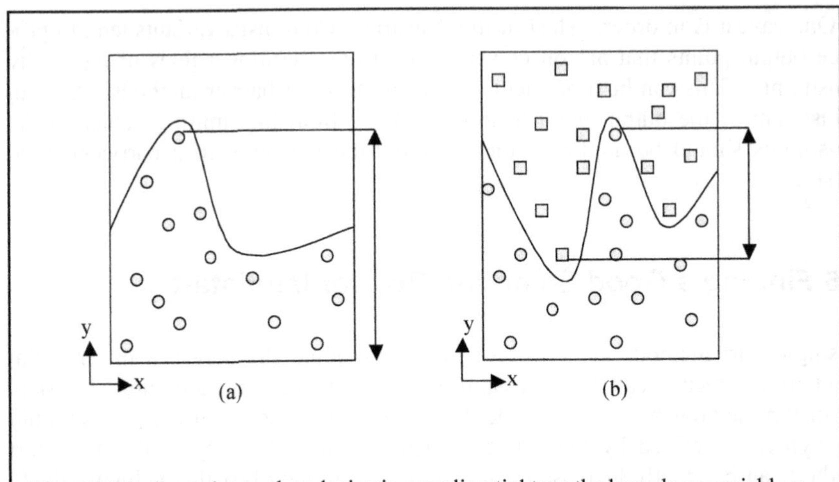

Fig. 5.5. Nonlinear interval analysis via sampling tightens the bounds on variable y. (a) inequality constraint. (b) equality constraint (Chinneck 2002)

ranges are (over)tightened, it may happen that the range returned for a variable by one constraint does not overlap with the range returned for the same variable by a different constraint. Because of the overtightening, it is possible that a feasible value for the variable exists in the gap between the two non-overlapping intervals. For this reason, when two intervals do not overlap, the interval that is returned consists of the gap between those intervals. Infeasibility is not assumed.

Nonlinear equality constraints can also be handled this way. The equality is first treated as a $\leq$ constraint and appropriate variable ranges are constructed, then it is treated as a $\geq$ constraint and a second set of variable ranges is constructed. The intersection of the two ranges is returned as the final tightened range after sampling the equality constraint. Fig. 5.5b illustrates this process.

Unbounded variables can cause difficulties for nonlinear interval analysis via sampling. This is because the feasible region for the constraint may comprise only a tiny portion of the variable box when the constituent variables are unconstrained, even when "unbounded" is numerically defined as $\pm 1 \times 10^{20}$, for example. Sampling then turns up no feasible points for the constraint, so that the variable ranges cannot be tightened.

When the range tightening methods described above are unsuccessful, the opposite approach may work: expanding boxes instead of shrinking them. The main idea is to start with a very small sampling box and to expand the bounds in several stages, looking for the approximate scale of the variables. The initial small box is centered on the origin. If the variable is unbounded in both directions, then the initial range extends from -1 to $+1$. If the variable is nonnegative unbounded, then the initial range extends from 0 to $+1$. This temporary sampling box is called a *nucleus box*.

For each nonlinear constraint for which no feasible points have been found, various nucleus boxes are sampled, usually at the scale of 1 as above, then increasing powers of 10 (10^1, 10^2, …10^5). The largest box that registers any feasible sample points for the constraint is used to reset the bounds on the variables in the constraint. In the case of equality constraints, the nucleus box is accepted if it registers at least one point at which the functional value is less than the constant, and at least one point at which the functional value is greater than the constant (or the low probability event that the point satisfies the equality constraint).

It is possible that the bounds will be overtightened by this procedure, but they could also be undertightened. Some manual adjustment may be desirable.

Nonlinear range cutting is another sampling heuristic that works by examining possible cuts on the edges of the variable range. For a given variable, a cut is proposed (as large as 90% of the total variable range if the range is very large, more commonly 30% of the variable range). The cut is accepted if, after sufficient samples, at least one of the constraints that use the variable has not been satisfied at any of the sample points. This method is similar to the range analysis described above, but operates on the infeasible zone of the constraint rather than the feasible zone. Equality constraints are considered satisfied in the test cut zone if they register at least one point at which the functional value is greater than the constant and at least one point at which the functional value is less than the constant, or the low probability event that a sample point satisfies the equality constraint. See Fig. 5.6 for an illustration.

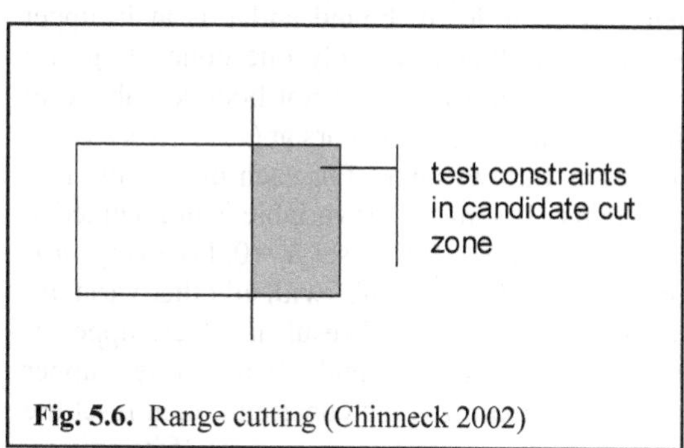

Fig. 5.6. Range cutting (Chinneck 2002)

Thus far, the sampling methods operate by examining a single constraint at a time. Bound tightening is more effective when the feasibility of all of the conastraints is considered simultaneously at a given point. This idea is difficult to incorporate in a random sampling approach given the presence of equality constraints. However, when a general convex sampling enclosure (see Sec. 5.2) is in use, the hit-and-run method guarantees that all of the sampling enclosure inequalities are simultaneously feasible. It is then simple to record the minimum and maximum values of the variables over all of the points sampled by the hit-and-run method.

Accuracy is improved by using the end points of the spanning line segments since these are at the extreme edges of the enclosure-feasible region. Overtightening of the variable bounds remains a possibility, but it is much less likely because of the use of the spanning line segment endpoints. All of the sample points used in resetting the maximum and minimum values are on the boundary of the enclosure-feasible region.

The MProbe software combines the various sampling techniques as follows:

1. One pass of presolving-style range tightening
2. Range sampling, one constraint at a time.
3. For those constraints which registered no feasible points during nonlinear range sampling, find a nucleus box.
4. Nonlinear range cutting.

MProbe also permits the use of three other techniques under user control: manual adjustment of variable bounds, the classic presolving techniques available in AMPL (Fourer et al. 2003), and convex enclosure sampling (Sec. 5.2).

Amarger et al. (1992) describe a method for tightening the variable bounds that makes use of the algebra defining the constraint functions. Each constraint is first analyzed for *monotonicity* in each variable it contains by inspection of the expression tree for the constraint. A variable is monotonic in the constraint if the constraint body is monotonically increasing or decreasing as the variable increases in value. The expression tree is a graph with variables and constants at the leaves and mathematical operators (+, –, /, ≤, etc.) at the nodes; this permits a systematic analysis of the variable interactions within the constraint.

The analysis then concentrates on the subset of constraints that contain only variables that are monotonic in the constraint, called monotonic constraints. Monotonic equality constraints are converted to a pair of oppositely-oriented inequalities. The procedure then finds the points that minimize and maximize the value of each constraint body. For example, if increasing x causes a monotonic increase in $g(x, y)$ while increasing y causes a monotonic decrease, then the minimum value of $g(x, y)$ occurs when x is at its lower bound and y is at its upper bound. However it is usual that many variables have only one bound (e.g. are nonnegative). If the variables are simply nonnegative but not bounded above in our example $g(x, y)$, then the minimum value of $g(x, y)$ occurs at $(x, y) = (0,\infty)$.

A series of bound-tightening iterations follows next. For each monotonic inequality, inspect each variable x_j it contains. If some other variable is unbounded at the minimum point for the constraint, then set $N_j = 1$, else set $N_j = 0$. For every variable for which $N_j = 0$, solve the constraint for the value of x_j with all other variables set at the value that minimizes the constraint. This will result in a new upper or lower bound on x_j, possibly tighter than the existing bound. If it is a new upper bound and is less than the value of x_j that maximizes the constraint, then reset the upper bound on x_j and the value of x_j that maximizes the constraint. If it is a new lower bound and is greater than the value of x_j that maximizes the constraint, then reset the lower bound on x_j and the value of x_j that maximizes the constraint. A small epsilon is used to ignore bound updates that are too small.

If at least one bound is updated by this procedure, then the process is repeated. The set of bounds is gradually tightened in this way. This process is similar to the usual cascade of bound tightening that takes place during presolving (see Sec. 6.1.1); the major difference is the pre-selection of the subset of monotonic constraints on which to operate due to their favourable properties for this operation.

5.6.2 Best Heuristic Sampling Box

Even when tightened as much as possible, the variable bounds may still be quite large, so it is reasonable to look for an even smaller sampling box. As for heuristic initial-point placement, there are certain regions that are more likely to yield a feasible point than others. For example, Lasdon and Plummer (2006) showed empirically that limiting the sampling box to $\pm1\times10^2$ or $\pm1\times10^4$ around the origin raised solver success rates.

The success of the randomized standard initial point heuristic shows that the small random region around the point supplied by the standard initial-point placement heuristic (Sec. 5.4) often provides a good starting point. Hence it is reasonable to expect that a larger box around that point will give good results for multistart. MacLeod and Chinneck (2007) investigated the size of the smallest box around the standard heuristic point that frequently includes the closest feasible point. Over a wide selection of models from the CUTE set, they determined that the average distance from the standard heuristic point to the closest feasible point was less than 1×10^4 per dimension for 98.3% of the models, and a maximum distance to the closest feasible point of less than 1×10^4 was recorded over all the dimensions for 97.4% of the models. Increasing the box beyond a distance of 1×10^4 from the standard heuristic initial point in each dimension showed very slow improvement per order of magnitude.

Given these results, the following initial multistart sampling box is recommended:

- If a variable is not bounded, bounds are $\pm1\times10^4$,
- If there is a single lower bound L, bounds are $(L, L+2\times10^4)$,
- If there is a single upper bound U, bounds are $(U-2\times10^4, U)$,
- If there are two bounds:
  - If separated by more than 2×10^4 with centre C, bounds are $(C-1\times10^4, C+1\times10^4)$
  - If separated by less than 2×10^4, bounds are (L, U).

MacLeod and Chinneck (2007) use this initial set of bounds in a multistart sampling procedure described in Sec. 5.7. The full procedure includes a method for expanding and shifting the sampling box as conditions warrant.

5.7 Multistart Methods

Single-start methods select a single initial point to submit directly to a nonlinear solver, or to a Constraint Consensus method whose output is then passed along to the nonlinear solver. For very difficult NLPs, the feasibility-seeking process may fail when started at any given initial point, hence another one must be chosen, perhaps several more, before feasibility can be attained. This is the premise of multistart methods. In *naïve multistart*, the nonlinear solver is launched at random starting points within the variable bounds. For difficult problems, the solver terminates unsuccessfully when started at most points in space, so this is inefficient and time-consuming. Newer multistart methods, in contrast, try to estimate which areas are most likely to contain a feasible point, and launch points there. As described in Sec. 5.6, one approach is to try sampling within a known promising sampling box; another approach is to learn where the promising launch areas are as the multistart process proceeds.

There are a variety of multistart methods for NLPs, but few that focus specifically on finding a feasible solution quickly. Most multistart methods focus on finding an optimum solution to the NLP, though this can be useful for finding a feasible solution when a phase-one type objective function is substituted in place of the original objective. A brief summary of optimum-seeking multistart NLP methods follows.

Variations on genetic algorithms (Michalewicz et al. 1994, Michalewicz and Nazhiyath 1995) are available, but can only reliably handle nonlinear constraints if the feasible region is known to be convex; this cannot be guaranteed in practice. In fact, the most difficult NLPs are likely to be nonconvex. Further, genetic algorithms can handle only very small models.

Scatter Search (Glover et al. 2000, 2003) uses an initial set of reference solutions, often generated randomly. Members of the reference set are combined in various nonconvex ways to create new solutions, which may be heuristically improved before being considered for inclusion in the updated reference set. Scatter search has been empirically tested on bound-constrained multimodal NLPs (Laguna and Martí 2005) with good results, but only for very small models. Path relinking (Glover et al. 2000, 2003) adds the exploration of trajectories between good solutions, on the theory that even better solutions may be located in between. Elements of one solution are progressively added to the other, with optional local improvement searches every few steps.

Glover, Laguna and Martí (2004) describe OptQuest, a commercial implementation of scatter search and path relinking that can handle nonlinear inequalities in conjunction with a local nonlinear solver. Ugray et al. (2006) provide the results of empirical tests of OptQuest for global optimization problems of moderate size.

The Greedy Randomized Adaptive Search Procedure (GRASP) (Resende and Ribeiro 2003a) keeps a restricted list of the best known candidate solution components, and then randomly selects among them to build a possible solution. Local searches are generally performed around candidate solutions. GRASP can also be combined with path relinking (Resende and Ribeiro 2003b). Meneses et al. (2005) apply GRASP to box-constrained nonlinear objective functions with good results.

The Efficient Global Optimization (EGO) method (Jones et al. 1998) samples a select few points of an expensive function and fits a surface to them. Successive samples are biased towards two kinds of areas: those that are strongly predicted to have better function values and those where the uncertainty of the function is so large that a better function value is quite possible, regardless of the predicted value. One of the key insights is the use of an initial Latin hypercube sample in order to start with a minimal amount of uncertainty. An initial sample of around 10 times as many points as the number of dimensions is recommended, which may be impractical to evaluate for truly expensive functions or very large models.

Some ideas from global optimization are useful in a multistart approach. Elwakeil and Arora (1996) define a clustering method as one in which local search is (ideally) only used once per local optimum. Clusters of points believed to be in the same region of attraction (i.e. near the same optimum) are identified, and only one point per cluster is used to initialize an expensive local search. Results show that the local search performance tends to depend heavily on dimension. Many global optimization methods (e.g. (Tu and Mayne 2002a, 2002b)) solve a series of quadratic sub-problems at each step to identify these clusters, which can be quite expensive, and is not practical for very large models. Tu and Mayne (2002a) report that multistart with clustering outperforms non-clustering multistart in terms of the number of local searches conducted, the number of minima found, the identification of the global minimum and the number of the function evaluations required.

Invoking a local solver is expensive, so it is important for multistart methods to avoid unnecessary invocations by filtering points before they are passed to the solver. An acceptance-rejection technique can be used to launch a local solver only when the initial point is thought to be promising. In general, the filtering becomes more severe as the algorithm progresses. Extensions of this general approach are used in the Zooming and Domain Elimination method of Elwakeil and Arora (1996) and the "multistart nonlinear programming" (MSNLP) software (Lasdon et al. 2004).

In addition to using acceptance-rejection filters, MSNLP incorporates a method of spherical approximation of the regions of attraction of the local minima. The distance filter compares each generated trial point to the set of best known solutions, and rejects it if it is within a solution's estimated basin of attraction (analogous to the clusters in clustering methods). It obtains an estimate of the radius of the basin of attraction by using the distance from the starting point for each returned solution. This radius is dynamically increased if another starting point returns the same solution from a greater distance, and decreased if the radius around this solution has rejected too many trial points. Further extensions initialize the radius to be slightly smaller than the calculated radius, and dynamically adjust the basins so they do not overlap. The general idea is similar to the repulsion algorithm (Sepulveda and Epstein 1996), except that with repulsion every new point is made use of by 'pushing it away' from the previously solved points in whose basins it lies. This is perhaps more computationally intensive than outright rejecting a point and moving on, depending on the problem characteristics.

The MSNLP merit filter, similar to the classic acceptance-rejection technique, compares the penalty function value of a candidate launch point to a certain threshold. Whenever a point with the new lowest value of the penalty function is discovered, the threshold is set to this value. If the threshold has rejected too many points in a row (the default value is 20) it is relaxed by a certain fraction. After a stage 1 in which *n1* points are generated and their penalty function values evaluated, stage 2 begins with the best point found in stage 1. This point is passed to the local solver, and the filters initialized with the result. In this stage *n2* points are generated one by one. If any point passes the two filters, the local solver is launched from it. The filters are updated based on the result and the next iteration begins. Note that the point returned by the local solver may not actually be locally optimal (or even feasible). Iterations are halted when a certain target value of the objective function is obtained, or when progress has slowed.

We now turn our attention to two multistart methods designed specifically for finding feasible points quickly in NLPs.

5.7.1 MSNLP Feasibility Mode

Lasdon and Plummer (2006) adapt the MSNLP concepts for the specific task of reaching a feasible point in a nonlinear program. The original optimization form of MSNLP uses the L1 exact penalty function as a quality metric for evaluating candidate starting points, i.e.

$$P(\boldsymbol{x},\boldsymbol{w}) = f(\boldsymbol{x}) + \sum_{i=1}^{m} w_i viol(g_i(\boldsymbol{x}))$$

where $f(\boldsymbol{x})$ is the original objective function, $viol(g_i(\boldsymbol{x}))$ is defined as the absolute violation of the ith constraint, and w_i is a positive penalty weights for the ith constraint. Variable bounds are never violated.

The main adaptation in the new feasibility mode is the replacement of the L1 exact penalty function by a new measure of infeasibility that is designed to be nearly invariant under changes in scaling of the constraint functions. Constraints have the form $l_i \leq g_i(\boldsymbol{x}) \leq u_i$. The new measure is the sum of the constraint violations over all of the constraints, $sinf(\boldsymbol{x})$, where the violation of an individual constraint $viol(g_i(\boldsymbol{x}))$ is defined as follows:

$$viol(g_i(\boldsymbol{x})) = \begin{cases} (l_i - g_i(\boldsymbol{x}))/(1 + abs(l_i)) & \text{if } g_i(\boldsymbol{x}) < l_i \\ (g_i(\boldsymbol{x}) - u_i)/(1 + abs(u_i)) & \text{if } g_i(\boldsymbol{x}) > u_i \\ 0 & \text{otherwise} \end{cases}$$

The feasibility mode of MSNLP bases all of its merit evaluations on this new $sinf(\boldsymbol{x})$, keeps the points with the best values of $sinf(\boldsymbol{x})$ in the list of local solutions, and terminates when the first feasible solution is found. Lasdon and Plummer (2006) report good results using MSNLP modified in this way to seek feasible solutions to a number of difficult NLPs. They also empirically evaluate a number of methods for generating initial points; a uniform sampling within the variable bounds does very well on most problems, in fact better than more advanced sampling techniques. Of course the point filtering algorithms severely restrict the number of starting points from which a full-scale nonlinear solver is launched.

5.7.2 Multistart Constraint Consensus

Constraint Consensus methods (Sec. 5.5) are ideal for use in a multistart algorithm. They are inexpensive and relatively quick at returning a point that is approximately feasible, hence they are well suited for the space exploration portion of a multistart algorithm. MacLeod and Chinneck (2007) develop this idea in their Multistart Constraint Consensus (MCC) algorithm.

MCC uses a phased approach to generating initial points for the local solver. In each phase a particular method is used to generate initial points which are then improved via the application of a Constraint Consensus method. If the point output by the Constraint Consensus method meets certain criteria, then a local solver is launched. If the local solver reaches a feasible solution, then the process is halted successfully. The ordered phases are:

1. The randomized standard initial point.
2. Latin hypercube sampling in the heuristic initial sampling box defined in Sec. 5.6.2.
3. Weighted random multistart.

Given its high rate of success in previous testing (Ibrahim and Chinneck 2005), the randomized standard initial point heuristic (Sec. 5.4) is a natural choice to generate the first initial point during phase 1. If that is not successful, then the second phase uses Latin hypercube sampling in the heuristic box defined around the standard heuristic point (see Sec. 5.6.2). If that is also unsuccessful, then Constraint Consensus is used in a phase 3 scheme which samples the variable space to continually update weights that control where subsequent samples will be placed.

The main innovation is in the phase 3 system for sampling within the solution space and updating the probability map for the placement of the next initial trial point. A Constraint Consensus algorithm is started at each new sample point. The final consensus vector in a particular solution sequence is not used to update the current point, but is instead used just to indicate the quadrant that the updating process wishes to move into. The overlap of the quadrants indicated by several different final consensus vectors provides the weighted probability map to guide the placement of the next initial point. Fig. 5.7 illustrates this process. There are three discontiguous feasible regions in Fig. 5.7, shown by the elliptical shapes. The last two consensus vectors for 5 different Constraint Consensus invocations are shown. The tail of the last consensus vector is called a *marker point*, which is used to divide the plane into quadrants; the last consensus vector in each Constraint Consensus sequence, called the *pointer* (shown as an arrow in the figure), indicates the quadrant which is most likely to contain a feasible point. In Fig. 5.7, all of the final consensus vectors indicate quadrants that contain a feasible point.

For k marker points there will be k+1 zones (or *bins*) in each dimension, and if there are n dimensions, then there will be $(k+1)^n$ boxes. It is obviously impractical to assign a probability to each individual box due to the combinatorial explosion as the number of dimensions increases, so the MCC algorithm takes a different tack: it assigns probabilities to each bin along each axis. In this way there are just $n(k+1)$ probabilities to assign. A new sampling point is built up by determining the new value in the x_1 dimension using the weighted probabilities for the bins along the x_1 axis, then the new value in the x_2 dimension using the weighted probabilities for the bins along the x_2 axis, etc.

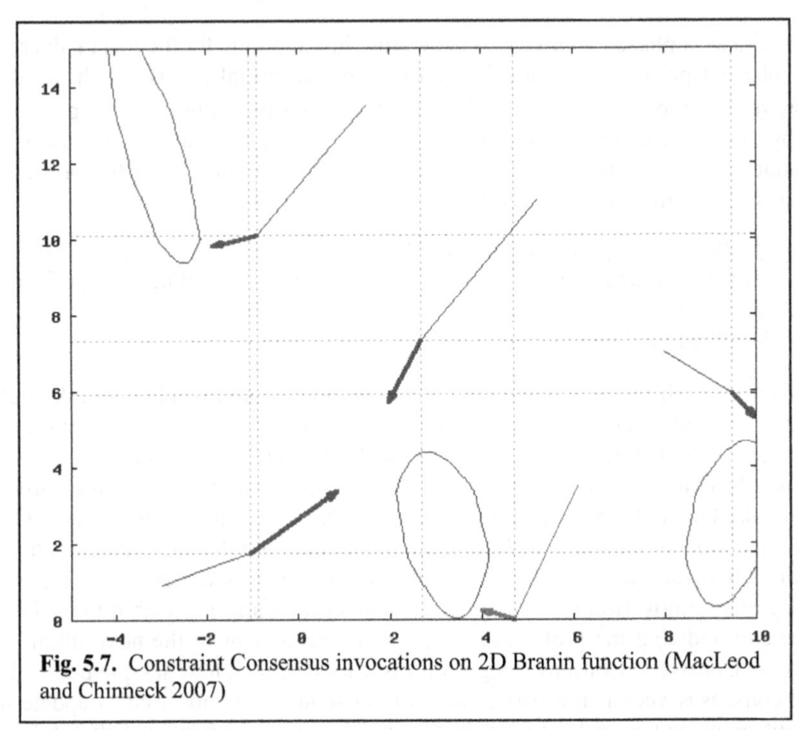

Fig. 5.7. Constraint Consensus invocations on 2D Branin function (MacLeod and Chinneck 2007)

The weight of each bin in each dimension is based on the number of pointers that cross that bin. This is illustrated in Fig. 5.8 which depicts the bins, marker points (thick dots) and pointers (thick arrows on the axis) for a single dimension. Each pointer indicates either a positive or a negative direction. A pointer adds a *vote* to every bin in the direction it indicates, until it encounters an oppositely-oriented marker. The long arrows above the axis line in Fig. 5.8 indicate the extent of the votes made by each pointer, and the numbers beneath the axis line indicate the vote totals for each bin.

The votes can be used to establish the weights for the bins in several ways. The simplest approach is to set the weight w_i for bin i as $w_i=v_i/V$, where v_i is the number of votes for bin i and $V=\Sigma v_i$ over all i. Once the bins and weights are known, the next point is chosen using the weighted probabilities. For each axis, choose bin i with probability w_i, then choose a uniformly distributed random point in the chosen bin. As implemented (MacLeod and Chinneck 2007), MCC uses a rela-

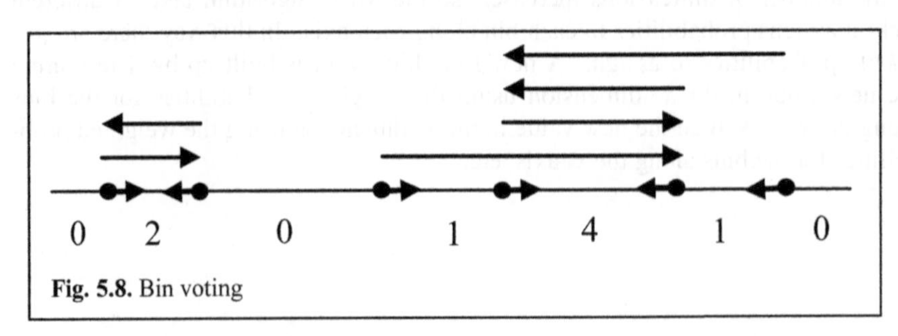

Fig. 5.8. Bin voting

tively small number of marker points and hence a small number of votes and bins. To compensate, exaggerating the differences between the weights in different bins gives generally better results. This is done by setting v_i to the square of the number of votes in the bin, but otherwise following the procedure above.

There are two special cases. First, it may happen that the pointer indicates neither the negative nor the positive direction in a certain dimension. In this case, a vote is placed in each of the two bins immediately bordering the marker point on either side. The rationale is that a zero pointer indicates that the immediate area around the marker point meets the criteria for feasibility for the current dimension, so sampling nearby is to be encouraged. Pointers voting towards the marker point extend their votes only to the marker point and not past it.

The second special case is when Constraint Consensus fails, or the local solver has been launched and fails. It is obviously desirable to promote less sampling in this region, so a negative vote is placed in each of the two bins immediately bordering the marker point on either side. This raises the possibility of a negative vote total for a bin. If the largest negative vote total is $-d$ in some bin i, then the vote total in every bin is adjusted upward by d. This sets the vote total in bin i to zero, meaning that it will not be chosen for sampling in the next iteration. It is also possible to adjust the vote totals so the minimum number of votes in any bin is 1 in order to retain a small possibility of sampling in low-probability spaces.

Note the contrast between the MCC weighted sampling scheme and that used by MSNLP (Lasdon and Plummer 2006). MSNLP concentrates on excluding areas that have already been searched, while MCC concentrates on identifying attractive areas for sampling that are likely to contain a feasible point. MSNLP uses hyperspheres for subdividing space, which assumes that all dimensions can be treated equally, which is not appropriate for models whose dimensions have different scales or different size intervals between upper and lower bounds. Hyperspheres are also unable to cover space completely without overlapping, an issue that becomes more pronounced at higher dimensions. The rectangular areas used in MCC do not have this problem.

MCC maintains a list of exactly N marker points at all times; a new point may replace an existing point if it meets these criteria: (i) fewer constraints evaluated at that point result in errors (e.g. overflow), and (ii) if two points have the same number of errors, the point whose longest feasibility vector is the shortest is better. The second criterion is equivalent to the smallest α that this point satisfies (the magnitude of all feasibility vectors will be less than or equal to this value). When a new point is better than an existing marker point, then it replaces the worst marker point and the bins and vote totals are recalculated. If the new point has the same longest feasibility length as the current worst point, it replaces it to avoid stagnation of the point set. MacLeod (2006) concluded that N=5 works well after a small study.

Sampling actually takes place inside the *sampling box* which is normally smaller than the box defined by the variable bounds. It is initially set to the box around the standard heuristic point as defined in Sec. 5.6.2. The sampling box is adjusted as evidence accumulates that it may be useful to sample outside the current box. The initial heuristic sampling box is unlikely to be the best choice for the sampling activities over the entire running time of the algorithm: it may be too big for some models while for others it may not actually encompass a feasible point.

The core region in the sampling box extends from the lowest marker point to the highest marker point in each dimension, but the sampling box should extend above and below these boundaries to some extent to handle the case where pointers indicate votes beyond the extreme upper or lower marker point. The sampling box is extended out from the core region by a fixed amount below the lowest marker point and above the highest marker point in each dimension. The fixed size assigned to the bin at each extreme edge in each dimension is found as follows. *Mag* is defined as the magnitude of the initial starting box (normally 10^4). *Edgewidth* is defined as $2 \times mag/N$. At any given iteration, the lowest bin in dimension x extends from d_1–*edgewidth* to d_1 and the highest bin extends from d_N to d_N+*edgewidth*, where d_1 and d_N are the lowest and highest marker values in that dimension.

If a newly generated and accepted point constitutes a new extreme marker in any dimension, the sample box expands accordingly in that dimension. Similarly, if the highest or lowest marker value in a dimension is eliminated when its point is replaced, the sample box will shrink in that dimension. The sample box is appropriately reduced if necessary to avoid sampling outside the variable bounds, but otherwise *edgewidth* remains a constant size to mitigate focusing too tightly on one area. This helps to avoid getting trapped in infeasible local minima.

An adaptive procedure is also used to dynamically adjust the feasibility distance tolerance α if the Constraint Consensus invocations frequently fail using the existing tolerance. If there are more than ρ consecutive Constraint Consensus invocations that do not trigger a local solver launch, then α is increased. The new value of α is set to the length of the longest feasibility vector in the best of the N marker points. Recall that the best marker point is the one whose longest feasibility vector is the shortest, hence this is equivalent to the smallest α that would trigger any of the current marker points to launch the local solver. There is no mechanism for reducing α.

An empirical study (MacLeod and Chinneck 2007) shows that Multistart Constraint Consensus is very effective in reaching feasibility for very difficult models. It is far more effective than naïve multistart. Simply adding a Constraint Consensus improvement phase to the points selected by a naïve multistart greatly improves its success rate, as expected (see Sec. 5.5). However Multistart Constraint Consensus is needed for the most difficult models. In a sample of 151 models, phase 1 of MCC solves 70.6% of the models, phase 2 solves 21.0%, and phase 3 solves the remaining 8.4%. As expected, the more complex phase 3 is needed only for the more complex models.

5.8 Bootstrapping Method of Debrosse and Westerberg

Bootstrapping methods (see Sec. 5.3) find a feasible point by generating an initial point that satisfies some subset of the constraints, and then adjusting the current point such that additional constraints are satisfied. Once a constraint has been satisfied, it is not violated in any subsequent step. In an ideal algorithm, the number of satisfied constraints increases monotonically until all are satisfied.

Debrosse and Westerberg (1973) provide some useful theorems concerning the feasibility of nonlinear systems of constraints and describe a bootstrapping method which produces one of two possible outcomes: (i) a feasible point or (ii) a minimal infeasible set of constraints (later termed an *IIS*: see Chap. 6). The method they propose is suitable for highly structured models in which each constraint involves only a few variables, which makes the set of constraints easily ordered by precedence. They consider a system consisting of equality constraints $f_i(\boldsymbol{x}) = 0$ and inequality constraints $g_j(\boldsymbol{x}) \leq 0$ in multiple dimensions.

The Debrosse and Westerberg bootstrapping method relies on identifying points at the intersection of subsets of the constraints. When dealing with inequalities, this means points at the intersection of the limiting values of the inequalities. A main technique is finding intersections of subsets of constraints that do not intersect with any other constraints. Some of their important theorems follow.

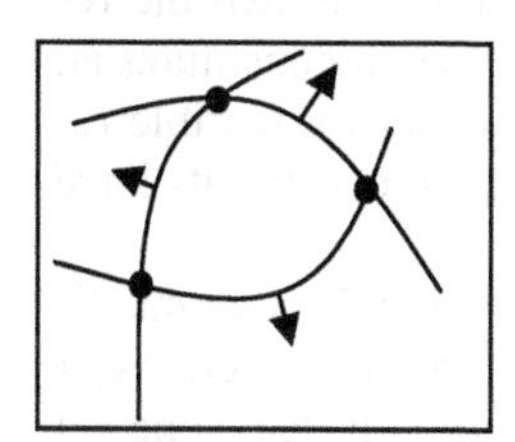

Fig. 5.9. Theorem 5.1

Theorem 5.1: *Intersecting surfaces* (Debrosse and Westerberg 1973, Theorem 1). For a set $\boldsymbol{P}$ of $p \geq 2$ inequality constraints, each proper subset $\boldsymbol{P}_i$ of constraints whose limiting surfaces intersect simultaneously determines a surface. Find a point on each such surface such that the surface intersects none of the constraints in $\boldsymbol{P} \backslash \boldsymbol{P}_i$ (such a surface is called a *surface of maximal intersection*). The system is infeasible if and only if none of the points satisfies all p constraints.■

Note that there is the possibility of multiple intersections of a given subset of constraints, in which case more than one surface is determined. In an n-dimensional system, some of these surfaces may be redundant (dimensionality is greater than $n-m$) or degenerate (dimensionality is less than $n-m$).

The general idea of Theorem 5.1 is illustrated in Fig. 5.9, which depicts an infeasible system of 3 inequalities. In the absence of multiple intersections due to curving and recrossing of the constraints, each of the dots represents a point on a surface of maximal intersection. Each of those points is defined by the intersection of two of the inequalities, but does not intersect the third. In addition, each of the points violates that third inequality. Hence the system is infeasible.

Consider the opposite case, in which one of the inequalities has the opposite sense. We would have the same points on the surfaces of maximal intersection, but now not all of the points will violate the third inequality. Hence the system is feasible.

Theorem 5.2: *Nonlinear turnabout* (Debrosse and Westerberg 1973, Lemma 1). Consider a set ***M*** of m constraints and a feasible proper subset ***P*** of p constraints. If every constraint p forms some part of the boundary of ***P***, and ***P*** is connected and intersects none of the constraints in ***M\P***, then either ***M*** is feasible, or there exists at least one constraint $i \in$ ***M\P*** such that ***P*** $\cup\, i$ is infeasible.■

A *multiple constraint* is a constraint that has multiple intersections and thereby creates more than one limiting surface.

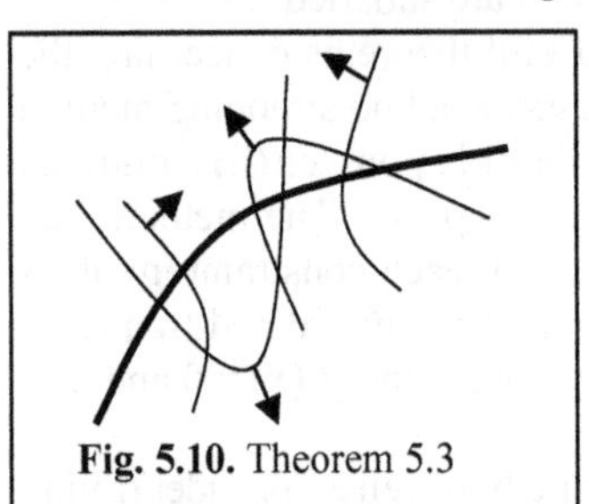

Fig. 5.10. Theorem 5.3

Theorem 5.3: *IIS along a line* (Debrosse and Westerberg 1973, Lemma 2). Given a line, if a set of $p > 2$ inequalities creates an IIS in conjunction with the line, then there are at least $p-2$ multiple constraints on the line.■

Debrosse and Westerberg consider a "line" to include a curve. If you are restricted to staying on the line, and there are no multiple constraints, then it should require just two intersecting constraints in addition to the line itself to cause infeasibility, one with a feasible region to the "left" and another with a feasible region to the "right", with no overlap of the two feasible regions. If an IIS has more than two constraints in addition to the line, then it implies that there are constraints that have multiple intersections with the line, creating discontiguous feasible regions. See Fig. 5.10 for an example of a line plus 4 inequalities that constitutes an IIS. By inspection, removing any one of the 4 inequalities creates a feasible region on the darker line. As required by the theorem, two of the constraints have multiple intersections with the line.

Theorem 5.4: *IIS in n dimensions* (Debrosse and Westerberg 1973, Lemma 3). Given a set of constraints ***P*** having p members and no multiple intersections or multiple constraints, and that constitutes an IIS, then (a) there is at least one set $\boldsymbol{P}_j = \boldsymbol{P}$-$\{j\}$ such that the feasible region of $\boldsymbol{P}_j$ is bounded by all $p-1$ constraints in $\boldsymbol{P}_j$, (b) the feasible region of $\boldsymbol{P}_j$ is connected, and (c) the feasible region of $\boldsymbol{P}_j$ is not intersected by constraint j.■

The various restrictions on ***P*** mean that a linear system qualifies, and the theorem is easily understood for a linear system.

Theorem 5.5: *IIS cardinality in n dimensions* (Debrosse and Westerberg 1973, Theorem 2). In an n-dimensional space, if a set of $n+p$ constraints forms an IIS where $p>1$, then there is at least one multiple intersection or multiple constraints in the set.■

This is a close cousin of Thm. 6.14 which states that the maximum cardinality of an IIS in a linear system is $n+1$. For more constraints to be involved there must be multiple intersections or multiple constraints to form the "extra" feasible regions, which is possible only with nonlinear constraints.

Theorem 5.6: *Number of point evaluations to establish infeasibility* (Debrosse and Westerberg 1973, Theorem 3). We are given a set ***P*** of p constraints and wish to determine whether ***P*** is infeasible. Consider a proper subset ***M*** of ***P*** which has m members, and which defines a surface S by the intersection of all m members.

We must evaluate a point on S if and only if all $p - m$ subsets determined by $\boldsymbol{M}$ plus one constraint from $\boldsymbol{P} \backslash \boldsymbol{M}$ have no simultaneous intersections.■

The point of this theorem is simply to reduce the number of simultaneous equation solutions that must be carried out. It follows from Thm. 5.1.

The authors also define *structural infeasibility* of a set of equations. This occurs when some subset of the equations has fewer variables than equations, and none of the equations are redundant.

Debrosse and Westerberg construct an algorithm based on these theorems. A simplified version for inequality constraints only that ignores multiple intersections and constraints (and the cycles they can cause) is shown in Alg. 5.7. The bootstrapping characteristic is apparent in Step 2 in which further constraints are added to the set of constraints already satisfied. The algorithm then attempts to satisfy the added constraints or to identify an IIS from among the constraints in the current hypothesis set $\boldsymbol{H}$.

No instructions are given as to the best way to construct the initial point in Step 0, or which subset of violated constraints to incorporate into the hypothesis set in Step 2. A further difficulty is that an enumeration of subsets is required in Steps 8 and 11, of which there are potentially very many.

The algorithm can be modified to handle equality constraints. The main difference is that all of the equality constraints are included in the hypothesis set at all times. This affects the results in that the set of constraints output in Step 14.1 of Alg. 5.7 may not be an IIS.

5.9 Global Optimization

Global optimization methods are designed to reliably reach the proven global optimum of a nonlinear function. Such methods generally use some form of branch and bound, which subdivides the solution space in an exhaustive search (Pintér 1998). Areas of the solution space that cannot contain a feasible solution are gradually eliminated, while more promising areas are subdivided further for closer examination. This is a time-consuming process, though provably correct, hence not especially suited for speedy identification of feasible points in NLPs. It is generally practical only for relatively small models. A widely available implementation is the BARON solver (Sahinidis 1996, 2000).

Global optimization methods are particularly useful when other methods, including extensive application of multistart methods, are unable to locate a feasible point. Global optimization can then be applied to make a definite determination as to whether the model is feasible or infeasible.

H is a set of constraints called the hypothesis.

GIVEN: a set of inequality constraints, some or all of which may be nonlinear

0. Generate an initial point $\boldsymbol{x}$.
1. Determine the set $\boldsymbol{V}$ of constraints violated at $\boldsymbol{x}$, and the set $\boldsymbol{S}$ of constraints satisfied at $\boldsymbol{x}$.
2. $\boldsymbol{H} \leftarrow \boldsymbol{S} \cup \{\text{subset of } \boldsymbol{V}\}$.
3. IF $|\boldsymbol{H}| \neq n + 1$ THEN:
   - 3.1 Try to find a new point $\boldsymbol{x}$ that satisfies all constraints in $\boldsymbol{H}$.
   - 3.2 IF there is no feasible point $\boldsymbol{x}$ THEN go to Step 4.
   - 3.3 Determine the set $\boldsymbol{V}$ of constraints violated at $\boldsymbol{x}$, and the set $\boldsymbol{S}$ of constraints satisfied at $\boldsymbol{x}$.
   - 3.4 IF $\boldsymbol{V} = \varnothing$ THEN exit with $\boldsymbol{x}$ as a feasible solution.
   - 3.5 Go to Step 2.
4. IF there exists a proper subset $\boldsymbol{H'}$ of $\boldsymbol{H}$ that is structurally infeasible THEN:
   - 4.1 $\boldsymbol{H} \leftarrow \boldsymbol{H'}$; go to Step 4.
5. Find n, the number of dimensions in $\boldsymbol{H}$.
6. IF $|\boldsymbol{H}| > n$ THEN $r = n$ ELSE $r = |\boldsymbol{H}| - 1$.
7. Let $l = 1$ and establish empty list L_l.
8. $L_l = \{\text{all subsets of } \boldsymbol{H} \text{ of size } r\}$; label subsets $E_r(k)$, $k = 1,2,\ldots |\boldsymbol{H}|!/r!(|\boldsymbol{H}|-r)!$
9. FOR each $E_r(k)$ on list L_l:
   - 9.1 IF $E_r(k)$ feasible at some point $\boldsymbol{x}$ THEN:
     - 9.1.1 Delete $E_r(k)$ from L_l.
     - 9.1.2 IF $\boldsymbol{x}$ satisfies all constraints in $\boldsymbol{H} \backslash E_r(k)$ THEN:
       - 9.1.2.1 $\boldsymbol{H}$ is disproved; go to Step 15.
10. IF $|L_l| < |\boldsymbol{H}| - r + 1$ THEN $\boldsymbol{H}$ is proved so go to Step 13.
11. Let $l \leftarrow l+1$ and establish new list L_l by filling it with all set intersections possible for entries on list L_{l-1} taken $|\boldsymbol{H}| - r + 1$ at a time. Place each resulting set intersection containing exactly $r - 1$ constraints into list L_l.
12. $r \leftarrow r - 1$; go to Step 9.
13. Form a list $\boldsymbol{T}$ of infeasible subsets of $\boldsymbol{H}$ in the order found in the last list L_q, then L_{q-1}, L_{q-2} etc., plus $\boldsymbol{H}$ itself.
14. For each set $\boldsymbol{T}_i$ on the list $\boldsymbol{T}$:
    - 14.1 Test feasibility of $\boldsymbol{T}_i$ by testing it against all of the points evaluated during Step 9. IF infeasible, THEN exit with $\boldsymbol{T}_i$ as an IIS.
15. IF all constraints satisfied at point $\boldsymbol{x}$, THEN exit with $\boldsymbol{x}$ as a feasible solution.
16. Go to Step 1.

Alg. 5.7. Bootstrapping method by Debrosse and Westerberg (1973)

PART II: ANALYZING INFEASIBILITY

As mathematical models grow larger and more complex, infeasibility happens more often during the process of model formulation, and is harder to diagnose than ever before. A linear program may have hundreds of thousands or even millions of constraints: which of these are causing the infeasibility and how should the problem be repaired? In nonlinear programs the issue is even more vexed: the model may be truly infeasible, or the solver may just have been given a poor starting point from which it is unable to reach feasibility.

Some form of automated or semi-automated assistance in diagnosing and repairing infeasibility is necessary in the face of the scale and complexity of modern optimization models. Fortunately, algorithmic tools have been developed in recent years. There are three main approaches. The first is the identification of an *Irreducible Infeasible Subset (IIS)* of constraints within the larger set of constraints defining the model. An IIS has the property that it is infeasible, but becomes feasible if any one or more of its constraints are removed; it is irreducible in that sense. Identifying an IIS allows the modeler to focus attention on a small set of conflicts within the larger model. Further refinements of the base algorithms try to return IISs that are of small cardinality, or that are easier for humans to understand. Other issues include trade-offs between the speed of identifying an IIS and the cardinality of the IIS that is returned.

The second main approach to analyzing infeasibility is to identify a *Maximum Feasible Subset* (MAX FS) of constraints within the larger set of constraints defining the model. This naturally focuses the analysis on the constraints that do not appear in this subset, i.e. the minimum cardinality set of constraints that must be removed so that the remainder constitutes a feasible set. Identifying a maximum feasible subset is an NP-hard problem, so the methods for doing so are clever heuristics.

Both of these approaches to analyzing infeasibility focus attention on a small part of a large model so that the modeler can determine how to repair it using human understanding of the meanings of the constraints. However the third approach seeks to suggest the best repair for the model, where "best" can be defined in various ways that can be handled algorithmically, e.g. the fewest changes to constraint right hand side values. The suggested repair can of course be accepted, modified, or rejected by the human modeler.

Many of the methods for analyzing infeasibility that are described in Part II depend on the ability of a solver to determine the feasibility status (feasible or infeasible) of an arbitrary set of constraints with very high accuracy. This ability is by and large available for linear programs, but it is much more problematic for nonlinear programs and for mixed-integer programs. In those cases, we may have to settle for the identification of an infeasible subset that is not irreducible (but is significantly smaller than the original set of constraints one hopes), or a *minimal intractable subset (MIS)* which is a minimal subset of constraints that causes a solver to report infeasibility under stated solver parameter settings (initial point, tolerances, etc.). For this reason, there are differing expectations for the success of the general analysis algorithms depending on the type of optimization model. However there are methods that are special to each type of model.

As you will see in Part II, effective algorithms for the analysis of infeasibility in linear programs and linear networks exist, and have been implemented in most commercial LP solvers. The situation is not yet so positive for nonlinear and mixed-integer programs, but research in this area is active and ongoing, with frequent new developments. Significant breakthroughs are likely in the relatively near future, especially as improved algorithms for reaching feasibility quickly reach maturity (see Part I).

Some of the algorithmic tools are best integrated directly into the solvers rather than into separate analysis software. This is the case for many of the infeasibility analysis algorithms since they make use of data that is available during the solution or re-solution of the problem. This is certainly the case for several of the algorithms for analyzing infeasible linear programs that use information from the final phase 1 basis, and thereafter from bases produced by repeated solutions of slightly differing versions of the model. Algorithms in this class benefit from the use of hot-starts based on the immediately previous solution.

Infeasibility analysis is part of larger efforts in *computer-assisted analysis* of complex optimization models originated by Greenberg (1981a, 1981b, 1983). He developed software such as ANALYZE (Greenberg 1983) which provides tools for the manipulation and analysis of linear programs. PERUSE (Kurator and O'Neill 1980) was another early system which permitted interactive query of the LP matrix and solution. MProbe (Chinneck 2001) is a more recent system that provides various tools for general probing of optimization models, particularly nonlinear programs. Practical approaches to infeasibility analysis in particular date to the 1970s, notably in the *Refinery and Petrochemical Modeling System* by Bonner & Moore (1979). Other model-specific approaches for infeasibility analysis were developed by Harvey Greenberg as part of his *Intelligent Mathematical Programming System* (IMPS) project originating in the early 1980s (see Greenberg (1987c, 1991) for a description of the IMPS project and Greenberg (1996b) for a summary of relevant literature). In contrast, the IIS isolation approach developed in Chap. 6 is independent of the particular model and has been developed for general solvers applicable to any LP.

The eventual goal in computer-assisted analysis is the development of a complete environment supporting optimization modeling, similar to the environments enjoyed by general software developers that include debuggers, profilers and other useful tools. This is an essential part of the verification and validation of optimization models. There has been significant progress in the development of techniques and tools supporting the debugging of complex optimization models. Most modern LP solvers now include routines for isolating IISs for instance, and model debugging is now included as a topic in modern textbooks on optimization. For example, see the textbook by Pannell (1997) for an excellent discussion of how to debug a linear program, including infeasibility analysis.

6 Isolating Infeasibility

When faced with infeasibility in a very large optimization model, such as a linear program containing thousands of constraints, it is immensely helpful to be able to narrow the focus of the diagnostic effort. The focus is narrowed as much as possible if you are able to isolate an *irreducible infeasible set (IIS)* of constraints from among the larger set defining the model. An IIS has this property: it is itself infeasible, but any proper subset is feasible. It is irreducible in the sense that every member contributes to the infeasibility. IISs are also known as *irreducible inconsistent sets* or *minimal infeasible subsystems*. Where the subset is infeasible, but is reducible, it is simply called an *infeasible subset (IS)* of constraints. A simple IIS consisting of three linear inequalities is shown in Fig. 6.1.

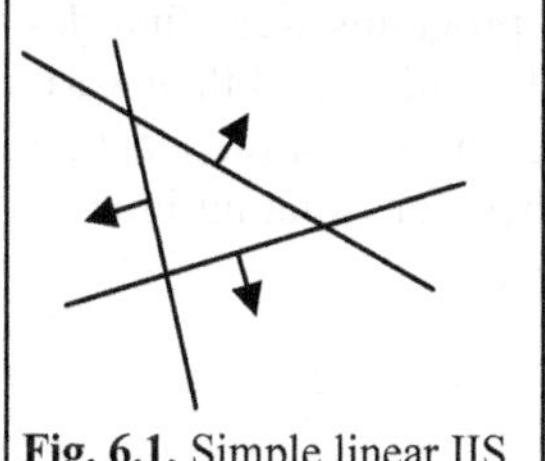

Fig. 6.1. Simple linear IIS

Greenberg (1992) performed an empirical comparison of three methods of LP infeasibility analysis and concluded that the isolation of IISs "performed consistently above midrange, and it never failed to provide useful information. It frequently gave an immediate diagnostic." See also Greenberg (1993) for further study of the value of isolating IISs during the diagnostic process.

After isolating an IIS, it can then be examined to see whether the model really is infeasible, or to determine which of the IIS members must be repaired. Human understanding of the model is necessary to make this decision. It may be that there are multiple infeasibilities in a model, hence IIS isolation is typically used in a cyclic manner: (1) isolate an IIS, (2) determine a repair for this IIS, (3) if the model is still infeasible, go to step (1).

Practical methods for isolating IISs in linear programs were developed during the 1990s and are now included in most commercial LP solvers. However it is still difficult to reliably isolate IISs in MIPs, NLPs, and other optimization model forms, mainly due to the difficulty in ascertaining the feasibility status of a set of constraints with perfect accuracy.

There are several practical issues related to IIS isolation. Many isolation methods require multiple solutions of slight variations of the original model, so speed can be an issue. For linear programs, basis re-use alleviates this problem to a great extent, fortunately, but it continues as a problem in other model forms. A second practical issue is how easy it is for the human modeler to understand the IIS that is isolated. The same infeasibility can sometimes be reflected in multiple different IISs, some of which are easier to understand than others. Heuristic methods are available to return IISs that are generally easier to understand, e.g. that have fewer complex row constraints and more column bounds.

IISs may also *overlap*, i.e. share at least one common constraint, or be organized into distinct *clusters* (Chinneck and Dravnieks 1991), i.e. maximal sets of IISs such that each IIS overlaps at least one IIS of the cluster. Where there are many overlapping IISs and clusters, it may prove useful to identify a minimal cardinality IIS set cover (i.e. the smallest set of constraints to remove such that the remainder constitute a feasible set), as described in Sec. 7.

The concept of infeasibility isolation is relatively old, though it was not developed to any extent until recently. Carver (1921) first mentions irreducibly inconsistent systems of linear constraints, but the first theorem on infeasible systems of linear constraints dates to Fourier (1827). Motzkin (1936) and Fan (1956) developed additional useful theorems for linear systems. Debrosse and Westerberg (1973) describe a procedure that can find minimal sets of infeasible constraints for nonlinear constraint sets under specific conditions. Van Loon (1981) describes a way to recognize minimal infeasible sets of linear constraints, but does not provide a method for finding them in an efficient manner. Greenberg (1987a) addressed the idea of searching for minimal substructures in infeasible LPs, but noted that how to find them was unclear.

Practical methods for isolating infeasibilities in linear programs were first developed by Chinneck and Dravnieks (1991). It is these algorithms and their variants that appear in modern LP solvers. Research continues at present on adapting these methods, and developing entirely different approaches, for isolating infeasibilities in other optimization forms such as MIPs and NLPs.

6.1 General Logical Methods

A number of the basic methods for isolating IISs do not depend on any properties of the optimization model itself. Instead they are purely logical, requiring nothing more than the ability to evaluate constraints or to determine whether a set or subset of constraints is feasible or infeasible. It is quite difficult to accurately determine the feasibility status of an arbitrary set of constraints in some model forms. In practice, we can rely on this ability only for sets of linear constraints. Numerical difficulties, usually related to the feasibility tolerances, can arise even for linear constraints, but this is fortunately relatively rare. Accurate assessment of the feasibility status for nonlinear programs can be quite difficult, and can even be problematic for mixed-integer programs.

For this reason, many of the general logical algorithms described in this Section are currently applicable only to linear systems. However there is hope that they will eventually be applicable to other classes of optimization models as better algorithms for accurately determining feasibility status in those other classes appear.

6.1.1 Logical Reduction of Models and Presolving

The usual goal of presolving or pre-processing an optimization model is to simplify it prior to applying a solution algorithm (Holder 2006). In the process of simplification, infeasibility may be discovered; hence this technique is of interest here. Standard presolving techniques include the removal of null constraints or variables, and the conversion of constraints containing a single variable into variable bounds that are possibly tighter than those currently in effect. This is usually followed by a series of bound tightening actions and simple tests to detect infeasibility.

The main idea in bound tightening is the detection and propagation of simple reductions, such as replacing a fixed variable with its value everywhere in the model, which may give rise to a cascade of further simplifications. Most bound tightening procedures are based on work by Brearly, Mitra, and Williams (1975). Consider the following simple example:

constraint row: $x_1 + x_2 \leq 10$

variable bounds: $4 \leq x_1 \leq 12,\ 7 \leq x_2 \leq 20$

The row bound is first tightened by substituting the variable bounds into it: the lower limit on the constraint row left-hand side is obtained when both of the variables are at their lower bounds: (4 + 7) = 11. This contradicts the constraint row right-hand side, so the conclusion is that the LP is infeasible. In general, the logical reduction can also work in the other direction, i.e. the tightening of the bound. Consider the upper bound on x_2 implied by the constraint row and the bounds on x_1: an upper bound on x_2 is obtained when x_1 is at its lower bound in the constraint row, implying an upper bound on x_2 of 6 (i.e. $4 + x_2 \leq 10 \Rightarrow x_2 \leq 6$). In general, this tightened bound might then propagate to other constraints, causing a sequence of further bound reductions.

Andersen and Andersen (1995) formalize bounds reduction by defining the upper and lower bounds on constraints as follows:

$$g_i = \sum_{j \in P_i} a_{ij} l_j + \sum_{j \in M_i} a_{ij} u_j$$

and

$$h_i = \sum_{j \in M_i} a_{ij} l_j + \sum_{j \in P_i} a_{ij} u_j$$

where $P_i = \{j | a_{ij} > 0\}$ (i.e. "plus" signs on the coefficients) and $M_i = \{j | a_{ij} < 0\}$ (i.e. "minus" signs on the coefficients). Given these definitions, we have

$$g_i \leq \sum_j a_{ij} x_j \leq h_i$$

for every constraint i.

The logical reduction of the model by such techniques is not limited to linear constraints: it can be used on nonlinear constraints as well. For example, the AMPL modeling system (Fourer et al. 2003) has a presolver that it applies to all model forms. Presolving can be especially effective in MIP models, in which case the bounds on integer variables can be rounded to integer values, introducing further tightening.

Bound tightening is not specifically designed for the analysis of infeasibility, so it can detect infeasibility only some of the time. Chinneck (1996a) shows that the presolver in a leading commercial LP solver is able to detect infeasibility in only 3 of 19 infeasible models tested. A similar conclusion about presolve procedures for analyzing infeasibility is reached by Andersen and Andersen (1995). When presolving does detect infeasibility, but provides a poor explanation of its cause, it is often a good idea to re-run the solution with the presolver turned off so that one of the more advanced IIS isolation routines described later in this chapter can run instead.

If infeasibility is detected during presolving, diagnosis may not be easy because a very long chain of reductions leading to the infeasibility may be reported. At a minimum, the rows and columns mentioned in the trace of the reductions can be used to provide an infeasible isolation, but it is often too large to be useful. However, bound reduction is available in many commercial solvers and modeling systems and may occasionally provide a useful analysis of infeasibility.

Here are the presolving inspections that can detect infeasibility in linear programs outlined by Andersen and Andersen (1995):

- If a row is empty (i.e. $a_{ij} = 0$ for $j = 1...n$ for some consraint i) but b_i is nonzero, then the row cannot be satisfied.
- If the bounds on a variable conflict (i.e. $l_j > u_j$) then these bounds cannot be satisfied.
- As described above, the bounds on the variables in a constraint may imply bounds on the constraint that conflict with the specified constraint bounds, in which case the constraint cannot be satisfied. As the model is reduced by various logical presolve operations, these revised bounds are checked for conflicts with the original bounds. The constraint cannot be satisfied if $h_i < b_i$ and the constraint is of the form $\boldsymbol{a}_i\boldsymbol{x} \geq b_i$ or $\boldsymbol{a}_i\boldsymbol{x} = b_i$, or if $b_i < g_i$ and the constraint is of the form $\boldsymbol{a}_i\boldsymbol{x} \leq b_i$ or $\boldsymbol{a}_i\boldsymbol{x} = b_i$.

A number of researchers have developed presolving techniques that can detect dual infeasibility (e.g. Andersen and Andersen (1995), Mészáros and Suhl (2003)). This is equivalent to primal unboundedness and hence is not considered further here.

As shown in Chap. 4, logical reduction is a main theme in constraint programming, under the name *constraint propagation*. The detection and analysis of infeasibility is also an important theme in constraint programming, a topic we will revisit in Sec. 6.5.

6.1.2 The Deletion Filter

Chinneck and Dravnieks (1991) introduced the *deletion filter*, shown in Algorithm 6.1. If the solver is able to accurately determine the feasibility status of an arbitrary set of constraints, then the deletion filter guarantees the identification of exactly one IIS after a single pass through the set of constraints. This is an essential property possessed by very few of the IIS isolation methods.

INPUT: an infeasible set of constraints.
FOR each constraint in the set:
1. Temporarily drop the constraint from the set.
2. Test the feasibility of the reduced set:
   IF feasible THEN return dropped constraint to the set.
   ELSE (infeasible) drop the constraint permanently.

OUTPUT: constraints constituting a single IIS.

Alg. 6.1. The deletion filter

Theorem 6.1: *Deletion filter functionality* (Chinneck 1997a). The deletion filter returns exactly one IIS.

Proof: The initial set of constraints input to the deletion filter is infeasible, and constraints are removed only when the reduced set of constraints remains infeasible without them. The only constraints retained in the set are those whose removal renders the set feasible. Hence these must be members of an IIS, by definition. There is only a single IIS because if there were two or more IISs, you would be able to remove at least one constraint from the set and it would remain infeasible. ■

The main idea of the deletion filter is to remove constraints from the set one at a time. If a constraint is removed, and the remainder of the model is still infeasible, then the constraint is not necessary to the infeasibility and can be removed permanently. On the other hand, if a constraint is removed and the model becomes feasible, then that constraint is necessary to the infeasibility and is replaced. A simple example illustrates the action of the algorithm: consider a set of constraints {A,B,C,D,E,F,G} which contains the embedded IIS consisting of the constraints {B,D,F}. The members of the IIS are shown in boldface below. The deletion filter considers the effect of dropping each constraint:

1. Remove A: {**B**,C,**D**,E,**F**,G} infeasible. A deleted permanently.
2. Remove **B**: {C,**D**,E,**F**,G} feasible. **B** reinstated.
3. Remove C: {**B**,**D**,E,**F**,G} infeasible. C deleted permanently.
4. Remove **D**: {**B**,E,**F**,G} feasible. **D** reinstated.
5. Remove E: {**B**,**D**,**F**,G} infeasible. E deleted permanently.
6. Remove **F**: {**B**,**D**,G} feasible. **F** reinstated.
7. Remove G: {**B**,**D**,**F**} infeasible. G deleted permanently.
8. Output: the IIS {**B**,**D**,**F**}

Where there are several IISs in the model, exactly one is returned because the testing set will remain infeasible when elements of any other IISs are removed.

Which IIS is returned depends on the order in which the constraints are tested. The deletion filter always returns the IIS whose *first* member is tested *last* (because the testing set remains infeasible to this point).

Theorem 6.2: *Deletion filter IIS selection* (Chinneck 1997a). The deletion filter returns the IIS whose first member is tested last.

Proof: As long as one IIS remains intact in the set of constraints, then test constraints are dropped permanently because the set remains infeasible. If there are several IISs in the model, then the IIS whose first member is tested last remains intact while members of the other IISs are tested. Since the set remains infeasible while constraints from other IISs are tested, those constraints are eliminated. Thus the IIS whose first member is tested last is isolated. ■

6.1.3 The Additive Method

The additive method is the opposite of the deletion filter: starting with an empty set of constraints, constraints are added until infeasibility is triggered, which shows that the last added constraint is involved in the infeasibility. The testing set of constraints is then primed by emptying it of all constraints except those that have been implicated in this manner and the process repeats.

Tamiz et al. (1995, 1996) introduced the additive method to the optimization community, though it was discovered earlier in the constraint programming community (see Sec. 6.5). Tamiz et al. named it GPIIS because they conceived it based on methods from LP goal programming. Their development uses "deviational variables" (equivalent to elastic variables, see Sec. 6.1.4) and an elastic objective function to decide feasibility status of the intermediate test sets of constraints. This apparatus is not necessary, though, because the main feature of the method is the adding in of constraints as the algorithm proceeds and the testing of the feasibility of the resulting set. Alg. 6.2 shows the simpler version of the algorithm without the elastic variables and elastic objective function.

C: ordered set of constraints in the infeasible model.
T: the current test set of constraints.
I: the set of IIS members identified so far.

INPUT: an infeasible set of constraints C.
0. $T \leftarrow \varnothing, I \leftarrow \varnothing$.
1. $T \leftarrow I$.
    FOR each constraint c_i in C:
        $T \leftarrow T \cup c_i$.
        IF T infeasible THEN
            $I \leftarrow I \cup c_i$.
            Go to Step 2.
2. IF I feasible THEN go to Step 1.
    Exit.
OUTPUT: I is an IIS.

Alg. 6.2. The additive method

A simple example will illustrate the workings of the algorithm. Consider an IIS {B,D,F} embedded in {A,B,C,D,E,F,G}. The members of the IIS are shown in boldface:

1. {A}, {A,**B**}, {A,**B**,C}, {A,**B**,C,**D**}, {A,**B**,C,**D**,E} all feasible.
2. {A,**B**,C,**D**,E,**F**} infeasible: I = {**F**} is feasible.
3. {**F**,A}, {**F**,A,**B**}, {**F**,A,**B**,C} all feasible.
4. {**F**,A,**B**,C,**D**} infeasible: I = {**F**,**D**} is feasible.
5. {**F**,**D**,A} feasible.
6. {**F**,**D**,A,**B**} infeasible: I = {**F**,**D**,**B**} is infeasible. Stop.
7. Output: the IIS {**F**,**B**,**D**}

The additive method also guarantees the identification of a single IIS, even when several are present in the original infeasible set.

Theorem 6.3: *Additive algorithm* (Chinneck 1997a). The additive algorithm returns a single IIS.

Proof: Constraints are added to I only when their addition to T changes its status from feasible to infeasible (i.e. at least one complete IIS is in T), thus each constraint added to I must be part of all of the IISs just created in T. Assume c_k has just been added to I. Because T is "primed" with I at the beginning of Step 1, the FOR loop in Step 1 will not proceed beyond c_k-1 because at that point T would be identical to the previous T and hence infeasible, causing exit from the loop (and the addition of c_k-1 to I). Thus the maximum number of iterations of the FOR loop decreases by 1 each time it is entered. In fact, the FOR loop in Step 1 is exited as soon as a complete IIS is in T, which may happen well before c_k-1.

When there is a single IIS in T, each subsequent constraint added to I is a member of the same IIS, hence Step 2 will eventually cause the algorithm to exit with I containing a single IIS. When there is more than one IIS in the current T, the subsequent pass through Step 1 will either (a) identify another element common to all of the multiple IISs in T, or (b) complete one of the IISs before the others, eliminating some of the elements of the other IISs from further consideration because they are past the current c_i in the list, thereby also eliminating the other IISs themselves from eventual output. Thus T eventually contains only a single IIS, which will be recognized and output by Step 2. ∎

Theorem 6.4: *Additive method IIS selection* (Chinneck 1997a). The additive method isolates the IIS whose last member is tested first.

Proof: The loop in Step 1 of Alg. 6.2 is exited the first time that T becomes infeasible or equivalently, the first time that a complete IIS is in T. Thus, while parts of various IISs may be added to T as it builds up, the process exits only when the last member of any IIS is added to T. Therefore the IIS whose last member is tested first is isolated. ∎

Theorem 6.4 shows that, as for the deletion filter, which IIS is isolated by the additive method is affected by the ordering of the constraints. The advantage of the additive method is that it may require fewer tests of feasibility, and tests of smaller sets of constraints, especially when the IIS that is isolated is small compared to the cardinality of C.

Guieu and Chinneck (1999) introduced an improved variant of the basic additive method, called the *dynamic reordering additive method.* The main insight is that some of the feasibility tests can be avoided as follows: if an intermediate test subproblem is feasible, then scan all of the constraints past the current constraint just added, and add to T all constraints that are satisfied by the current solution point. The original version by Guieu and Chinneck (1999) was specific to the analysis of infeasible MIPs. A generalized and improved version of the algorithm appears in Alg. 6.3.

First note that the main effect of the algorithm is nothing more than a dynamic reordering of the constraints. If the constraints had been ordered originally as they are after the dynamic reordering (i.e. if the constraints had been ordered as $\ldots c_i$, *temp*..., then obviously all of the sets tested from c_i through the last constraint in *temp* would have been feasible (we know this because the current point is feasible for all of the constraints in *temp*). Thus Theorem 6.3 continues to hold.

There are two main efficiency improvements in Alg. 6.3. First, the dynamic reordering eliminates some feasibility tests, and makes it much more probable that a given feasibility test will result in infeasibility. This is good because every infeasible outcome identifies another member of the IIS. The second efficiency improvement comes from the truncation of the set C that occurs after an infeasible

C: ordered set of constraints in the infeasible model.
T: the current test set of constraints.
I: the set of IIS members identified so far.

INPUT: an infeasible set of constraints C.

0. $T \leftarrow \varnothing$, $I \leftarrow \varnothing$.
1. $T \leftarrow I$.
   FOR each constraint c_i in C:
      $T \leftarrow T \cup c_i$.
      IF T infeasible THEN
         $I \leftarrow I \cup c_i$.
         $C \leftarrow C \backslash \{c_k \mid k > i\}$
         Go to Step 2.
      ELSE
         $temp \leftarrow \{c_k \mid k > i, c_k \text{ satisfied at current point}\}$
         Reorder C by inserting temp just after c_k.
         $T \leftarrow T \cup temp$.
         $i \leftarrow i + |temp| + 1$
2. IF I feasible THEN go to Step 1.
   Exit.

OUTPUT: I is an IIS.

Alg. 6.3. The dynamic reordering additive method

outcome of a feasibility test. The truncation means that the algorithm does not have to check constraints for satisfaction beyond the constraint that just triggered feasibility. Constraints beyond the c_i that triggered infeasibility are obviously not part of the IIS: it is completely contained in the current infeasible T, hence those later constraints can be safely ignored. Note that the truncation is not necessary in the basic additive method in Alg. 6.2 because no constraints beyond the c_i that triggered infeasibility are ever added to T.

While there is some cost associated with checking whether constraints farther along in the list are satisfied at the current solution point, this is usually negligible compared to the cost of conducting another feasibility test.

Finally, there is one additional small efficiency improvement that is omitted from the algorithm statements for clarity. Under most circumstances, the first feasibility test is only conducted when two constraints have been added since at least two are needed to cause infeasibility. However this may not be true for certain nonlinear constraints.

6.1.4 The Elastic Filter

Useful information about an infeasible model can be obtained if the constraints can be violated in a graceful manner. For example, in the familiar linear programming Phase 1 procedure, nonnegative *artificial variables* (a_i) are added to all equality and ≥ constraints (see e.g. Winston and Venkataramanan 2003), which allows those constraints to be violated so that an initial basic "feasible" solution can be established. This initial solution is feasible in the space consisting of the original plus artificial variables, but not in the space consisting of just the original variables. The LP Phase 1 objective is to minimize the sum of the artificial variables, i.e. minimize $W = \sum a_i$, via standard linear programming (see Sec. 2.1).

If W reaches a minimum value of zero, then all of the artificial variables are themselves zero, hence a feasible solution has been found for the original model, and the LP solution now proceeds to Phase 2, the solution of that original model. If the minimum value of W is not zero, then at least one of the artificial variables cannot be forced to zero, so the corresponding constraint remains violated in the original variable space, and the LP as a whole is determined to be infeasible.

Viewed in the space of the original variables, the linear equality and ≥ constraints are able to *stretch*, or violate their original bounds: the value of the associated artificial variable corresponds directly to the size of the adjustment of the right hand side needed to provide a feasible solution in the original variable space.

This idea can be extended to allow all forms of constraints to adjust in all directions, as originally described by Brown and Graves (1975). A fully elastic program adds a nonnegative elastic variable (or variables) s_i (or s_i' and s_i'') to every constraint. This allows a solver to find a "feasible" solution for the original infeasible model. The rules for adding elastic variables are as follows:

nonelastic constraint	*elastic version*
$\Sigma_j a_{ij} x_j \geq b_i$	$\Sigma_j a_{ij} x_j + s_i \geq b_i$
$\Sigma_j a_{ij} x_j \leq b_i$	$\Sigma_j a_{ij} x_j - s_i \leq b_i$
$\Sigma_j a_{ij} x_j = b_i$	$\Sigma_j a_{ij} x_j + s_i' - s_i'' = b_i$

An elastic constraint "stretches" (violates its original bounds) when one of its elastic variables takes on a positive value. Stretching is resisted by the elastic objective function (minimize $\Sigma_i s_i$) which replaces the original objective function. This is similar to a textbook phase 1, except that elastic variables are added to all constraints, and equality constraints are elasticized in both directions. Note that integer restrictions cannot be elasticized, so elastic filtering can be applied only to LPs, NLPs, and the linear part of MIPs. In this sense it is slightly less general than the deletion filter and the additive method.

The elastic filter (Chinneck and Dravnieks 1991) makes extensive use of elastic programming. All constraints are initially elasticized, but since the original model is infeasible, at least one constraint must stretch to achieve a feasible solution for the elastic program. The elastic variables are removed from any constraints that stretch; this *enforces* the constraint in the next round. The cycle repeats until enough elastic variables have been removed that the partly-elastic model becomes infeasible. At this point the de-elasticized constraints constitute a small infeasible set that is not necessarily an IIS, but that has some very desirable properties. The details of the algorithm are shown in Alg. 6.4.

INPUT: an infeasible set of constraints.
1. Make all constraints elastic by incorporating nonnegative elastic variables s_i.
2. Solve the model using the elastic objective function.
3. IF feasible THEN
    Enforce the constraints in which any $s_i > 0$ by permanently removing their elastic variable(s).
    Go to step 2.
ELSE (infeasible)
    Exit.
OUTPUT: the set of de-elasticized enforced constraints contains at least one IIS.

Alg. 6.4. The elastic filter

As described in Alg. 6.4, the elastic filter identifies constraints which must be part of some IIS because they have stretched. Because these stretched constraints are then de-elasticized, some other member of the IIS must stretch in the next iteration. The process halts when all of the members of at least one IIS have been enforced, which renders the partly-elastic model infeasible. The deletion filter or additive method can then be applied to the output set to identify a single IIS. We assume that the solver is perfectly accurate in minimizing the elastic objective function. In practice this currently limits the application of the elastic filter to linear programs since we cannot guarantee to find the global minimum of an NLP and we cannot elasticize the integer restrictions in a MIP.

Consider again the small example in which the IIS {B,D,F} appears in the set of constraints {A,B,C,D,E,F,G}. Let us assume that our hypothetical solver stretches just one constraint in the IIS at each iteration. The members of the IIS are shown in boldface for clarity, and elasticized constraints are underscored:

1. {A,**B**,C,**D**,E,**F**,G} is feasible, B stretches, so is de-elasticized.
2. {A,**B**,C,**D**,E,**F**,G} is feasible, F stretches, so is de-elasticized.
3. {A,**B**,C,**D**,E,**F**,G} is feasible, D stretches, so is de-elasticized.
4. {A,**B**,C,**D**,E,**F**,G} is infeasible.
5. Output: the set {**B**,**F**,**D**}.

Note that the output is not necessarily an IIS; a single IIS must be identified by applying the deletion filter or the additive method to the output.

Lemma 6.5: *Elastic stretching 1* (Chinneck 1997a). Each elastic program (or partly elastic program) in Alg. 6.4 stretches only elastic constraints which belong to an IIS.

Proof: The cost of stretching a constraint is strictly positive when the elastic objective function is used, and only IIS constraints need to be stretched to achieve a feasible solution for any elastic (or partly elastic) program, hence only constraints belonging to an IIS will stretch. ■

Lemma 6.6: *Elastic stretching 2* (Chinneck 1997a). The elastic filter will stretch at least one previously unstretched elastic constraint from each IIS in the current constraint set at each iteration of Step 2 of Alg. 6.4.

Proof: In Step 2 of Alg. 6.4, the solver must stretch at least one elastic member of each IIS to achieve a feasible solution, otherwise the algorithm exits during Step 3. By Lemma 6.5, any stretched constraint will be a member of an IIS, and because any stretched constraints are enforced during Step 3, the stretched constraint will not have stretched previously. ■

Theorem 6.7: *Elastic filter termination* (Chinneck 1997a). The output set of Alg. 6.4 will contain at least one IIS.

Proof: By Lemma 6.6, Alg. 6.4 will not terminate prematurely and will add at least one constraint to the output set at each iteration. Since each IIS is composed of a finite number of members, Alg. 6.4 will terminate in a finite number of steps, when all of the members of at least one IIS have been enforced, creating an infeasible LP which will be detected during Step 3, causing exit. Because this output set is infeasible, it must contain at least one IIS. ■

Note that the output set may contain more than one IIS, and may also contain partial IISs. The deletion filter or the additive method must be applied to the output set to guarantee the isolation of a single IIS.

Theorem 6.8: *Elastic filter iterations* (Chinneck 1997a). The number of elastic filter iterations (i.e. elastic programs solved in Step 2 in Alg. 6.4) is at most equal to the cardinality of the smallest-cardinality IIS in the input set.

Proof: By Lemma 6.6, at least one constraint from each IIS is stretched at each iteration of Step 2 of Alg. 6.4. Let k be the cardinality of the smallest-cardinality IIS in the input set. Then in at most k iterations of Step 2, all members of the smallest-cardinality IIS will have been enforced, rendering the partly-elastic model infeasible and causing exit during Step 3. ■

The practical significance of the elastic filter derives from Theorem 6.8, which provides a lower bound on the size of the smallest IIS in the input set. Assuming that exactly one member of each IIS is stretched during each iteration of the elastic filter, its output set will contain a smallest-cardinality IIS. This is a very desirable property because small cardinality IISs are much easier for humans to diagnose. While this assumption does not hold in general, it does hold quite often in practice, so the elastic filter provides a very good heuristic for isolating small-cardinality IISs. We will return to the use of the elastic filter to find useful infeasibility isolations in a later chapter.

6.1.5 Speed-ups: Treating Constraints in Groups

The basic versions of both the deletion filter and the additive method treat the constraints in the model one by one. However Chinneck (1995) suggested that the speeds of both algorithms can be improved by considering constraints in groups and this idea was implemented for the analysis of infeasible MIPs by Guieu and Chinneck (1999). For example, during deletion filtering, constraints could be dropped in groups of size k. If the reduced model remains infeasible, then there is a savings of $k-1$ feasibility tests. If dropping the group causes feasibility, then reinstate all k constraints and repeat the deletion filter over that set of k constraints dropping individual constraints one by one; this results in 1 extra feasibility test compared to the basic method. The efficiency of the method depends on how often the model remains infeasible after dropping a group vs. how often dropping a group results in feasibility. The choice of group size is obviously important.

A similar idea applies in the case of the additive method, except that the trigger for repeating the analysis with a group size of 1 is that the addition of a group causes infeasibility.

Guieu and Chinneck (1999) looked at several variations of grouping, as listed below for the case of deletion filtering of MIPs. In all cases below, when the test subset is feasible, the algorithm backtracks and re-tests the individual constraints in the group. The group size is then reset as shown for the next group test. k is the group size.

- *Fixed Group Size.* k is fixed by user.
- *Additive Adaptive Grouping A.*
  - Set $k = 2$.
  - IF test subset is infeasible THEN $k \leftarrow k + 2$.
  - ELSE $k = \text{maximum}[k-2, 1]$
- *Additive Adaptive Grouping B.*
  - Set $k = 2$.
  - IF test subset is infeasible THEN $k \leftarrow k + 2$.
  - ELSE $k = 2$.
- *Multiplicative Adaptive Grouping A.*
  - Set $k = 1$.
  - IF test subset is infeasible THEN $k \leftarrow k \times 2$.
  - ELSE $k = \text{maximum}[\text{integer}(k/2), 1]$.

- *Multiplicative Adaptive Grouping B.*
  - Set $k = 1$.
  - IF test subset is infeasible THEN $k \leftarrow k \times 2$.
  - ELSE $k = 1$.

For the specific MIP experiments in (Guieu and Chinneck 1999), the most efficient grouping algorithm proved to be a fixed group size of $k = 4$. Further experimentation is needed to determine whether this choice is a good general choice, or whether other grouping algorithms or sizes are better for models of other optimization types or with specific characteristics.

As suggested by Guieu and Chinneck (1999), Atlihan and Schrage (2006) use binary search to generalize the grouping idea. M is the original set of inconsistent constraints. As the algorithms proceed, the constraints are divided into several subsets:

- I : set of constraints already shown to be in the IIS.
- D : set of constraints that contains at least one IIS member.
- S : set of constraints that is likely to contain an IIS member (though S may contain no IIS members).
- R : set of removed constraints. The constraints in R are definitely not in the IIS.

Elastic programming is used extensively to assign the constraints to the different subsets. For example, the set of stretched constraints after an elastic solution helps identify constraints that are definitely or likely to be part of the IIS that is eventually isolated.

INPUT: an infeasible set of constraints M.
0. $T = M; I = R = S = \varnothing$.
1. IF $|T| \leq 1$ THEN:
   1.1. $I \leftarrow I + T$.
   1.2. IF I is infeasible THEN exit.
   1.3. $T \leftarrow S; S \leftarrow \varnothing$.
   1.4. IF $|T| \geq 2$ THEN go to Step 1.
   1.5. $T_2 \leftarrow T; T_1 \leftarrow \varnothing$.
   ELSE
   1.6. Split T into T_1 and T_2.
2. IF $\{I + S + T_1\}$ is feasible THEN:
   2.1. $S \leftarrow S + T_1$
   2.2. $T \leftarrow T_2$
   ELSE
   2.3. $R \leftarrow R + T_2$
   2.4. $T \leftarrow T_1$
3. Go to Step 1.

OUTPUT: I is an IIS.

Alg. 6.5. The depth first binary search filter

Atlihan and Schrage's *Depth First Binary Search Filter* (*DFBS*) uses a dynamic group size, the simplest form of which drops half of the constraints remaining in the set that is known to contain at least one constraint belonging to the IIS that is being isolated. The general DFBS algorithm is given in Alg. 6.5.

The crux of the algorithm is Step 1.6 where the testing set is subdivided. The simplest subdivision algorithm is to divide the set in half. We will illustrate the working of the algorithm using binary subdivision (where the set cannot be equally subdivided we make T_1 the larger set), and a small example in which the IIS {B,D,F} appears in the set of constraints {A,B,C,D,E,F,G,H}. The members of the IIS are shown in boldface:

- T= {A,**B**,C,**D**,E,**F**,G,H}; $I = R = S = \varnothing$.
- Split T into T_1= {A,**B**,C,**D**} and T_2= {E,**F**,G,H}.
- $\{I+S+T_1\}$ = {A,**B**,C,**D**} is feasible, so $S\leftarrow${A,**B**,C,**D**}, and $T\leftarrow${E,**F**,G,H}.
- Split T into T_1= {E,**F**} and T_2= {G,H}.
- $\{I+S+T_1\}$ = {A,**B**,C,**D**,E,**F**} is infeasible, so $R\leftarrow${G,H}, and $T\leftarrow${E,**F**}.
- Split T into T_1= {E} and T_2= {**F**}.
- $\{I+S+T_1\}$ = {A,**B**,C,**D**,E} is feasible, so $S\leftarrow${A,**B**,C,**D**,E}, and $T\leftarrow${**F**}.
- $|T| \leq 1$, so $I\leftarrow${**F**}.
- I feasible, so $T\leftarrow${A,**B**,C,**D**,E}; $S\leftarrow\varnothing$.
- Split T into T_1= {A,**B**,C} and T_2= {**D**,E}.
- $\{I+S+T_1\}$={**F**,A,**B**,C} is feasible, so $S\leftarrow${A,**B**,C}, and $T\leftarrow${**D**,E}.
- Split T into T_1= {**D**} and T_2= {E}.
- $\{I+S+T_1\}$={**F**,A,**B**,C,**D**} is infeasible, so $R\leftarrow${G,H,E}, and $T\leftarrow${**D**}.
- $|T| \leq 1$, so $I\leftarrow${**F**,**D**}.
- I feasible, so $T\leftarrow${A,**B**,C}; $S\leftarrow\varnothing$.
- Split T into T_1= {A,**B**} and T_2= {C}.
- $\{I+S+T_1\}$={**F**,**D**,A,**B**} is infeasible, so $R\leftarrow${G,H,E,C}, and $T\leftarrow${A,**B**}.
- Split T into T_1= {A} and T_2= {**B**}.
- $\{I+S+T_1\}$ = {**F**,**D**,A} is feasible, so $S\leftarrow${A}, and $T\leftarrow${**B**}.
- $|T| \leq 1$, so $I\leftarrow${**F**,**D**,**B**}.
- I is infeasible, so exit with I = {**F**,**D**,**B**} as the output IIS.

As this example shows, the DFBS algorithm has characteristics of both the deletion filter and the additive method, both with grouping. If the subset $\{I+S+T_1\}$ tested in Step 2 is feasible, then constraints are added to the testing set S in Step 2.1, but if $\{I+S+T_1\}$ is infeasible, then constraints are permanently deleted in Step 2.3. In both cases, half of the constraints in the testing set T are either added or deleted.

Note that the IIS isolation via the DFBS algorithm requires 10 feasibility tests in this small example vs. the 8 that would be required by a straight deletion filter or the 15 required by a straight additive method. This is because the IIS constitutes a large portion of the model and is well distributed within the list of constraints. In a larger model where the proportion of IIS members is smaller and the members closer together in the list, the DFBS algorithm can be much more efficient. Atlihan and Schrage (2006) show that if there are k constraints in the IIS that is isolated, then DFBS requires fewer feasibility tests than a deletion filter if $k \cdot \log_2(|M|) < |M|$, or $k < |M| / \log_2(|M|)$, though of course k cannot be known in advance. They also introduce a few simple modifications to the splitting rules based on the relative size of $|I|$ and $|M|$ to improve this value somewhat.

Atlihan and Schrage (2006) also introduce another constraint grouping algorithm called the *Generalized Binary Search Filter* (*GBF*). The main idea here is that the stretched constraints after an elastic solution are known to be involved in some IIS, hence it is better to focus attention on those stretched constraints while deleting other constraints if possible. A subset of constraints that includes a stretched constraint then becomes the focus of a binary search. The details are shown in Alg. 6.6.

D_i: set that contains at least one IIS member.

INPUT : an infeasible set of constraints M.
0. $k = 0$; $S = M$; $R = \emptyset$; $D_i = \emptyset$; for all $i \leq k$.
1. IF $S = \emptyset$ and $|D_i| = 1$ for all $i \leq k$ THEN
    1.1. $I = D_1 \cup D_2 \cup \ldots D_k$; exit.
2. IF $|S| = 1$ and $|D_i| = 1$ for all $i \leq k$ THEN
    2.1. $T \leftarrow S$; $T_2 \leftarrow T$; $T_1 \leftarrow \emptyset$.
   ELSE
    2.2. $T \leftarrow S$ or $T \leftarrow D_i \in \{D_1, \ldots, D_k\}$ such that $|T| \geq 2$.
    2.3. Split T into T_1 and T_2.
3. Solve elastic program consisting of the constraints $M \backslash R$ in which the constraints in T_2 are elasticized.
4. IF feasible THEN
    4.1. Form the set T_3 as a subset of the set of stretched constraints.
    4.2. IF $T = S$ THEN
        4.2.1. $k \leftarrow k + 1$; $D_k \leftarrow T_3$; $S \leftarrow S \backslash T_3$.
      ELSE
        4.2.2. $D_k \leftarrow T_3$; $S \leftarrow S \cup T \backslash T_3$.
   ELSE
    4.3 IF $T = S$ THEN
        4.3.1. $R \leftarrow R \cup T_2$; $S \leftarrow S \backslash T_2$.
      ELSE
        4.3.2 $D_k \leftarrow T_1$; $R \leftarrow R \cup T_2$.
5. Go to Step 1.
OUTPUT: I is an IIS.

Alg. 6.6. The generalized binary search filter

To illustrate the workings of the algorithm, consider the IIS {B,D,F} that appears in the set of constraints {A,B,C,D,E,F,G,H}. As before, when called to split the set T into two subsets T_1 and T_2, we do so equally, or give T_1 the extra member in case $|T|$ is odd. The members of the IIS are shown in boldface and elasticized constraints are underlined:

- $M = \{A,\mathbf{B},C,\mathbf{D},E,\mathbf{F},G,H\}$.
- $k = 0$; $S = \{A,\mathbf{B},C,\mathbf{D},E,\mathbf{F},G,H\}$; $R = \varnothing$; $D_i = \varnothing$ for all $i \le k$.
- $T = \{A,\mathbf{B},C,\mathbf{D},E,\mathbf{F},G,H\}$; $T_1 = \{A,\mathbf{B},C,\mathbf{D}\}$; $T_2 = \{E,\mathbf{F},G,H\}$.
- Solve elastic program $\{A,\mathbf{B},C,\mathbf{D},\underline{E},\underline{\mathbf{F}},\underline{G},\underline{H}\}$: feasible, **F** stretches.
- $T_3 = \{\mathbf{F}\}$.
- $T = S$ so $k=1$; $D_1 = \{\mathbf{F}\}$; $S = \{A,\mathbf{B},C,\mathbf{D},E,G,H\}$.
- $T = \{A,\mathbf{B},C,\mathbf{D},E,G,H\}$; $T_1 = \{A,\mathbf{B},C,\mathbf{D}\}$; $T_2 = \{E,G,H\}$.
- Solve elastic program $\{A,\mathbf{B},C,\mathbf{D},\underline{E},\mathbf{F},\underline{G},\underline{H}\}$: infeasible.
- $T = S$ so $R = \{E,G,H\}$; $S = \{A,\mathbf{B},C,\mathbf{D}\}$.
- $T = \{A,\mathbf{B},C,\mathbf{D}\}$; $T_1 = \{A,\mathbf{B}\}$; $T_2 = \{C,\mathbf{D}\}$.
- Solve elastic program $\{A,\mathbf{B},\underline{C},\underline{\mathbf{D}},\mathbf{F}\}$: feasible, **D** stretches.
- $T_3 = \{\mathbf{D}\}$.
- $T = S$ so $k = 2$; $D_2 = \{\mathbf{D}\}$; $S = \{A,\mathbf{B},C\}$.
- $T = \{A,\mathbf{B},C\}$; $T_1 = \{A,\mathbf{B}\}$; $T_2 = \{C\}$.
- Solve elastic program $\{A,\mathbf{B},\underline{C},\mathbf{D},\mathbf{F}\}$: infeasible.
- $T = S$ so $R = \{C,E,G,H\}$; $S = \{A,\mathbf{B}\}$.
- $T = \{A,\mathbf{B}\}$; $T_1 = \{A\}$; $T_2 = \{\mathbf{B}\}$.
- Solve elastic program $\{A,\underline{\mathbf{B}},\mathbf{D},\mathbf{F}\}$: feasible, **B** stretches.
- $T_3 = \{\mathbf{B}\}$.
- $T = S$ so $k = 3$; $D_3 = \{\mathbf{B}\}$; $S = \{A\}$.
- $|S| = |D_1| = |D_2| = |D_3| = 1$ so $T = \{A\}$; $T_2 = \{A\}$; $T_1 = \varnothing$.
- Solve elastic program $\{\underline{A},\mathbf{B},\mathbf{D},\mathbf{F}\}$: infeasible.
- $T = S$ so $R = \{A,C,E,G,H\}$; $S = \varnothing$.
- $S = \varnothing$ and $|D_1| = |D_2| = |D_3| = 1$ so $I = \{\mathbf{F},\mathbf{D},\mathbf{B}\}$ and exit.

The GBS algorithm solves 6 elastic programs en route to finding the IIS, as compared to a maximum of 3 elastic programs and 3 deletion filter iterations for the usual elastic filter followed by deletion filter. Again, the relative efficiency depends on the size and placement of the IIS in the set of constraints.

To better understand how the GBS algorithm operates, look at the sequence of elastic programs solved in the small example. This shows how the binary search gradually identifies elements that must be part of the IIS.

There are various possibilities for selecting T in step 2.2. T could be selected as the subset of largest cardinality, which leads to many subsets D_i. Because the number of subsets k is a lower bound on the cardinality of the IIS being isolated, this gives a more accurate (i.e. higher) lower bound earlier.

In both the DFBS and GBS algorithms there are also various ways to subdivide T into T_1 and T_2 where required. This can be done in a straightforward binary manner as in the worked examples, or randomly. In GBS it can also be done based on criteria such as the number of times a constraint has been previously stretched (e.g. if a constraint has stretched relatively frequently, then we may assign it to T_2 to encourage an infeasible result and the augmentation of the set R of removed constraints).

Atlihan and Schrage (2006) show empirically that the DFBS and especially the GBS grouping strategies can be very effective, particularly for nonlinear and mixed-integer programs. They are less advantageous for linear programming where basis re-use is more of a factor in overall speed; in general they require fewer feasibility tests, but they take longer because the next feasibility test is sufficiently different from the last one that the previous basis is not very useful in providing an advanced start.

6.1.6 Speed-ups: Combining the Additive Method and the Deletion Filter

The three basic methods described so far can be combined in several ways to produce faster and more effective isolation methods. We have already seen how all three can be combined in the Generalized Binary Search algorithm (Alg. 6.6). The same principle will hold later when we introduce methods specialized for different classes of mathematical programs such as LPs.

As suggested by Guieu and Chinneck (1999), the additive and deletion methods are very easily combined: simply run the additive method until feasibility is first detected and then change to the deletion filter for the final IIS isolation. Details of the additive/deletion algorithm are given in Alg. 6.7.

C: ordered set of constraints in the infeasible model.
T: test set of constraints.

INPUT: an infeasible set of constraints C.
0. Set $T = \varnothing$.
1. FOR each constraint c_i in C:
    Set $T = T \cup c_i$.
    IF T infeasible THEN go to Step 2.
2. FOR each constraint t_i to $t_{|T|-1}$ in T:
    Temporarily drop the constraint t_i:
    Test the feasibility of the reduced set:
        IF feasible THEN return dropped constraint to T.
        ELSE (infeasible) $T \leftarrow T \setminus t_i$.
OUTPUT: T is an IIS.

Alg. 6.7. The additive/deletion method

Note that the deletion filter in Step 2 of Alg. 6.7 does not test the final constraint in *T*. It is already known that this constraint is part of the IIS being isolated since it triggered infeasibility during the additive method hence there is no need to test it. The worst-case time complexity of the additive/deletion method occurs when *C* itself constitutes an IIS. In this case there will be m additive method feasibility tests followed by $m-1$ deletion filter feasibility tests for a total of $2m-1$ feasibility tests. In practice, efficiency is affected by the location of the last IIS constraint in *C*: if it occurs early in *C* then the additive/deletion method can be significantly more efficient than either the deletion filter or the additive method by itself.

6.1.7 Sampling Methods

Feng (1999) describes a method for isolating IISs based on sampling the solution space randomly. A simple example in Fig. 6.2 illustrates the concept. There are 4 linear inequalities A through D, and a set of 10 sample points *a* through *j*. A 4-tuple is associated with each sample point. Each entry in the 4-tuple is a binary value that indicates whether the associated constraint is satisfied (value is 0) or violated (value is 1). A set covering matrix is constructed from the complete set of tuples associated with the sample points.

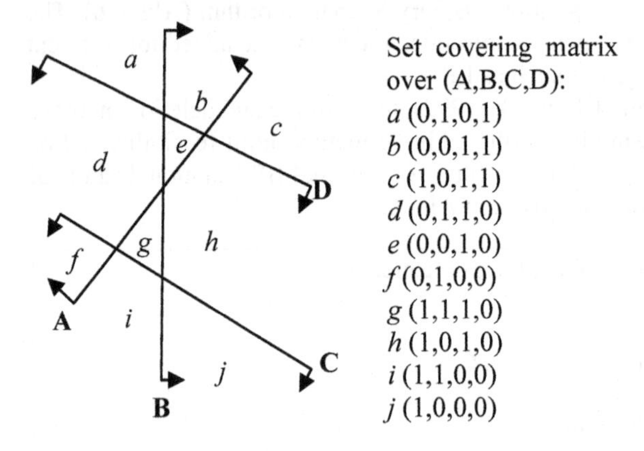

Fig. 6.2. Identifying an IIS by sampling

An IIS is found by solving the associated set covering problem: find the smallest cardinality set of constraints (columns in the set covering matrix) such that all of the tuples are covered. The solution to this set covering problem yields the smallest cardinality IIS under certain conditions. It is easy to verify by inspection in this small example that (i) the smallest cardinality set cover is {A,B,C}, and that (ii) this is the only IIS in the model. If the sampling points provide good coverage of the sample space, then the associated set covering problem is really answering this question: what is the smallest set of constraints such that at least one of them is violated at every point in the solution space? The effectiveness of the method is of course very dependent on how well the sample points cover the solution space.

Sampling methods have a number of major limitations that restrict their use in practice. First, the result returned very much depends on whether a suitable subset of the relevant subspaces was sampled. For instance, if the sample point *j* were missed in the example, then the smallest cardinality set cover returned would be {B,C}, which is clearly not an IIS. This implies that a great number of sample points are needed, though steps can be taken to reduce the number. Second, the method is restricted to models composed entirely of inequalities and cannot handle equality constraints, or even implied equalities generated by inequalities. Third, the method is restricted to convex constraints, a condition that may be difficult to verify in practice. Third, the resulting set covering matrix may be very large, and hence difficult to solve. For this reason, the method has not been used in practice, though it is of theoretical interest.

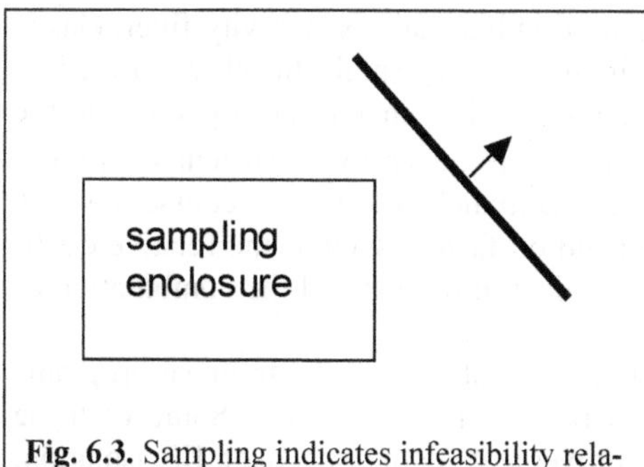

Fig. 6.3. Sampling indicates infeasibility relative to constraint

As described in Sec. 5.2, the MProbe software (Chinneck 2001, 2002) samples the functions defining an optimization model within either a box enclosure, or a subset of the constraints that defines a convex envelope. One of the measures returned for each constraint is its *effectiveness*, defined as the fraction of sample points that violate the constraint. If an inequality constraint reports an effectiveness of 1.0 then no sample points satisfied the constraint, and the model is most likely infeasible; see Fig. 6.3. For equality constraints, a constraint effectiveness of 1.0 is returned in either of two cases: (i) if the function value is greater than the right-hand side constant at all sample points, or (ii) the function value is less than the right-hand side constant at all sample points.

Note that an IIS is not isolated by this sampling method. However, the constraint reporting an effectiveness of 1.0 is isolated as a constraint that cannot be satisfied relative to the bounds and constraints that define the sampling enclosure, and hence is a good candidate for further analysis.

6.2 Methods Specific to Linear Programs

Analysis of infeasibility is easiest when the constraints are restricted to linear forms and associated variable bounds in real-valued variables, i.e. linear programs. For one thing, it is known that the maximum cardinality of any IIS is $n + 1$ when there are n variables in the linear model (Chvátal 1983, p. 146). We are also able to make use of sensitivity analysis, alternative forms of the LP, and information provided both by pivoting and interior point methods. All of these elements have a role to play in the development of efficient and effective methods for the analysis of infeasible LPs.

All of the methods described in Sec. 6.1 are also effective for linear programs. On first glance it may seem that these methods are relatively slow, given that they may require the solution of a significant number of LPs. However they can be surprisingly fast due to the advanced start that each LP solution provides for the next.

Chinneck (1994) compares (*a*) the computer time needed to find an IIS (after phase 1 has completed) to (*b*) the computer time needed to identify infeasibility in the first place (i.e. the phase 1 time) for his modified version of the MINOS LP solver. For the combination of methods used in his study (sensitivity filter, elastic filter, deletion filter), the ratio a/b is frequently very small. In other words, it is often much faster to isolate an IIS after phase 1 has ended than to complete the tial phase 1 itself. The ratios seen in practice are generally consistent with theoretical analyses presented by Chinneck and Dravnieks (1991). Of course the a / b ratio achieved for a particular model depend on factors such as the relative cardinality of the IIS vs. the entire set of constraints in the LP and the combination of IIS isolation algorithms applied.

There are a number of IIS isolation methods that are specific to linear programs in that they take advantage of the properties of linear systems. Some of these methods provide a very significant improvement in the speed of IIS isolation. Such methods are the subject of this section.

We will refer several times to a handy collection of infeasible linear programs that is available online in the netlib collection (Chinneck 1993). The basic characteristics of these models are shown in Table 6.1. The models cover a range of sizes, characteristics, origins, and difficulty (both to solve and to analyze). The infeasibility is original in a number of the cases, but is introduced in many of the models by adjusting a constraint in a feasible model to cause infeasibility. Some details are available in the online *readme* file.

Table 6.1. Characteristics of the netlib infeasible LPs

Model	Rows	Columns	Nonzeroes
bgdbg1	349	407	1485
bgetam	401	688	2489
bgindy	2672	10116	75019
bgprtr	21	34	90
box1	232	261	912
ceria3d	3577	824	17604
chemcom	289	720	2190
cplex1	3006	3221	10664
cplex2	225	221	1059
ex72a	198	215	682
ex73a	194	211	668
forest6	67	95	270
galenet	9	8	16
gosh	3793	10733	97257
gran	2569	2520	20151
greenbea	2505	5405	35159
itest2	10	4	17
itest6	12	8	23
klein1	55	54	696
klein2	478	54	4585
klein3	995	88	12107
mondou2	313	604	1623
pang	362	460	2666
pilot4i	411	1000	5145
qual	324	464	1714
reactor	319	637	2995
refinery	324	464	1694
vol1	324	464	1714
woodinfe	36	89	209

6.2.1 The Reciprocal Filter

Chinneck (1997b) defines the reciprocal filter, which applies when a variable or a row constraint has a pair of distinct upper and lower bounds.

Theorem 6.9: *The reciprocal filter (Chinneck 1997b).* In the absence of simple upper and lower bound reversal, if a variable or row constraint has distinct upper and lower bounds and one of the bounds is involved in an IIS, then the other bound cannot be involved in the same IIS.

Proof: An IIS can be rendered feasible by stretching one of its members until a feasible point is reached. The constraint stretching creates at least one point that satisfies the stretched version of the constraint and all of the other members of the IIS, thereby rendering the IIS feasible. Since this new point already satisfies the other bound on the constraint, there is no need to stretch the other bound in order to satisfy the other members of the IIS, hence it cannot be a member of the IIS. ■

The reciprocal filter depends on the fact that bound constraints are linear and parallel. It can be used to eliminate the second bound on a constraint as soon as the first bound is identified as being part of the IIS being isolated. This may reduce the number of feasibility tests in the deletion filter and the additive method.

6.2.2 The Sensitivity Filter

Chinneck and Dravnieks (1991) presented the sensitivity filter as a way of quickly eliminating many constraints that are not involved in the infeasibility detected by the phase 1 LP solution. It uses the fact that a phase 1 solution of an infeasible LP is a partially elastic program, and will of necessity stretch a constraint by assigning nonzero values to one or more of the artificial variables, in the same way that an elastic program stretches a constraint by assigning nonzero values to one or more elastic variables. A phase 1 (or elastic) solution of an infeasible LP will thus be sensitive to an infinitesimal adjustment of the RHSs of the stretched constraints or the constraints that oppose them to cause the stretching. But the phase 1 objective function will never be sensitive to an infinitesimal adjustment of the RHS of a constraint that is *not* in the IIS(s) detected by the phase 1 solution. The sensitivity filter is summarized in Alg. 6.8.

C: ordered set of constraints in the infeasible model (includes both functional constraints and variable bounds.

INPUT: an infeasible set of constraints C.

1. Solve the phase 1 LP.
2. For every c_i in C:
   2.1 If the reduced cost of c_i is 0, then $C = C \setminus c_i$.

OUTPUT: C contains at least one IIS.

Alg. 6.8. The sensitivity filter

Theorem 6.10: *Functional constraints found by sensitivity filter (Murty 1983).* The set of functional constraints having nonzero shadow prices in the optimal tableau of a phase 1 LP which reports infeasibility contains all of the functional constraints in a least one IIS. ■

Roodman (1979) provides a similar argument and is listed as a reference by Murty.

Theorem 6.11: *Nonnegativity constraints in IISs (Murty 1983).* Original variables having nonzero reduced costs in the optimal tableau of a phase 1 LP solution for an infeasible LP identify nonnegativity constraints that are involved in IISs. ■

The important fact resulting from Theorems 6.10 and 6.11 is that the output set of constraints following a sensitivity filter is still infeasible because it still contains one or more IISs. Further it is usually of much reduced size since all of the constraints that are not involved in the detected infeasibility will have been removed. However the output is not guaranteed to include a single IIS: it must be further processed by the deletion filter or the additive method to guarantee this.

The sensitivity filter has several important properties. First, it is very inexpensive, involving only the inspection of the results of a phase 1 LP solution. Very large numbers of constraints are immediately eliminated from further consideration. Second, the output set is not guaranteed to include all of the constraints in all of the IISs in the model (Chinneck and Dravnieks 1991, Observation 7). This happens because the phase 1 LP is partly elastic, and stretches constraints to reach the optimum phase 1 solution in the expanded space created by adding artificial variables. See the example in Fig. 6.4, in which constraint B stretches from its original position in the left diagram to its final position at B' after the phase 1 solution as shown in the right diagram. Constraint A, which is in the IIS {A,B}, has a reduced cost of zero after the phase 1 solution and so is removed by the sensitivity filter.

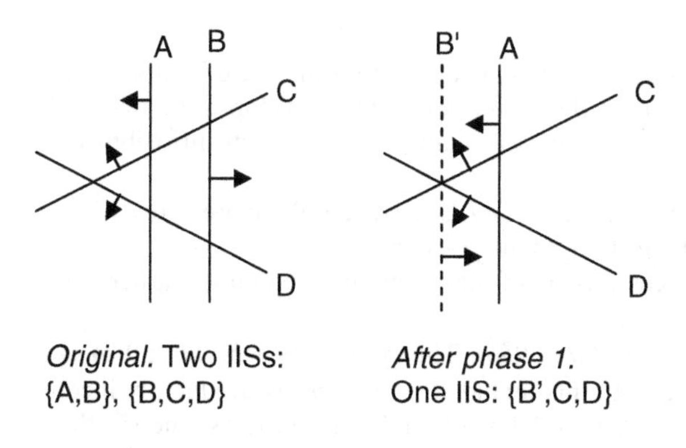

Fig. 6.4. The sensitivity filter

It is possible to identify more constraints that are part of some IIS by using an extension of the sensitivity filter. Additional implicated constraints can be found by examining the LP basis inverse matrix $\boldsymbol{B}^{-1}$. The nonzero elements in a row of $\boldsymbol{B}^{-1}$ corresponding to a basic artificial/elastic variable index the constraints whose RHSs affect the final value of the corresponding artificial or elastic variable. Some of the constraints so indexed may not have been identified by the sensitivity filter, and hence can be added to the output set. Constraints may be in this situation because the effect of increasing a certain elastic variable is exactly counterbalanced by a decreasing effect on another elastic variable, hence the net reduced cost is zero.

Where there is degeneracy in the solution of the phase 1 elastic LP, as indicated by a basic variable with a value of zero, there may be additional constraints that are tight, and which form part of an IIS, but which are not included in the output set produced by the sensitivity filter. These constraints are indexed by the nonzero elements in the row of $\boldsymbol{B}^{-1}$belonging to the basic variable with value zero. These constraints can also be added to the output set of the sensitivity filter (Chinneck and Dravnieks 1991).

The sensitivity filter is easily combined with other IIS isolation procedures to produce faster hybrid methods, as we will see later in this chapter. Further, it has the property that it tends to isolate larger IISs when applied immediately following the initial phase 1 that recognizes infeasibility. Fig. 6.4 is an example of this phenomenon: the smallest cardinality IIS is {A,B}, and yet the larger IIS {B,C,D} appears in the output set and will be positively identified after further analysis. This probably happens because the overlapped constraints in a cluster are the cheapest to stretch because they eliminate more than one IIS at a time, and the larger IISs require a larger stretch on average, so the smaller IISs are bypassed. We return to this characteristic in Sec. 6.2.7 when we are concerned with finding IISs that are easiest to understand.

6.2.3 Pivoting Methods

Pivoting methods rely on a series of theorems that permit the use of simple pivoting to identify the constraints in an IIS. Van Loon (1981) developed the first such pivoting method. His method relies on two main theorems about infeasibility in systems of linear inequalities:

Theorem 6.12: *IIS matrix rank (Motzkin 1936).* Where there are p rows in an IIS, the coefficient matrix of the IIS has rank $p - 1$. ■

Van Loon presents a stronger version of this theorem based on an earlier theorem by Fan (1956):

Theorem 6.13: *IIS conditions (Fan 1956).* The system $\boldsymbol{Ax} \le \boldsymbol{b}$, $\boldsymbol{x},\boldsymbol{b} \ge \boldsymbol{0}$, $\boldsymbol{A}$ is a $p \times n$ matrix (nonnegativity constraints included in $\boldsymbol{Ax} \le \boldsymbol{b}$), is irreducibly inconsistent if and only if (i) there exist $p - 1$ linearly independent rows, and (ii) there exist numbers $\lambda_i > \boldsymbol{0}$ such that $\sum_{i=1}^{p} \lambda_i a_{ij} = 0$ and $\sum_{i=1}^{p} \lambda_i b_i < 0$. ■

Van Loon also notes the following result derived from these theorems (Chvátal (1983) subsequently provides a proof). This simplifies the search for a set of constraints that meet the conditions described in the previous two theorems.

Theorem 6.14: *IIS dimension (Chvátal 1983, p. 146).* Every unsolvable system of linear inequalities in n variables contains an unsolvable subsystem of at most $n + 1$ inequalities. ■

Van Loon (1981) introduces a further theorem that establishes conditions under which the tableau developed by various simplex variants will recognize the conditions described in Theorem 6.13. For the system $\boldsymbol{Ax} + \boldsymbol{y} = \boldsymbol{b}$ with $\boldsymbol{y} \ge \boldsymbol{0}$, solve the system in terms of a single slack variable y_1, thus treating the corresponding row as if it were the objective function of an LP. We use this notation in the following theorem: $\boldsymbol{y}^1$ is the vector $\boldsymbol{y}$ without y_1, $\boldsymbol{x}_B$ is the vector of $m-1$ basic variables, $\boldsymbol{x}_N$ is the vector of $n - m + 1$ nonbasic variables, $\boldsymbol{A}^1$ is the remaining $(m - 1) \times n$ submatrix of $\boldsymbol{A}$ after the removal of row 1 (the row for which s_1 is the slack variable), $\boldsymbol{b}^1$ is the vector $\boldsymbol{b}$ without b^1, $\boldsymbol{B}$ is the columns of $\boldsymbol{A}^1$ corresponding to the basic variables, and $\boldsymbol{N}$ is the columns of $\boldsymbol{A}^1$ corresponding to the nonbasic variables.

Theorem 6.15: *Recognizing an IIS (Van Loon 1981).* The system $\boldsymbol{Ax} + \boldsymbol{y} = \boldsymbol{b}$, $\boldsymbol{y} \geq \boldsymbol{0}$ is an IIS if and only if there is a slack variable, say y_1, such that the system can be solved with respect to y_1 and a set $\boldsymbol{x}_B$ of basic variables as follows: (1) $y_1 = u - (\boldsymbol{w}^1)^T \boldsymbol{y}^1$, (2) $\boldsymbol{x}_B = \boldsymbol{B}^{-1}\boldsymbol{b}^1 - \boldsymbol{B}^{-1}\boldsymbol{N}\boldsymbol{x}_N - \boldsymbol{B}^{-1}\boldsymbol{y}^1$, with $u < 0$ and $\boldsymbol{w}^1 > \boldsymbol{0}$. ■

Thm 6.15 allows an LP solver to recognize an IIS. Van Loon's search for tableaux that meet these conditions is undirected, and will in general enumerate many bases that do not provide any information about the cause of the infeasibility. Greenberg and Murphy (1991) point out that his method could be extended to find IISs more efficiently by pivoting through alternative bases.

Gleeson and Ryan (1990) improve on Van Loon's approach by developing a method that avoids uninformative bases and enumerates only those bases that correspond to IISs (in the absence of degeneracy). Their method rests on Thm. 6.16, a variant of Farkas' Theorem of the Alternative, and polyhedral theory.

Theorem 6.16: *Efficient IIS pivoting (Gleeson and Ryan 1990).* Let $\boldsymbol{A}$ be a rational $m \times n$ matrix and let $\boldsymbol{b}$ be a rational m-vector. Then the indices of the IISs of the system $\boldsymbol{Ax} \leq \boldsymbol{b}$ are exactly the supports of the vertices of the polyhedron $P = \{\boldsymbol{y} \in R^m \mid \boldsymbol{y}^T\boldsymbol{A} = \boldsymbol{0}, \boldsymbol{y}^T\boldsymbol{b} \leq -1, \boldsymbol{y} \geq \boldsymbol{0}\}$. ■

Gleeson and Ryan apply Dyer's method (Dyer 1983) to efficiently enumerate all of the bases of the system established in Thm. 6.16, and an IIS is identified at each basis (though the same IIS may be identified multiple times if there is degeneracy). All of the IISs in a model are identified in this manner. A very similar result is reported by De Backer and Beringer (1991) based on Fourier's theorem (Fourier 1827).

While Gleeson and Ryan's method is much more efficient than van Loon's, theoretical comparisons (Chinneck and Dravnieks 1991) show that it is likely to be much slower than the filtering methods when only the first IIS is desired. Much of the speed disadvantage is due to the necessity of converting equality constraints to oppositely signed pairs of inequalities, which causes a blow-up in model size. In comparison to the filtering methods, Gleeson and Ryan's method operates at a disadvantage in any system that has numerous nonnegativity constraints and equality constraints. This is true of many general LPs, and especially true for network LPs.

Parker and Ryan (1996) modify Gleeson and Ryan's method slightly by showing that you can construct a cone instead of a polyhedron and identify IISs based on the extreme rays of the cone:

Theorem 6.17: *IIS cone (Parker and Ryan 1996).* Let $\boldsymbol{Ax} \leq \boldsymbol{b}$ denote an inconsistent set of inequalities. Then the IISs are in 1 – 1 correspondence with the extreme rays of the cone $P' = \{\boldsymbol{y} \in R^m \mid \boldsymbol{y}^T\boldsymbol{A} = \boldsymbol{0}, \boldsymbol{y}^T\boldsymbol{b} < 0, \boldsymbol{y} \geq \boldsymbol{0}\}$. In particular, the nonzero components of any extreme ray of P' index an IIS. ■

The motivation for the work by Gleeson and Ryan (1990) and Parker and Ryan (1996) is not to identify a first IIS quickly: it is instead to identify a minimum-weight cover of the set of IISs. When the weights are all identical, this is the same as the minimum-cardinality IIS set cover, i.e. the smallest set of constraints to remove from the LP such all of the remaining constraints constitute a feasible set.

This is an important question that we will return to in Chapter 7. Still, the pivoting methods can be used to isolate individual IISs.

6.2.4 Interior Point Methods

Greenberg (1996a) shows how to use an interior point method solution of an infeasible LP as a filter that separates constraints into two sets: (i) those that might be part of some IIS, and (ii) those that cannot be part of any IIS. This is an improvement over the sensitivity filter which cannot always identify all of the constraints that are part of some IIS.

Following Greenberg (1996a), let $S = \{\boldsymbol{Ax} \geq \boldsymbol{b}\}$ be a finite collection of inequalities. $S(I) = \{\boldsymbol{A}_i\boldsymbol{x} \geq b_i \text{ for } i \in I\}$. $X(S) = \{\boldsymbol{x} | \ \boldsymbol{x} \text{ is feasible in } S\}$.The dual system is $S^d = \{\boldsymbol{\pi} \geq \boldsymbol{0}, \boldsymbol{\pi A} = \boldsymbol{0}, \boldsymbol{\pi b} \geq 0\}$. Define *LP*: max $\boldsymbol{\pi b}$ subject to $\boldsymbol{\pi A} = \boldsymbol{0}$, $\boldsymbol{\pi} \geq \boldsymbol{0}$, $\boldsymbol{\pi b} \leq 1$, $\pi_i = 0$ for $i \in$ I. The support set $\sigma(v)$ of a nonnegative vector v is the set of indices for which the coordinate is positive. A solution in $X(S^d)$ has the support set $\sigma(\boldsymbol{\pi}) = \{i | \ \pi_i > 0\}$. If $\boldsymbol{x} \in X(S)$ and $\boldsymbol{\pi} \in S^d$ then we have complementary slackness, i.e. $\boldsymbol{A}_i\boldsymbol{x} = b_i$ for all $i \in \sigma(\boldsymbol{\pi})$. The solutions are strictly complementary if $\boldsymbol{A}_i\boldsymbol{x} > b_i$ for all $i \notin \sigma(\boldsymbol{\pi})$. A strictly complementary solution induces a support partition, $\sigma(\boldsymbol{\pi}) \cup \sigma(\boldsymbol{Ax} - \boldsymbol{b})$ on the indices of the inequalities.

Theorem 6.18: *Strictly complementary partition (Greenberg 1996a).* If S is consistent, there exists a strictly complementary solution, $(\boldsymbol{x}, \boldsymbol{\pi}) \in X(S) \times X(S^d)$. Further, the support partition is the same for all strictly complementary solutions.■

If the optimal solution to *LP* is obtained by an interior point method, then the optimal partition, say $\boldsymbol{\pi}^0$, is strictly complementary. Now $\sigma(\boldsymbol{\pi}^0) = \{i | \ \boldsymbol{A}_i\boldsymbol{x} \geq b_i$ is in some IIS of $S \backslash S(I)\}$. $S \backslash S(I)$ is the set of violated constraints at the interior point solution, and it is separated into two parts by the strictly complementary solution: those that might be part of some IIS, and those that are not part of any IIS. This partition can be used to eliminate the inequalities that are not part of any IIS.

6.2.5 Speed-ups: Combining Methods

As we saw in Section 6.1.6, the deletion filter and the additive method can be combined to create a hybrid method that may be faster than either method by itself. The opportunity for combining methods is even greater when the IIS isolation methods that are specific to LP are considered.

Combining the deletion and sensitivity filters results in the *deletion/sensitivity* filter, one of the quickest ways to isolate an IIS. As shown in Alg. 6.9,the sensitivity filter is applied whenever an intermediate deletion filter test proves infeasible. The deletion/sensitivity filter has the useful property given in Thm. 6.19 that can be used in a backtracking scheme to find other IISs.

INPUT: an infeasible set of linear constraints.
1. Solve the phase 1 LP.
2. Sensitivity filter the phase 1 result.
3. FOR each constraint in the set:
   3.1 Temporarily drop the constraint from the set.
   3.2 Test the feasibility of the reduced set:
      3.2.1 IF feasible THEN return dropped constraint to the set.
      3.2.2 ELSE (infeasible)
         3.2.2.1 Drop the constraint permanently.
         3.2.2.2 Apply the sensitivity filter.

OUTPUT: constraints constituting a single IIS.

Alg. 6.9. The deletion/sensitivity filter

Theorem 6.19: *Deletion/sensitivity filter (Chinneck 1994).* Assume a sensitivity filter is applied to the phase 1 final basis which originally signals infeasibility. During a subsequent deletion/sensitivity filtering, any constraint removed by the deletion filter, along with any constraints removed by the sensitivity filter in the same iteration, are part of a different IIS than the output IIS.

Proof: The initial sensitivity filter following the phase 1 solution which signals model infeasibility retains only constraints which are part of some complete IIS in the sensitivity filter output. If a constraint is subsequently removed by the deletion filter, then it must be part of a different IIS than the final IIS eventually isolated by the deletion/sensitivity filter. Any constraint removed by the sensitivity filter in the same iteration must be part of an IIS with the deletion filtered constraint, otherwise the phase 1 objective function would continue to be sensitive to it. ■

The sensitivity filter can also be combined with the additive method to yield the *additive/sensitivity* method. This has a very beneficial effect on the speed of the algorithm. The sensitivity filter is simply applied each time the additive method discovers an infeasible set, and is applied to the current test set T. Any constraints that can be eliminated from T are also eliminated from C, and hence are not included in the additive testing on the next round. This is especially effective the first time that the additive method detects infeasibility in T because numerous non-IIS constraints may have been added to T prior to the infeasible outcome. See Alg. 6.10 for details.

The sensitivity filtering step continues to be useful even after the first infeasible outcome in the additive method. The final constraint that triggered the original infeasibility may be a member of several overlapped IISs, all of which appear in T. If infeasibility is encountered a second or subsequent time, the algorithm may be homing in on a particular IIS in the IIS cluster, meaning that there now exist members of partial IISs in T, which will then be eliminated by the sensitivity filter. After the first sensitivity filter, a result similar to Thm. 6.19 is also available for the additive/sensitivity method.

C: ordered set of constraints in the infeasible model.
T: the current test set of constraints.
I: the set of IIS members identified so far.

INPUT: an infeasible set of constraints C.
0. Set $T = I = \varnothing$.
1. Set $T = I$.
2. FOR each constraint c_i in C:
    Set $T = T \cup c_i$.
    IF T infeasible THEN
        Apply the sensitivity filter to T. Constraints dropped from T are likewise dropped from C.
        Set $I = I \cup c_i$.
        Go to Step 2.
3. IF I feasible THEN go to Step 1.
    Exit.
OUTPUT: I is an IIS.

Alg. 6.10. The additive/sensitivity method

The various independent methods can be combined in a variety of ways to improve overall speed, or to provide improved characteristics such as the identification of smaller IISs. Some possible combinations, including those mentioned so far, are:

- Additive/deletion method
- Deletion/sensitivity filter
- Additive/sensitivity method
- Combine reciprocal filter with any method suitable for LP
- Elastic filter followed by deletion or deletion/sensitivity filter
- Elastic filter followed by additive or additive/sensitivity method
- Etc.

6.2.6 Guiding the Isolation

The modeler normally brings additional knowledge to the task of identifying the cause of infeasibility. He may know, for example, that one part of the LP has been well-tested and running reliably for a long time and hence is a very unlikely source of difficulties, and hence would like to steer the IIS isolation away from that part of the model. Or he may know that a complex new portion has been recently added and so is the most likely source of infeasibility, and so would like to steer the IIS isolation towards that part of the model.

Fortunately it is straightforward to guide the model in various ways. The experimental code MINOS(IIS) (Chinneck 1990, 1996a) demonstrates this: it allows the user to tag individual constraints with codes that influence how the constraint is treated during a deletion or deletion/sensitivity filter. Codes include (i) eliminate immediately, before the IIS search begins, (ii) encourage elimination from IIS, (iii) discourage elimination from IIS, and (iv) never eliminate. The "encourage" and "discourage" codes are most useful in guiding the IIS search away from reliable portions of the model and towards suspect portions. Name masking can be used to apply guide codes to many similarly named constraints simultaneously. This is useful, e.g. in protecting large well-trusted portions of the model.

The guide codes influence the deletion or deletion/sensitivity filter as follows:

1. Remove all constraints coded for immediate removal. If the model becomes feasible, issue a message and exit.
2. Deletion filter all constraints coded as "encourage elimination". If the sensitivity filter is applied, removed only the constraints coded as "encourage elimination" or not specially coded.
3. Deletion filter all constraints not specially coded. If the sensitivity filter is applied, remove only constraints coded as "encourage elimination" or not specially coded.
4. Deletion filter all constraints coded as "discourage elimination". Do not apply the sensitivity filter.
5. Do not deletion filter the constraints coded as "never eliminate". Instead, run a sensitivity filter on these constraints simply to identify the constraints that could possibly be removed; alert the modeler about these.

Note that the output may not be an IIS under two conditions, both of which are brought to the user's attention. In Step 1, constraints that are essential to the IIS may be removed immediately, rendering the model feasible. In Step 5, constraints that should be dropped may be protected from doing so, so the output is not an IIS. The sensitivity filter in Step 5 is only a partial solution to this difficulty. If there is a single IIS in the constraints remaining after Step 4, then it will reliably indicate which protected constraints could be dropped to yield that IIS. On the other hand, the protected constraints could contain additional complete IISs which will not be identified as candidates for elimination by the final sensitivity filter. It is for this reason that the "never eliminate" code should normally only be applied to well-tested parts of the model.

As we will see in the next section, there is an important distinction between the functional constraints (or rows) and variable (or column) bounds in isolating IISs in LPs. It is usually much easier to understand an IIS that has few rows, regardless of the number of column bounds involved. For this reason, MINOS(IIS) allows special guidance for the treatment of the column bounds via the IIS PROTECTION parameter. In the first mode, column bounds are protected as much as possible; they can be eliminated only by a deletion test. In the second mode, column bounds are protected until all of the rows have been deletion filtered, then the columns are deletion/sensitivity filtered (even if sensitivity filtering was not initially enabled).

In MINOS(IIS), the use of guide codes conflicts with some of the other IIS isolation approaches, such as the elastic filter. Hence there is a strict hierarchy of methods as follows:

1. IIS set covering (see Chapter 7).
2. Elastic filtering.
3. Column protection options.
4. General guide codes.

Methods nearer the top of the list take precedence over those lower in the list. For example, the column protection options operate by making wholesale settings of guide codes on all of the constraints in the model to encourage the removal of row constraints and discourage the removal of column bounds. Implementation details make it difficult for the elastic filter to follow the guide codes, though guiding the elastic filter is certainly possible to some extent.

Guiding the elastic filter would use the fact that a constraint that is never de-elasticized will definitely *not* appear in the output IIS; constraints that *are* de-elasticized *may* appear in the output IIS, though this depends on the final filtering by the deletion filter or the additive method. This implies a rank-ordering of the stretched constraints to de-elasticize after each elastic program solution: (1) constraints coded "discourage removal", (2) constraints not specially coded, (3) constraints coded "encourage removal". At each opportunity to de-elasticize constraints, choose all of those from the non-empty group that is highest in this list. Constraints coded "never remove" will of course simply be included in the output of the elastic filter.

Weighting approaches can also be used to guide the isolation in various ways. For example, the usual elastic objective of minimizing the sum of the elastic variables can be replaced by the objective of minimizing a weighted sum of the elastic variables. The elastic variables associated with constraints that should be honoured as much as possible can be given higher weights. In the same way, weights c_i can be applied to an objective function introduced in the alternative system in Thm. 6.16 as suggested by Bruni (2005). The new objective to minimize $\Sigma c_j y_j$ influences which set of constraints is returned as the initial IIS.

6.2.7 Finding Useful Isolations

A modeling error resulting in infeasibility is often reflected in several different alternative IISs, not all of which are equally easy for humans to understand. Which of the several IISs is reported to the user can have a major impact on the speed of diagnosis. Experiments with users show clearly that the IIS having the fewest row constraints is the easiest to understand, and hence the most useful. For example, one model generated two IISs: one involving 12 rows (of 2393 bounded rows) and

68 columns, and another involving 1 row and 93 columns. Although the first IIS is smaller in terms of the total number of constraints involved, the second is much easier to interpret and to diagnose. In another model, one IIS involved all 323 of the bounded rows, while another involved only 76, effectively confining further analysis to about one quarter of the original model.

It is not surprising that analysts prefer IISs having few rows. Column bounds are easy to understand, but rows tie together both variables and other rows in complicated ways. Limiting the number of rows reduces the complexity of the subsequent human analysis. A small number of rows helps to pinpoint part of the model or a class of constraints for further analysis, e.g. the blending units, or perhaps the crude oil supply limits in a refinery model. Variable bounds are rapidly verified and are therefore of less consequence to the analysis process. The analyst would rather accept more variable bounds in the IIS in return for fewer rows. In general, minimizing the number of rows in an IIS also tends to reduce the number of variable bounds involved because the smaller number of rows interacts with fewer variables. A side effect of minimizing the number of rows is usually a reduction in the total size of the IIS.

Chinneck (1996a, 1997b) raised the issue of finding IISs that have few rows in infeasible LPs, and discovered that combining and controlling the IIS isolation algorithms can result in methods that usually return IISs that have few rows. One interesting result of the analysis was the observation that the sensitivity filter, which greatly increases the speed of IIS isolation, tends to return IISs of larger cardinality.

One way to find the minimum row-cardinality IIS is to enumerate all of the IISs in the model, and then choose the one having the fewest rows. Unfortunately, Chakravarti (1994) showed that the number of IISs in an infeasible LP could be exponential in the worst case. This means that, in general, the minimum row-cardinality IIS cannot be found in polynomial time by enumeration methods. However Parker and Ryan show that their method of generating IISs while solving the IIS set covering problem (see Sec. 7.2) can indeed identify small cardinality IISs in reasonable amounts of time, and hence that this approach has some value in practice. However, their empirical results show that their enumeration method finds a smallest IIS that is larger than the IIS found by the heuristic methods developed below in about two-thirds of the cases studied, and never returns an IIS having fewer rows than the heuristic methods. Further, it requires more time.

Mindful of the difficulties in enumerating IISs, Chinneck (1997b) instead takes a heuristic approach which does not guarantee to find the minimum row-cardinality IIS, but often finds IISs with a small row-cardinality. The heuristic method makes use of the filtering algorithms described earlier. Their inherent characteristics affect their ability to isolate IISs having few rows. We will review the relevant characteristics of each method next.

As shown in Thm. 6.2, the deletion filter returns the IIS whose first member is tested last. Obviously, it is not possible to predict in advance the ordering of the constraints that will return the IIS having the fewest rows. However, a general heuristic can be formulated based on this behaviour: order the constraints so that rows are eliminated before column bounds in an attempt to eliminate as many rows as possible. In the MINOS linear programming code (Murtagh and Saunders 1987) that underlies Chinneck's MINOS(IIS) code (Chinneck 1994), variables are ordered as follows: (i) the original model variables, then (ii) the slack/surplus variables for the rows. Deletion testing proceeds by removing and reinstating the bounds on these variables as appropriate. Hence proceeding in the natural order of the variables is detrimental to isolating IISs that have few rows. It is better to proceed by first deletion testing the bounds on the slack/surplus variables for the rows, followed by the bounds on the original variables. Since this is the reverse of the natural ordering of the variables found in many LP solver codes (such as MINOS), it is referred to as *reverse deletion filtering*. Chinneck (1997b) demonstrates the superiority of reverse deletion filtering over the normal forward deletion filtering empirically: over 14 tested IIS isolation procedures, the deletion filter returns IISs having the most rows on average, while the reverse deletion filter returns IISs having the fewest rows on average.

Similar thinking applies to the additive method. To make sure that few rows are included in any output result, include all of the variable bounds in the testing set at all times, while rows are added one by one as usual. The algorithm will eventually terminate with the output of a minimal set of row constraints plus all of the column bounds. We will see how to deal with the excess column bounds presently.

As shown in Thm. 6.4, the additive method isolates the IIS whose last member is tested first. Hence, as for the deletion filter, exactly which IIS is returned depends on the ordering of the constraints. For both algorithms, assuming a random distribution of the members of the IISs through the list of constraints, the largest IIS is unlikely to be returned. For the deletion filter, this is because some members of the largest IIS are likely to be eliminated before the first member of a smaller IIS is encountered. For the additive method, this is because an entire small IIS is likely to appear in the testing set before an entire large IIS. Hence we expect that these methods will have reasonably good average case ability to identify IISs having few members (or having few rows when the algorithm is modified as described above).

The sensitivity filter greatly increases the speed of IIS isolation, but it has the unfortunate side effect of tending to find IISs that have larger numbers of rows. Why does this happen? It is mainly due to the operation of the phase 1 objective function, which normally attempts to minimize the sum of the artificial variables. The effect is to "stretch" some constraints by setting their artificial variables to positive values, in effect moving the constraints as shown in Fig. 6.4. The final positions of all of the constraints determine which ones the phase 1 objective is sensitive to, and hence which ones are retained by the sensitivity filter for further analysis: the phase 1 objective will always be sensitive to any stretched constraints (as it will be to all of the active constraints).

The important fact is that the phase 1 process tends to stretch the constraints that give the greatest reduction in the overall phase 1 objective function value per unit of stretch, i.e. the constraints that are involved in the most IISs. Consider two IISs that overlap on a single constraint: is it cheaper to stretch two constraints, one from each IIS, or just the single overlapped constraint? Fig. 6.4 illustrates this effect. The smallest value of the phase 1 objective is achieved at the point defined by the intersection of constraints B' (the stretched version of constraint B), C, and D: these are the constraints that will be retained by the sensitivity filter. Since the phase 1 process tends to stretch as few constraints as possible (since this is cheaper), this means that more members of the larger IISs tend to stay in place, with the overlapped constraints tending to move. The side effect is that the smaller IISs are bypassed as in Fig. 6.4. This effect is demonstrated empirically by Chinneck (1997b): while the reverse deletion filter is among the best methods for finding IISs having few rows, the reverse deletion/sensitivity filter is among the worst. The inclusion of the sensitivity filter has negative consequences.

The elastic filter has the especially useful property described in Thm. 6.8: the number of elastic filter iterations (i.e. elastic programs solved) is at most equal to the cardinality of the smallest-cardinality IIS in the input set. Assuming that exactly one member of each IIS is stretched during each iteration of the elastic filter, its output set will contain a smallest-cardinality IIS, plus parts of all of the larger IISs. The smallest cardinality IIS will be found when the elastic filter output set is subjected to the deletion filter or additive method for positive identification of a single IIS. There is no guarantee that exactly one member of each IIS is stretched during each elastic solution, but experimentation shows that this does happen very frequently.

If the goal is to find IISs having few rows, then a simple modification of the elastic filter can assist. Since enforced constraints appear in the output set, enforce only the variable bounds that stretch in each elastic program solution. Row constraints are enforced only when there are no stretched variable bounds in the elastic solution. This process can be speeded considerably simply by starting the entire elastic filter process with all of the variable bounds already enforced. This then produces an output set that has a small number of row constraints, plus all of the variable bounds. We now deletion filter just the row constraints in the output set.

At this point, the output set has a minimal set of row constraints, plus the full complement of variable bounds, exactly the same situation as in a reverse deletion filter before the variable bounds are tested or in the modified additive method. From here forward, no further row constraints will be removed since all are definitely part of the IIS being isolated. This is the usual outcome of the strategy of *column protection* (Chinneck 1997b), i.e. preserving all of the variable bounds until a minimal set of row constraints has been identified. Column protection is very helpful in identifying IISs that have few rows. Further, once a minimal set of row constraints has been identified, sensitivity filtering can be safely used to remove large numbers of variable bounds quickly, greatly speeding the overall process.

The modified elastic filter described above performs very well in empirical tests (Chinneck 1996a, 1997b), giving results about as good as those for the reverse deletion filter and the modified additive method. The modified elastic filter has a significant speed advantage over the other two methods for larger models, but is slower on the smaller models.

To summarize, the following two general principles assist in finding IISs that have few rows: (i) Protect the variable bounds from elimination until a minimal set of row constraints has been identified. This is the main principle in the reverse deletion filter, the modified additive method, and the modified elastic filter. (ii) Do not use the sensitivity filter until a minimal set of row constraints has been identified. It can be applied thereafter.

The best methods for finding IISs that have few rows are: (a) reverse deletion filter (with sensitivity filter enabled after the rows are deletion filtered), (b) modified additive method (with sensitivity filter enable after a minimal set of rows is returned), and (c) modified elastic filter, followed by the reverse deletion filter (with sensitivity filter enabled after the rows are deletion filtered). The modified elastic filter is best for large models, the other two are best for smaller models.

Another approach to reducing the number of rows in an IIS is to *aggregate* the rows by simply summing them to yield a single row (Chinneck 1996b). This can be especially effective in network LPs where the large number of row constraints obscures a very simple diagnosis: incompatible input and output restrictions linked by many flow conservation equations. Aggregation of the row constraints condenses the "bridge" of equations connecting the conflicting input and output restrictions. Consider the following network example (Chinneck 1996b):

Rows in the IIS:
c125: $-x50 + x379 - x380 = -1825$
c126: $-x379 + x380 - x382 = -2535$
c127: $-x381 + x382 + x383 - x384 = -1658$
c128: $-x30 - x383 + x384 + x387 - x459 = -15466$
c147: $-x69 + x435 - x437 = -338$
c148: $-x435 + x437 + x438 - x439 = -1037$
c149: $-x438 + x439 + x440 - x442 = -5713$
c150: $-x440 + x442 + x443 - x444 = -16$
c151: $-x443 + x444 + x446 - x448 = -1954$
c153: $-x446 + x448 + x449 - x450 = -4255$
c154: $-x449 + x450 + x451 - x453 = -5155$
c155: $-x451 + x453 + x454 - x455 = -1274$
c156: $-x454 + x455 + x456 + x457 - x458 - x463 = -1454$
c157: $-x387 - x456 + x458 + x459 = -6401$
c158: $-x457 + x463 + x464 - x491 = -14$
c165: $-x475 + x477 + x478 - x479 = -246$
c166: $-x478 + x479 + x480 - x482 = -232$
c167: $-x480 + x482 + x483 - x484 = -61$
c168: $-x483 + x484 + x485 - x486 = -1536$
c169: $-x485 + x486 + x487 - x488 = -3648$
c170: $-x487 + x488 + x489 - x490 = -3676$
c171: $-x464 - x489 + x490 + x491 = -1848$

Column Bounds in the IIS:
x30 <= 12509
x50 <= 12509
x69 <= 14434
x475 <= 14434
x477 >= 0
Aggregated IIS Rows:
– x30 – x50 – x69 – x475 + x477 = – 60342

While diagnosis of the infeasibility is difficult in the full IIS, it is straightforward using the aggregated row and the column bounds: the column bounds conflict with the aggregate effect of the rows.

Aggregation has been used for some time to analyze infeasibility in general LPs (e.g. (Murty 1983)), but it is especially useful for the pure portion of network models.

6.2.8 Analyzing Infeasible Network LPs

The most straightforward approach to analyzing an infeasible network LP is to simply treat it as you would an infeasible general LP, applying the various IIS isolation techniques described previously (Chinneck 1996b). Aggregation can be applied to the resulting IIS to improve ease of understanding. However a number of more specialized methods are also available for ordinary flow-conserving networks. These rely almost exclusively on the supply and demand balancing procedures by Gale (1957), Fulkerson (1959), Hoffman (1960), and Ford and Fulkerson (1962).

As an example, the main theorem by Gale states that the total demand over a network is feasible if and only if for every subset S of nodes, the total demand over the complement of S is less than the total capacity of the arcs that cross from S to its complement. The proof depends mostly on the minimum cut theorem. Note that it applies to individual nodes as well as any larger collection of nodes. These balancing rules are used to construct more sophisticated analysis procedures such as those by Greenberg (1987b, 1988) and by Aggarwal et al. (1988).

As Greenberg and Murphy (1991) point out, the guidance provided directly by the Gale-Fulkerson-Hoffman flow balancing algorithms is often insufficient to clearly identify the cause of the infeasibility. More exact localization is needed. Greenberg (1987a, 1988) combines the flow balancing results with logic about network behaviour to yield heuristics that give better localization of infeasibility. New specific tests such as path and cycle generation are combined with methods akin to bound reduction and similar techniques described in Sec. 6.1.1. These heuristics improve the usefulness of the base flow balancing techniques, but there is no guarantee that an IIS will be isolated, or that the resulting reductions will be helpful in understanding the infeasibility, as for all logical reduction/presolving methods. These techniques are available in the ANALYZE software (Greenberg 1993a).

Aggarwal et al. (1988) apply the Gale-Fulkerson-Hoffman theorems in a maximum flow algorithm to develop a method for identifying a *witness* set of nodes for which the net supply and the total outflow capacities conflict. They refine the procedure so that it is able to identify a minimal witness set. Note that this isolation is not as precise as an IIS since the LP constructed from the witness set of nodes and incident arcs will in general include constraints that do not appear in the associated IIS. Conversely, a minimal witness is easily obtained from an IIS simply by listing the nodes whose equations appear in the IIS.

Straightforward flow balancing is not effective for more advanced network forms such as generalized or processing networks in which flow conservation is not guaranteed. These are still LPs however, so the general IIS isolation techniques described earlier can be used.

A related modeling error for networks is *nonviability* (Chinneck 1990a, 1990b, 1992), a structural condition in which the only feasible flow for some of the arcs is zero. This condition can also be diagnosed using IIS isolation, as described in Sec. 9.2. Detecting and analyzing nonviability as well as infeasibility in an integrated method is the best approach for analyzing networks (Chinneck 1996b).

6.2.9 Software

Infeasibility isolation routines were rapidly adopted in both academic prototypes and commercial LP solvers after their introduction in the early 1990s. A brief survey of some of the noteworthy software follows below. See also the earlier survey by Chinneck (1997a). Note that presolving (which may occasionally detect infeasibility, but does not provide a useful analysis of the cause) is universally available in commercial LP solvers.

MINOS(IIS) (Chinneck 1994, 1996a; Chinneck and Saunders 1995) is Chinneck's research code that incorporates all of the IIS isolation filtering algorithms for LPs: the deletion, sensitivity, elastic and reciprocal filters, and their various combinations. It also includes routines for finding the minimum cardinality IIS set cover (see Chap. 7). It includes routines to guide the isolation as described in Sec. 6.2.6 and for finding IISs that have few rows as described in Sec. 6.2.7, and for producing output IIS files that can be read by the ANALYZE software described below. The The earliest version of MINOS(IIS) was produced around 1989.

CLAUDIA is a proprietary LP solver produced by BP Oil International (Main 1993a, 1993b). Earlier versions performed various analyses of infeasible LPs, and IIS isolation was added in the mid-1990s. CLAUDIA uses a nonstandard phase 1 procedure in which an infeasible row called the *control row* is selected as the objective function and pivoting is carried out to induce feasibility in the control row subject to the other nonviolated constraints. Infeasibility is recognized when the control row cannot be rendered feasible. This is similar to the elastic filter except that any row constraints that are initially satisfied are immediately enforced, whereas in the elastic filter constraints are enforced only after they have been violated once.

A sensitivity filter is applied to the final infeasible control row to yield a set of *mutually incompatible constraints (MIC)*, but note that this is not necessarily an IIS. Variable lower bounds of zero are ignored during processing and are not reported by CLAUDIA. A straight deletion filter can be applied to the MIC to isolate an "IIS". However, since the "IIS" may omit needed variable nonnegativity constraints, it may not be a true IIS.

Another routine examines the effect on the total infeasibility of dropping individual constraints. Constraints are then rank-ordered in terms of their impact on the total infeasibility and presented to the analyst. This idea is a precursor of Chinneck's minimum cardinality IIS set covering heuristic (see Chap. 7).

Some analysis of the MIC members can be done using agreed-upon naming conventions, providing added diagnostic power. For example, material flow balancing constraints, recognized by the naming convention, can be treated differently during the analysis than constraints in other classes.

LINDO is a commercial LP solver. Its *debug* command applies a deletion filter to infeasible LPs, though variable bounds are not tested, so the output is not a true IIS. It further tests each member of the output "IIS" to determine how much effect it has on removing all of the infeasibility in the entire model via the procedure outlined in Sec. 7.8.1. This results in a labeling of each IIS member as being "necessary" (i.e. necessary to that particular IIS) or "sufficient" (i.e. sufficient to remove all infeasibility in the model). The IIS isolation routines in LINDO date to about 1993, but binary search grouping strategies (see Sec. 6.1.5) have recently been added. Primal unboundedness is also analyzed in LINDO using IIS isolation applied to the dual. See Sec. 9.1 for further information.

CPLEX is a commercial LP solver that offers a choice of two methods of isolating IISs: (i) a deletion/sensitivity filter applied to the rows and then the columns, and (ii) an elastic filter followed by the first method. Method (i) is preferred if an IIS is needed quickly, and method (ii) is preferred if an IIS having fewer rows is needed. An aggregation of the rows can also be produced. These IIS isolation routines date to version 3.0, released in 1993. Cplex version 10.0 (released 2006) includes a *conflict refiner* that uses groups and preferences to allow some guidance of the IIS isolation process, similar to the guide codes described in Sec. 6.2.6.

IBM's *OSL* LP solver incorporated the IIS filtering routines in 1995, but is no longer available.

Tamiz et al. (1995, 1996) built *additive method* routines into the FortLP LP solver (Mitra and Tamiz 1988). There are several versions of the basic additive algorithm, differing mainly in implementation details.

PROFLOW (Chinneck 1996b) is a computer tool for formulating, analyzing, and solving network LPs of many forms, including processing networks. MINOS(IIS) is used as the solver, hence IISs can be isolated if the network is infeasible. PROFLOW uses infeasibility analysis to isolate the cause in the case of nonviability (see Sec. 9.2).

Xpress-Optimizer (Dash Optimization 2006) has included the ability to isolate IISs in LPs since 1997. It uses a combination of the filtering algorithms to isolate IISs and will also try to find IISs having a small number of rows. In addition, it can search for several IISs at once.

The *Frontline Systems solvers*, available as Microsoft Excel add-ins, have included the ability to find IISs in infeasible LPs via the filtering methods since 1997. See Fylstra et al. (1998).

The *XA* solver from Sunset Technologies finds IISs using the filtering methods; see e.g. Holmström et al. (2006).

ANALYZE (Greenberg 1993a) is a general purpose tool for manipulating and analyzing linear programs. It includes a number of routines that are helpful in analyzing infeasible LPs, including bound tightening, path and cycle tracing for infeasible networks, row aggregation, a form of sensitivity filtering, and tools for syntax-based explanation. While it is not able to isolate IISs directly, it can read IIS output files produced by MINOS(IIS) and apply the tools mentioned above to provide a deeper analysis of the infeasibility.

6.3 Methods Specific to Mixed-Integer Linear Programming

The sole work on methods for isolating IISs in mixed-integer linear programs is by Guieu and Chinneck (1999), hence this is the main subject of this section. We will use the term *MIP* to refer to mixed-integer linear programs as well as fully integer or binary programs, and any combinations thereof. The model must include at least one integer or binary variable, along with linear constraints and variable bounds.

Of course, the biggest difference between MIP and LP is the addition of the integer (or binary) restrictions on some or all of the variables. This has far-reaching effects, not only on the analysis of infeasibility, but on the very algorithms used to optimize the model. As shown in Chap. 3, MIP models are typically solved by a branch and bound procedure which has characteristics that make infeasibility analysis difficult. First, infeasibility is not detected until the branch and bound tree is fully expanded, with every leaf node reporting infeasibility. Little useful information is available when infeasibility is detected. There is no single overall LP solution that can be subjected to a sensitivity filter for example; instead there is a large set of infeasible LPs, one at each leaf. Second, if the model is insufficiently constrained, the branch and bound solution may not terminate. An example of nontermination is given in Fig. 6.5. These characteristics have a severe negative impact on the ability to isolate an IIS.

Note that these characteristics also exist if a branch and cut solution method is applied.

Another consequence of nontermination is that the branch and bound tree may grow very large, possibly exceeding the available memory. It may also happen that a model will eventually terminate, but requires an excessive number of iterations to do so. For example, imagine that the two parallel constraints in Fig. 6.5 are angled very slightly towards one another so that they eventually cross at a great distance from the origin. It may take a great number of iterations before

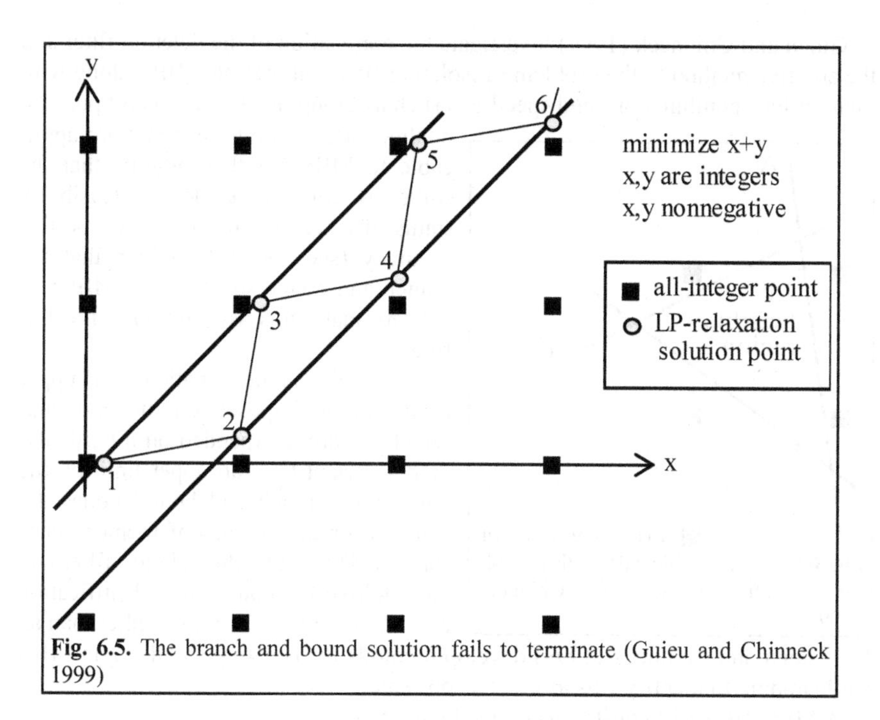

Fig. 6.5. The branch and bound solution fails to terminate (Guieu and Chinneck 1999)

infeasibility can be determined. If the constraints are angled very slightly away from each other, it may likewise require a great number of iterations before a MIP-feasible point is reached.

There is one degenerate case in which it is simple to identify an IIS: when the initial LP relaxation is itself infeasible, an IIS can be isolated simply by using the techniques for analyzing an infeasible LP. In the rest of this section we assume that the initial LP relaxation is feasible, but the entire MIP is not. An example of an infeasible MIP whose LP relaxation is feasible is shown in Fig. 6.6.

Few useful tools for analyzing infeasible MIPs are currently available. Savelsbergh (1994) describes a bound-tightening presolve procedure for MIPs (implemented in the MINTO solver (Nemhauser et al. 1994)) that may detect infeasibility as a side effect of the reformulation. Backtracking the complete set of reformulation operations may then isolate a set of constraints and integer restrictions that cause the infeasibility. However, there is no guarantee that the presolver will detect infeasibility, or that the backtrack of the reformulation operations will provide any useful information. Greenberg also uses related bound-tightening methods for dealing with binary variables in the reduce command of his ANALYZE software (Greenberg 1993a). These methods all fall into the class of general logical methods for model reduction discussed in Sec.6.1.1 and have the same drawbacks for diagnosing infeasibility.

Guieu and Chinneck (1999) investigate the application of the deletion filter and the additive method to the problem of isolating IISs in infeasible MIPs, along with their various combinations and speed-ups such as grouping. These general purpose methods are the only ones that are applicable to MIP, but they assume that the solver is able to decide the feasibility status of a set of constraints with perfect accuracy (see Sec. 6.1). Given that the branch and bound solution of a MIP may not terminate, this assumption is not fulfilled.

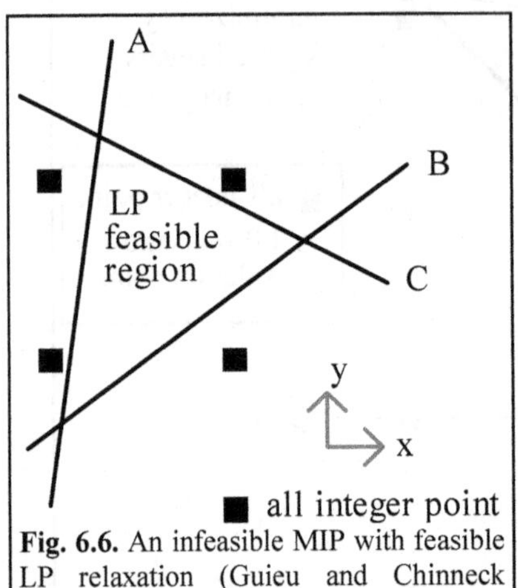

Fig. 6.6. An infeasible MIP with feasible LP relaxation (Guieu and Chinneck 1999)

To avoid nontermination, an upper limit can be imposed on the computational resources expended on a particular model variant (e.g. an upper limit on the number of branch and bound nodes developed, or the amount of memory consumed). This limits the deletion filter and the additive method to the identification of an *Infeasible Subset* (*IS*), rather than an IIS. Guieu and Chinneck (1999) develop methods for isolating small ISs in MIPs while hoping to identify IISs as often as possible.

A MIP consists of constraints divided into three sets:

- *LC*: the set of linear constraints (or rows),
- *BD*: the set of variable bounds (upper and lower bounds, if any),
- *IR*: the set of integer restrictions. Variables in *IR* are restricted to taking on integer values while variables not in *IR* are real-valued. Some integer variables may be binary, having a solution restricted to the set $\{0,1\}$.

We denote the presence of an integer restriction on a variable x_i by $[x_i]$. Binary variables are treated as integer variables with a lower bound of 0 and an upper bound of 1. An IIS for a MIP consists of a subset of the constraints in *LC*, *BD*, and *IR*. In the non-degenerate case the IIS must include at least one member of *IR*.

The entire MIP consists of a linear objective function plus the complete set of constraints $\{LC,BD,IR\}$. In an ordinary linear program, the set *IR* is empty. In an integer linear program, all of the variables are in *IR*. In a mixed integer program, at least one variable is in *IR*, and at least one variable is not in *IR*. The LP-relaxation of a MIP is created by considering only the objective function plus the subset of constraints $\{LC,BD\}$. Since the LP-relaxation has fewer restrictions, its feasible region is larger.

6.3.1 A Deletion Filter for MIPs

Applying a straightforward deletion filter to an infeasible MIP necessitates the solution of $|LC| + |BD| + |IR|$ MIPs, which is very time consuming, but guarantees the identification of a single IIS if no subproblem is aborted because it exceeds a time or memory limit. When the removal of a constraint generates a subproblem that does exceed a computation limit, that particular constraint is labeled *dubious* and is retained in the output set. This guarantees that the output set is infeasible. If there is at least one dubious constraint in the output set, then it is definitely an IS, but it is not known whether it is also an IIS. It may be possible to apply post-processing tests to the output IS to see whether the dubious constraints can be eliminated to yield a true IIS, but appropriate tests are not known at this time (other than increasing the computation limits and re-running the deletion tests on just the dubious constraints).

To avoid dubious constraints as much as possible it is preferred that all variables be both upper and lower bounded. For the same reason it is also preferred that the variable bounds stay in place as long as possible during deletion filtering. Because the size of the branch and bound tree tends to increase with the number of integer restrictions it is preferable to try to eliminate IRs before LCs or BDs. This helps in generating smaller branch and bound trees as the deletion filter proceeds.

LC_0, BD_0, IR_0 are the original sets of constraints.

INPUT: an infeasible MIP.

0. $status \leftarrow$ *"IIS"*.
   $T = LC_0 \cup BD_0$.
   IF T is infeasible, go to Step 2.
   $T \leftarrow T \cup IR_0$.
1. FOR each $ir_k \in IR_0$:
   IF $T \setminus \{ir_k\}$ is infeasible, $T \leftarrow T \setminus \{ir_k\}$.
   ELSE IF $T \setminus \{ir_k\}$ exceeds computation limit THEN
   $status \leftarrow$ *"IS"*, label ir_k *dubious*.
2. FOR each $lc_k \in LC_0$:
   IF $T \setminus \{lc_k\}$ infeasible, $T \leftarrow T \setminus \{lc_k\}$.
   ELSE IF $T \setminus \{lc_k\}$ exceeds computation limit THEN
   $status \leftarrow$ *"IS"*, label lc_k *dubious*.
3. $BD_1 \leftarrow BD_0 \setminus \{$BDs on variables not in $lc \in T\}$.
   $T \leftarrow (T \setminus BD_0) \cup BD_1$.
   FOR each $bd_k \in BD_1$:
   IF $T \setminus \{bd_k\}$ is infeasible, $T \leftarrow T \setminus \{bd_k\}$.
   ELSE IF $T \setminus \{bd_k\}$ exceeds computation limit THEN
   $status \leftarrow$ *"IS"*, label bd_k *dubious*.

OUTPUT: If $status =$ *"IIS"*, T is an IIS, else T is an IS.

Alg. 6.11. The (*IR-LC-BD*) deletion filter for MIPs

Given the preferences to eliminate IRs early and BDs late, this suggests that the deletion filter should operate on the constraint sets in the order IR-LC-BD. A deletion filter that uses this ordering of the constraints is shown in Alg. 6.11. Note that Step 3 avoids testing any bounds on variables that are not present in the set of LCs remaining after Step 2.

A similar algorithm can be constructed for the *LC-IR-BD* ordering during deletion filtering. Since linear constraints are removed first with this ordering, some variables will no longer be represented in the reduced set of linear constraints after they are deletion tested, and hence the bounds and integer restrictions on those variables can be removed before they are deletion tested.

6.3.2 Additive Methods for MIPs

In adapting the additive method for MIPs, there is again a choice of the order in which the classes of constraints are added. Given our assumption that the initial LP relaxation is feasible, it makes sense to proceed as though the sets *LC* and *BD* have already been added without causing infeasibility. This leaves only the members of *IR* to be tested. Hence the additive method for MIPs, shown in Alg. 6.12, begins by testing the addition of members of *IR* to $LC \cup BD$.

Unlike the deletion filter, the additive method is not able to directly identify dubious constraints. This is because the test set is maintained in a feasible or indeterminate state until sufficient constraints are added to render the test set infeasible. If no indeterminate subproblems are encountered in the course of the isolation, then it is known that the output set is an IIS, but if at least one subproblem exceeds a computation limit, then the output set is an IS (it may also be an IIS, but this is not known).

The worst case time complexity of the additive method occurs when the entire original problem is an IIS. In this case, the method solves n+1 MIPs during the first iteration, one for every constraint in the model plus one MIP for the test of I. During the next iteration, n MIPs are solved, etc. The overall worst case time complexity is then $\frac{1}{2}(|IR| + |LC| + |BD|)^2$ MIP solutions. However, by considering the model in stages as in Alg. 6.12, the worst-case time complexity is reduced to $\frac{1}{2}(|IR|^2 + |LC|^2 + |BD|^2)$ MIP solutions.

A dynamic reordering version of the additive method is also possible, as shown in Alg. 6.13. Note that the reordering applies only within the class of constraints currently being tested.

C: ordered set of constraints in the original infeasible MIP ($IR_0 \cup LC_0 \cup BD_0$).
T: the current test set of constraints.
I: the set of IS members identified so far.

INPUT: an infeasible MIP.

0. *status*← "*IIS*", $I \leftarrow \phi$.
   IF $LC_0 \cup BD_0$ is infeasible, go to Step 2b.
1. $T \leftarrow I \cup LC_0 \cup BD_0$.
   FOR each $ir_k \in IR_0$:
   - $T \leftarrow T \cup \{ir_k\}$.
   - IF *T* exceeds computation limit THEN *status*← "*IS*".
   - ELSE IF *T* infeasible THEN:
     - $I \leftarrow I \cup \{ir_k\}$.
     - IF $I \cup LC_0 \cup BD_0$ exceeds computation limit THEN *status*← "*IS*".
     - ELSE IF $I \cup LC_0 \cup BD_0$ infeasible, go to Step 2.
     - Go to Step 1.
2. a. IF $I \cup BD_0$ exceeds computation limit THEN *status*← "*IS*".
   ELSE IF $I \cup BD_0$ infeasible, go to Step 3.
   b. $T \leftarrow I \cup BD_0$.
   c. FOR each $lc_k \in LC_0$:
   - $T \leftarrow T \cup \{lc_k\}$.
   - IF *T* exceeds computation limit THEN *status*← "*IS*".
   - ELSE IF *T* infeasible THEN:
     - $I \leftarrow I \cup \{lc_k\}$.
     - IF $I \cup BD_0$ exceeds computation limit THEN *status*← "*IS*".
     - ELSE IF $I \cup BD_0$ infeasible, go to Step 3.
     - Go to Step 2b.
3. a. IF *I* exceeds computation limit, *status*← "*IS*".
   ELSE IF *I* inconsistent, exit.
   b. $BD_1 \leftarrow BD_0 \backslash \{$BDs on variables not in $lc \in I\}$.
   c. $T \leftarrow I$.
   d. FOR each $bd_k \in BD_1$:
   - $T \leftarrow T \cup \{bd_k\}$.
   - IF *T* exceeds computation limit THEN *status*←"*IS*".
   - ELSE IF *T* infeasible THEN:
     - $I \leftarrow I \cup \{bd_k\}$.
     - IF *I* exceeds computation limit THEN *status*←"*IS*".
     - ELSE IF *I* infeasible, exit.
     - Go to Step 3c.

OUTPUT: If *status* ="*IIS*", *I* is an IIS, else *I* is an IS.

Alg. 6.12. The basic additive method for MIPs

C: ordered set of constraints in the original infeasible MIP ($IR_0 \cup LC_0 \cup BD_0$).
T: the current test set of constraints. I: the set of IS members identified so far.
INPUT: an infeasible MIP.

0. $status \leftarrow$ "IIS"; $I \leftarrow \varnothing$.
   IF $LC_0 \cup BD_0$ infeasible, go to Step 2b.
1. $T \leftarrow I \cup LC_0 \cup BD_0$.
   FOR each $ir_k \in C$:
      IF ir_k unmarked, $T \leftarrow T \cup \{ir_k\}$, ELSE skip to next iteration.
      IF T exceeds computation limit THEN $status \leftarrow$ "IS".
      ELSE IF T infeasible THEN:
         $I \leftarrow I \cup \{ir_k\}$; $C \leftarrow C \setminus \{ir_j | j > k\}$.
         IF $I \cup LC_0 \cup BD_0$ exceeds computation limit THEN $status \leftarrow$ "IS".
         ELSE IF $I \cup LC_0 \cup BD_0$ infeasible, go to Step 2.
         Go to Step 1.
      ELSE $temp \leftarrow \{ir_j | j > k, ir_j \in C, ir_j$ satisfied$\}$.
         $T \leftarrow T \cup temp$; mark all members of $temp$.
2. a. IF $I \cup BD_0$ exceeds computation limit THEN $status \leftarrow$ "IS".
   ELSE IF $I \cup BD_0$ infeasible, go to Step 3.
   b. $T \leftarrow I \cup BD_0$.
   c. FOR each $lc_k \in C$:
      IF lc_k unmarked, $T \leftarrow T \cup \{lc_k\}$, ELSE skip to next iteration.
      IF T exceeds computation limit THEN $status \leftarrow$ "IS".
      ELSE IF T infeasible THEN:
         $I \leftarrow I \cup \{lc_k\}$; $C \leftarrow C \setminus \{lc_j | j > k\}$.
         IF $I \cup BD_0$ exceeds computation limit THEN $status \leftarrow$ "IS".
         ELSE IF $I \cup BD_0$ infeasible, go to Step 3.
         Go to Step 2b.
      ELSE $temp \leftarrow \{lc_j | j > k, lc_j \in C, lc_j$ satisfied$\}$.
         $T \leftarrow T \cup temp$; mark all members of $temp$.
3. a. IF I exceeds computation limit, $status \leftarrow$ "IS".
   ELSE IF I inconsistent, exit.
   b. $BD_I \leftarrow BD_0 \setminus \{$BDs on variables not in $lc \in I\}$.
   c. $T \leftarrow I$.
   d. FOR each $bd_k \in BD_I$:
      IF bd_k unmarked, $T \leftarrow T \cup \{bd_k\}$.
      IF T exceeds computation limit THEN $status \leftarrow$ "IS".
      ELSE IF T infeasible THEN:
         $I \leftarrow I \cup \{bd_k\}$; $BD_I \leftarrow BD_I \setminus \{bd_j | j > k\}$.
         IF I exceeds computation limit THEN $status \leftarrow$ "IS".
         ELSE IF I infeasible, exit.
         Go to Step 3c.
      ELSE $temp \leftarrow \{bd_j | j > k, bd_j \in BD_I, bd_j$ satisfied$\}$.
         $T \leftarrow T \cup temp$; mark all members of $temp$.

OUTPUT: If $status$ = "IIS", I is an IIS, else I is an IS.

Alg. 6.13. Dynamic reordering additive method for MIPs

6.3.3 An Additive/Deletion Method for MIPs

An additive/deletion method is also available for MIPs, as shown in Alg. 6.14. The basic additive/deletion method proceeds by adding IRs to $LC_0 \cup BD_0$ until infeasibility is triggered, and then switches to the deletion filter to complete the isolation of the infeasibility. The status of the output set as an IS or IIS is determined only during the deletion filtering portion of the algorithm, which is able to identify dubious constraints. During the additive portion of the algorithm, indeterminate subproblems are treated in the same manner as feasible subproblems. Alg. 6.14 is easily modified to incorporate the dynamic reordering version of the additive method in Step 1.

The time complexity of the additive/deletion method derives partly from the time complexity of the additive method as applied to the IRs, and to the time complexity of the deletion filter as applied to the LCs and BDs. The worst case time complexity is $O(|IR|^2+|LC|+|BD|)$ MIP solutions.

T: the current test set of constraints.
I: the set of IS members identified so far.

INPUT: an infeasible MIP.

0. $status \leftarrow$ *"IIS"*; $I \leftarrow \varnothing$.
   IF $LC_0 \cup BD_0$ infeasible THEN go to Step 2a.
1. $T \leftarrow I \cup LC_0 \cup BD_0$.
   FOR each $ir_k \in IR_0$:
      $T \leftarrow T \cup \{ir_k\}$.
      IF T infeasible THEN:
         $I \leftarrow I \cup \{ir_k\}$.
         IF $I \cup LC_0 \cup BD_0$ infeasible THEN go to Step 2.
         Go to Step 1.
2. a. $T \leftarrow I \cup LC_0 \cup BD_0$.
   b. FOR each $lc_k \in LC_0$:
      IF $T \setminus \{lc_k\}$ infeasible THEN $T \leftarrow T \setminus \{lc_k\}$.
      ELSE IF $T \setminus \{lc_k\}$ exceeds computation limit THEN
         $status \leftarrow$ *"IS"*, label lc_k *dubious*.
3. $BD_1 \leftarrow BD_0 \setminus \{$BDs on variables not in $lc \in T\}$.
   $T \leftarrow (T \setminus BD_0) \cup BD_1$.
   FOR each $bd_k \in BD_1$:
      IF $T \setminus \{bd_k\}$ infeasible THEN $T \leftarrow T \setminus \{bd_k\}$.
      ELSE IF $T \setminus \{bd_k\}$ exceeds computation limit THEN
         $status \leftarrow$ *"IS"*, label bd_k *dubious*.

OUTPUT: If $status =$ *"IIS"*, T is an IIS, else T is an IS.

Alg. 6.14. Basic additive/deletion method for MIPs

6.3.4 Using the Information in the Initial Branch and Bound Tree

A great deal of information is contained in the original branch and bound tree that initially discovers that the MIP is infeasible. Some of this information is useful in the subsequent IIS isolation. We develop three theorems in this regard.

Some initial definitions are needed. A *leaf node* of a branch and bound tree is either a node in which all of the IRs are satisfied (i.e. it is an integer-feasible solution), or one in which the LP-relaxation is infeasible. An *intermediate node* is a node that is not a leaf node. For an intermediate node K, IR_K is the set of all IRs satisfied by the LP-relaxation at that node. $BBBD_K$ is the set of BDs added by the branch and bound procedure at some node K (intermediate or final).

Theorem 6.20: *IRs at intermediate nodes (Guieu and Chinneck 1994).* An infeasible MIP does not have any IISs whose integer part is identical to the IR_K at any intermediate node.

Proof: At an intermediate node K, the current set of constraints is $LC \cup BD \cup IR \cup BBBD_K$. Since the node is intermediate, the LP-relaxation is feasible, or equivalently, $LC \cup BD \cup IR_K \cup BBBD_K$ is MIP feasible. An IIS having IR_K as its complete integer part must have as its linear part either $LC \cup BD \cup BBBD_K$ or some subset of it, but no such IIS can exist because it is already known that $LC \cup BD \cup IR_K \cup BBBD_K$ is MIP feasible. ■

Theorem 6.21: *Sensitivity filtering leaf nodes (Guieu and Chinneck 1999).* If a sensitivity filter is applied to every leaf node, and all original LCs and BDs having nonzero reduced costs are marked, then the set $IR \cup$\{marked LCs\}$\cup$\{marked BDs\} is infeasible.

Proof: The unmarked LCs and BDs are not marked because they are not tight in any of the leaf nodes. Hence those unmarked LCs and BDs could have been relaxed in the original MILP and the same branch and bound tree would still have proven infeasibility of the modified MILP. ■

Some further definitions are needed. A *path* in a branch and bound tree is a set of branches leading from the root to a leaf in which each branch is labeled with the name of the integer variable that was branched on. The set of *active IRs (A_T)* is the union of all of the IRs for the branched variables in any of the paths in a branch and bound tree.

Theorem 6.22: *Branched variables (Guieu and Chinneck 1999).* For an infeasible MIP, the set $LC \cup BD \cup A_T$ is infeasible.

Proof: Given the MIP $LC \cup BD \cup A_T$, a branch and bound tree identical to the original branch and bound tree can be generated, arriving at the conclusion that $LC \cup BD \cup A_T$ is infeasible. ■

Notice also that each path provides an interesting candidate for an IS: the constraint set $LC \cup BD \cup$\{IRs on variables in the path\}. This candidate for an IS is more likely to prove infeasible because the set of branches in the path terminates at an infeasible node. There is no guarantee that the candidate IS is actually infeasible, however, since the path may consist partly or entirely of one-sided branches (i.e. a particular variable is branched upon only in the higher-valued direction or only in the lower-valued direction).

Where the MIP has multiple IISs, it may be possible to develop a different branch and bound tree for the same model (perhaps by varying parameters such as the bounding rule or branching variable selection rule) in which different sets of LCs, BDs and IRs can be eliminated using these three theorems. This happens when a different IIS drives the development of the branch and bound tree.

Thms. 6.20 – 6.22 suggest a preprocessing of the MIP after it has been found infeasible but before the infeasibility isolation algorithms are applied. Thm. 6.21 allows the initial elimination of any unmarked LCs or BDs. Thm. 6.22 allows the initial elimination of any IRs that do not appear in A_T.

Each path in the original branch and bound tree provides a candidate for the set of IRs in an IS. This set can be pruned by comparing the sets of IRs associated with the paths with the sets of IRs associated with the nodes. Any IR set associated with a node (and any subset of such a set) cannot be the entire IR set in an IS in conjunction with $LC \cup BD$ by Thm. 6.20.

These ideas can be combined as shown in Alg. 6.15. For efficiency, as new IRP_i are discovered during the initial branch and bound solution, they can be checked against the current IRN^*. Similarly, as new IRN_i are discovered, the current members of IRP can be checked against it and its subsets. This would, however, slow the solution in the case of a feasible MIP.

IRP: IRP_i is the set of IRs defined by the variables in path i.
    $IRP = \{IRP_i | i = 1 \text{ to (number of paths)}\}$.
IRN: IRN_i is the set of IRs defined by the satisfied IRs at an intermediate node i.
    $IRN = \{IRN_i | i = 1 \text{ to (number of intermediate nodes)}\}$.
    $IRN^* = IRN \cup \{\text{all proper subsets of members of } IRN\}$.
LCM: the set of marked LCs.
BDM: the set of marked BDs.

INPUT: a MIP, feasibility status unknown.

1. Solve the MIP. Compile the sets A_T, *IRP*, *IRN*, *LCM*, *BDM* while solving. If feasible, exit.
2. $IRP \leftarrow IRP \backslash (IRP \cap IRN^*)$.
   Order *IRP* from smallest to largest cardinality.
3. FOR each $IRP_i \in IRP$:
        IF $LCM \cup BDM \cup IRP_i$ is infeasible THEN
            $IRP' \leftarrow IRP_i$.
            Go to Step 4.
   $IRP' \leftarrow A_T$.
4. Isolate an IIS or IS in $LCM \cup BDM \cup IRP'$ using any algorithm.

OUTPUT: an IIS or IS.

Alg. 6.15. Using information from the original branch and bound tree

6.3.5 Speed-ups

Grouping of constraints is a useful strategy for reducing the number of MIP solution needs for the deletion filter, the additive method and the additive/deletion method. This is an effective strategy in this case, as shown in Sec. 6.3.6.

The settings of the MIP solver have a great influence on the speed of the solution. Since the IIS isolation algorithms require the solution of numerous test MIPs, it is worthwhile determining the MIP solver settings that provide quick solutions for the intermediate test MIPs. Two solver settings have the most influence on the speed of the MIP solution: the method of node selection and the method of branching variable selection.

The two most common methods of node selection are best-bound and depth-first. Since determining that a MIP is infeasible requires a complete expansion of the branch and bound tree, neither method is likely to be faster for infeasible MIPs. However, when the MIP is feasible, it is likely that a depth-first node selection will reach feasibility faster. In addition, depth-first node selection allows re-use of the final LP basis from the parent node, which will be near-feasible for the child nodes. Since we need only to determine feasibility status when examining the test MIPs, depth-first node selection may be preferred.

A number of branching variable selection schemes are possible, including use of estimates of the branching variable impact on the objective function, a simple list ordering, user-defined priority weighted ordering, and the variable that is most infeasible. See Chap. 3 for details.

The original objective function does not play a useful role during infeasibility analysis. It may even slow the infeasibility isolation by the way in which it guides the development of the branch and bound tree. Speed improvement may be possible by replacing the original MIP objective by one that tends to decide feasibility status more rapidly.

When a subproblem is MIP-infeasible (but LP-relaxation feasible), two child nodes are generated, each having a new bound added based on the branching variable x_k, whose non-integer value in the parent node is α. The typical form of the added constraint (with nonnegative slack variable s_k included) is: $x_k+s_k=\lfloor\alpha\rfloor$ or $x_k-s_k=\lceil\alpha\rceil$. A new objective function can then be introduced: *minimize* Σs_k over all of the slack variables introduced during branching. The effect is to drive the MIP towards feasibility in a manner analogous to an ordinary LP phase 1, which should speed the decision of feasibility status in the test subproblems.

6.3.6 Conclusions from Empirical Studies

Guieu and Chinneck (1999) carried out an extensive study of the methods described above. Atlihan and Schrage (2006) studied their binary grouping strategies applied to the deletion filter (see Sec. 6.1.5) for infeasible MIPs. The conclusions arising from these studies are summarized here.

The test set collected by Guieu and Chinneck consists of 20 infeasible MIPs; Atlihan and Schrage used the same models. The infeasibility in two of the models is original, with unknown cause. The remaining 18 models were taken from the

MIPLIB 3.0 set (Bixby et al. 1996) and altered to be MIP-infeasible with a feasible initial LP relaxation. This was done by adding a constraint constructed from the objective function that requires it to take a value midway between the objective function value of the initial LP relaxation and the final MIP-feasible objective function value. Further details about the experimental setup are available in Gueiu and Chinneck (1999).

The results of the experiments by Guieu and Chinneck are summarized in Table 6.2. Methods in the Table are sorted by decreasing number of average LP iterations for the IS isolation process to run. Numbers are averages over the 20 test models. The column *#dubious LC|BD (#IISs)* reports the average number of dubious constraints in each category along with the number of cases having no dubious constraints (i.e. an IIS is reported); this is omitted in favour of comments on the results in the case of the additive methods which cannot detect dubious constraints. The columns *#IR*, *#LC* and *#BD* report the average number of each class of constraint in the output IS.

Some very general conclusions can be drawn from these small experiments. First, notice that the deletion filtering methods always complete: there are no cases that time out whereas the two additive-only methods time out on 3 models. Note that the much smaller numbers of LP iterations and nodes reported for the two additive-only methods results from the fact that the timed-out models are omitted from the averages. Second, the IR-LC-BD deletion filter is faster than the LC-IR-BD deletion filter, as expected: LP iterations are reduced by 44%. Third, grouping improves speed: LP iterations are reduced by 35% for the IR-LC-BD filter when a standard group size of 4 is used. Fourth, the dynamic reordering additive/deletion method is quick: LP iterations are reduced by 22% compared to the IR-LC-BD deletion filter. Based on these results, a grouped version of the dynamic reordering additive/deletion method should be tested.

Table 6.2. Summary of results for IS isolation methods for MIPs

Method	#dubious LC|BD (#IISs)	#IR	#LC	#BD	#B&B nodes	#LP iterations
LC-IR-BD deletion filter	17|181.9 (5)	16.1	131.8	289.6	499153.8	3401931.4
IR-LC-BD deletion filter	16.4|185.3 (5)	11.5	153.8	321.4	344796.8	1913248.4
dynamic reordering additive/ deletion method	0.3|8.1 (3)	7.9	135.4	308.6	124512.4	1487990.6
IR-LC-BD deletion filter with group size 4	16.4|186.1 (5)	11.5	153.4	311.6	189561.1	1246078.1
basic additive method	4 IISs, 3 models timed out	8.9	49.7	142.0	172687.8	982255.4
dynamic reordering additive method	4 IISs, 3 models timed out	7.6	40.5	145.2	130067.6	396176.2

Atlihan and Schrage (2006) tested a binary grouping deletion filter (see Sec. 6.1.5) applied to the same 20 models, but with the significant difference that they do not attempt to remove IRs or BDs. Only linear constraints are removed by their methods. This amounts to an LC-IR-BD deletion filter with binary grouping on the linear constraints, but halted after the linear constraints have been deletion tested. Their results show that the generalized binary search deletion filter (Alg. 6.6) and the basic deletion filter perform similarly in terms of CPU time, though the basic deletion filter requires fewer MIP solutions for 70% of the models. The depth-first binary search deletion filter performed relatively poorly.

While using different machines and solvers, the Atlihan and Schrage results can be roughly compared with those in Table 6.2 by looking at the average number of simplex iterations. Because Atlihan and Schrage deletion test only the linear constraints, their average simplex iterations for the basic deletion filter are lower (218644.5) than for the full LC-IR-BD deletion filter in Table 6.2 (3401931.4), as expected. The average iterations for their best method, the generalized binary search deletion filter, applied to only the linear constraints (1714413.9) is surprisingly worse than two methods in Table 6.2: the dynamic reordering additive/deletion method (1487990.6) and the IR-LC-BD deletion filter with group size 4 (1246078.1). This is likely due to the fact that both of these faster methods are geared to eliminating IRs early, which greatly speeds the subsequent MIP solutions.

Note that IISs are found only relatively rarely in Table 6.2, generally for only 3–5 of the 20 models. Most often the result is an IS. The methods do not vary significantly in the size or composition of the ISs returned, which are relatively large. This may be an artifact of the way in which the infeasibilities were constructed for these test models. Atlihan and Schrage report that all of their tested algorithms (basic deletion filter, DFBS and GBS) report an irreducible set of linear constraints in 18 of the 20 test cases (note that this is not the same as an IIS which would include irreducible numbers of BDs and IRs as well). The average number of LCs in their output sets is lower than in the four top rows of Table 6.2. Note that the original models have the following average statistics: 79.5 IRs, 237.6 LCs, and 518.0 BDs.

The use of the information in the branch and bound tree (as per Sec. 6.3.4) was also investigated. Results were mixed, with this approach proving extremely helpful for some models, but also worsening the results significantly for some other models. However the techniques can probably be implemented in a manner that will be significantly faster in practice.

Perhaps the most important conclusion is that these IS isolation techniques are extremely slow. Consider that the average number of simplex iterations needed for the initial solution of the models is just 1718.7 and the average number of branch and bound nodes is 436.5. The repeated solution of MIPs is very time consuming: there is as yet no hot start technique as there is in linear programming. Certainly the recent development of faster methods for reaching MIP feasibility (See Chap. 3) will help, but overall speed will continue to be a major stumbling block. Still, it is often useful to substitute machine time for human time: an IS analysis of an infeasible MIP can run overnight and provide a suitable reduction in the size of the problem that the analyst must deal with.

Given that the deletion filter based methods are the most robust, and that they produce IS isolations of about the same size, the best method for use in practice is simply the fastest one, in this case the IR-LC-BD deletion filter with group size 4. The dynamic reordering additive/deletion method with grouping may prove faster, but has not been tested.

The isolation of IISs in MIPs is fertile ground for further research with immense practical pay-offs.

6.3.7 Software Survey

As described in Sec. 6.3.6, Guieu and Chinneck (1999) built academic prototype software to carry out various combinations of the deletion filter and the additive method with several types of grouping. Their software used the Cplex 3.0 callable library to carry out the MIP solutions.

Two commercial solver systems claim infeasibility analysis systems for MIPs. The "conflict refiner" in Cplex 10.1 (Ilog 2006) apparently uses variations on the deletion filter to identify small conflict sets, and is described as a generalization and extension of the IIS finder. It includes some features for guiding the isolation (see Sec. 6.2.6) such as treating constraints in groups, assigning weights etc. which can be applied to LP models as well. The LINGO modeling system (Lindo Systems Inc. 2007) also claims the ability to isolate small infeasible subsets in MIPs, again using the deletion filter and additive method. See Atlihan and Schrage (2006) for further information.

6.4 Methods Specific to Nonlinear Programming

We are interested here in isolating infeasibility in models that include one or more nonlinear constraints, and consisting entirely of continuous variables, generically termed nonlinear programs or NLPs for ease of reference. Note that we are specifically excluding nonlinear systems in which only the objective function is nonlinear while all of the constraints are linear; infeasibility in such systems is easily isolated using the methods for linear systems.

Isolating infeasibility in nonlinear programs is significantly harder than for linear systems simply because it can be extremely difficult to determine the feasibility status of an arbitrary set of nonlinear constraints. A particular solver may determine that an NLP is infeasible when it is in fact feasible: the solver is just unable to find a feasible point. For local NLP solvers, the determination of infeasibility is always indefinite: you can only become more confident by trying different starting points or solver settings, but you can never be completely confident that the status is truly infeasible. Note, however, that if a feasible point is found, the status is definite.

There exist some special classes of nonlinear programs for which the feasibility status is frequently determined correctly (though this cannot be guaranteed due to the possibility of numerical difficulties). A *quadratically-constrained quadratic program (QCP)* is a nonlinear convex generalization of an LP and hence the determination of feasibility status is less difficult. This makes the subsets isolated by the deletion filter and the additive method more likely to be true IISs. Since an important class of solvers for general NLPs make use of successive quadratic approximations, this improved ability to isolate IISs is especially welcome.

We will concentrate here on local NLP solvers applied to general sets of constraints that include at least one nonlinear constraint. Feasibility status is more definitely determined by global optimizers, which normally use a space-covering strategy similar to branch and bound. The deletion filter and the additive method can be used with good success in conjunction with global optimizers; their use and related issues will be similar to their application to MIPs (see Sec. 6.4.4). However global optimizers are very slow and the state of the art in practice is still the use of local solvers. As we will see below the state of the art consists mostly of adapting the deletion filter from LP for use with local solvers for NLPs. Variations on the additive method are also used for Quadratically Constrained Quadratic programs. Curiously, the pure elastic filter has not yet been adapted for use with NLPs, though this would be straightforward development.

Debrosse and Westerberg (1973) develop a number of relevant theorems relating to IISs in systems of nonlinear constraints. These are given in Sec. 5.8 in the context of their bootstrapping method for reaching feasibility quickly.

6.4.1 Deletion Filtering

Chinneck (1995) studied the use of the deletion filter to isolate IISs in NLPs. Many factors affect the ability of a solver to decide feasibility status correctly, including the NLP algorithm and its implementation, the tolerances chosen, the initial point selected, the termination criteria, method of approximating derivatives, etc. The main difficulty in using the deletion filter for NLPs is the inability of local solvers to accurately determine the feasibility status of the NLP in all cases. However the decision error happens in one way only: a solver may declare that a feasible model is infeasible, but it will never declare that an infeasible model is feasible, issues of tolerance aside. The incorrect declaration of infeasibility generally happens when the solution process becomes trapped at an infeasible local minimum of a phase 1 penalty function. Incorrect declaration of feasibility does not happen because every candidate feasible point is checked against the constraints and feasibility is declared only if all constraints are satisfied.

The one-way error in deciding feasibility status has a particular effect on the deletion filter. The solver may decide that the reduced model remains infeasible when a certain constraint is temporarily removed, when in fact the model has become feasible but the solver is unable to detect this. The deletion filter will then incorrectly drop the constraint permanently when it should be reinstated. For this reason, the deletion filtering algorithm can only guarantee the identification of a *Minimal Intractable Subsystem (MIS)*, rather than an IIS.

An MIS is formally defined as follows. A Minimal Intractable Subsystem (MIS) of constraints in an NLP is a minimal set of constraints causing a given NLP solver to report infeasibility under a given set of parameter settings (including initial point, tolerances, termination conditions, etc.).

A further complication is that the removal of one or more constraints can result in a mathematical error, e.g. taking the square root of a negative number, or dividing by zero. This is especially serious as constraints are dropped during deletion filtering. Consider this infeasible set of constraints: $y-\sqrt{x}=0, x \geq 0, y \leq -1$. When the bound on x is dropped during deletion filtering, a mathematical error occurs. Strictly speaking, this is simply a special kind of infeasibility, but it does cause practical difficulties. Operational definitions are given below.

A model is judged *infeasible in the ordinary sense* when the solver decides that it is infeasible without encountering a mathematical error. A model is judged *infeasible due to mathematical error* when the solver decides that it is infeasible because it is requested to perform an illegal mathematical operation such as dividing by zero. Modern compilers are able to detect and recover from illegal operations without difficulty.

When a constraint is temporarily removed during deletion filtering, the solver will decide that the reduced model is in one of three states: (1) feasible, (2) infeasible in the ordinary sense, or (3) infeasible due to mathematical error. The deletion filter tries to keep the model in state (2) as constraints are removed one by one, always reinstating a constraint whose temporary removal causes the reduced model to enter states (1) or (3). Constraints whose removal places the reduced model in state (3) are called *guards* and are specially identified on output.

The modified deletion filter for NLPs is given in Alg. 6.16. In general, whether the isolated MIS is also an IIS is not known. However, if the constraints in the MIS are all linear, then it is easily shown that the MIS is also an IIS.

Elastic filtering is also easily adapted to the nonlinear case, but the reduced set of constraints that is output must still be deletion filtered to guarantee the identification of a single MIS. There is at present no nonlinear analog of sensitivity filtering. For these reasons, only the nonlinear deletion filter has been implemented.

INPUT: an infeasible set of constraints, at least one of which is nonlinear.
FOR each constraint in the set:
  1. Reset the initial point and solver parameters.
  2. Temporarily drop the constraint from the set.
  3. DO CASE:
     i. Solver reports feasibility:
        Return dropped constraint to the set.
     ii. Solver reports infeasibility in the ordinary sense:
        Drop the constraint permanently.
     iii. Solver reports infeasibility due to mathematical error:
        a. Mark dropped constraint as a guard.
        b. Return dropped constraint to the set.
OUTPUT: constraints constituting a single MIS (including guards).

Alg. 6.16. The deletion filter for NLPs

There are four possible cases when Alg. 6.16 is applied:

1. The model is feasible and this is correctly detected by the solver. Infeasibility analysis is not needed.
2. The model is feasible, but is reported infeasible by the solver and an MIS is isolated.
3. The model is infeasible, and an MIS is isolated which is also an IIS.
4. The model is infeasible, and an MIS is isolated which is not an IIS.

Case 1 is straightforward. The difficulty lies in distinguishing between cases 2–4. Knowledge of the physical meaning of the model is required to make a definite diagnosis, but the isolation provided in cases 2–4 is still very helpful. It is a subset of the model that is intractable to the solver in use with the current initial point and parameter settings. Assuming the modeller will continue using the same solver, the isolation provides a smaller subset of the model for further experimentation with other initial points, parameter settings, etc. Sometimes the problem is easily diagnosed by inspection, once the MIS shows where to look.

When the MIS contains at least one nonlinear constraint it is important to improve confidence about whether the model is really infeasible (cases 3 and 4), or is actually feasible (case 2). One simple approach is to submit several new initial points to the solver operating on the reduced problem formed by the MIS, as described in Alg. 6.17. This is faster than trying new initial points on the perhaps much larger original problem.

INPUT: An MIS from an infeasible NLP.
Make the MIS the current constraint set.
DO until satisfied:
  Postulate and record a new initial point.
  Apply solver to test feasibility of current constraint set using
    the new initial point; record final point.
  IF final point feasible THEN
    Apply solver to test feasibility of complete original constraint set
      using the feasible point as initial point.
    IF complete original constraint set feasible THEN
      Record the final feasible point and exit with appropriate message.
    ELSE (complete original constraint set infeasible)
      Find MIS in complete original constraint set and record it.
      Make current constraint set the union of the current constraint
        set and the new MIS.
OUTPUT: Either (1) a feasible point for the original problem, or
  (2) one or more MISs and a list of infeasible initial and final point pairs.

Alg. 6.17. Using the MIS

New initial points can be postulated in a number of ways, either by inspection of the current constraint set (note that the dimensionality is likely to be lower than in the original problem), or randomly. If output (2) of Alg. 6.17 is returned, then the sets of initial and final points can be used to form an idea of the regions of attraction of the MIS(s) in the model. The algorithm can then be restarted with initial points placed more appropriately. Human insight and knowledge of the physical meaning of the model are needed here.

If sufficient initial points are used, with the solver reporting infeasibility each time, then confidence is increased that the model really is infeasible. For further confidence, it may be appropriate to adjust the other solver parameters (e.g. tolerances, methods of approximating derivatives, etc.), or even to change solvers altogether and repeat the process.

Chinneck (1995) developed prototype software to implement Alg. 6.16 based on the local solvers MINOS (Murtagh and Saunders 1987) and LSGRG (Smith and Lasdon 1992). The modified version of LSGRG, named LSGRG(MIS), was used to analyze some small examples. It is instructive to consider the infeasible NLTEST1 model:

$$row1: -x_1^2 + 10x_1 + x_2 \leq 25$$

$$row2: -x_1^2 + 10x_1 - x_2 \leq 15$$

$$row3: x_1 + x_2 \leq 25$$

$$x_1 \; bounds: 4 \leq x_1 \leq 6$$

$$x_2 \; bounds: 0 \leq x_2 \leq 12$$

NLTEST1 is sketched in Fig. 6.7. Applying LSGRG(MIS) directly to NLTEST1 using the initial point (5.5,5.5) and the default settings of all parameters isolates MIS1:{*row1*, *row2*, $x_1 \leq 6$}. This is an example of case 4 in which the MIS is not an IIS. Fig. 6.8 shows the surface formed by the absolute sum of the constraint violations and explains why this happens: NLTEST1 has an infeasible local minimum at (7.236,5) which traps the phase 1 solution. The absolute sum of the constraint violations plotted for the MIS alone, shown in Fig. 6.9 shows a similar infeasible local minimum. From the given initial point, the MIS is a minimal set of constraints creating an infeasible local minimum which traps the solver.

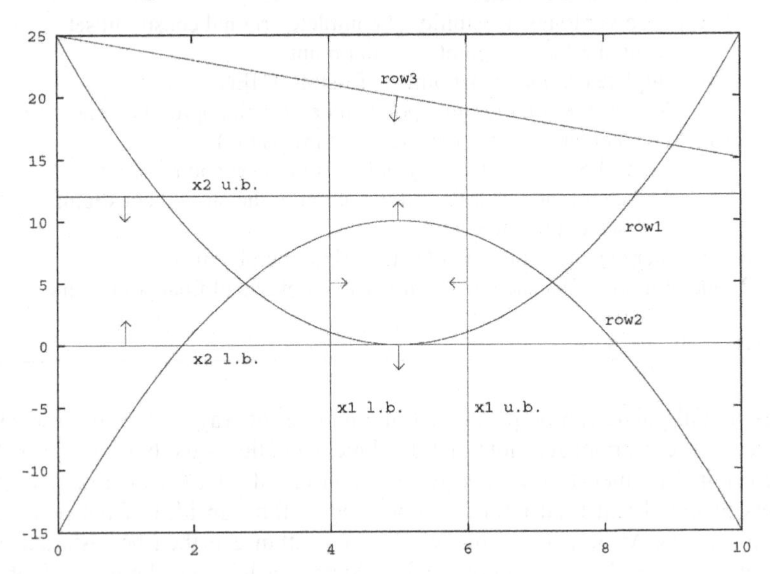

Fig. 6.7. Constraints in NLTEST1

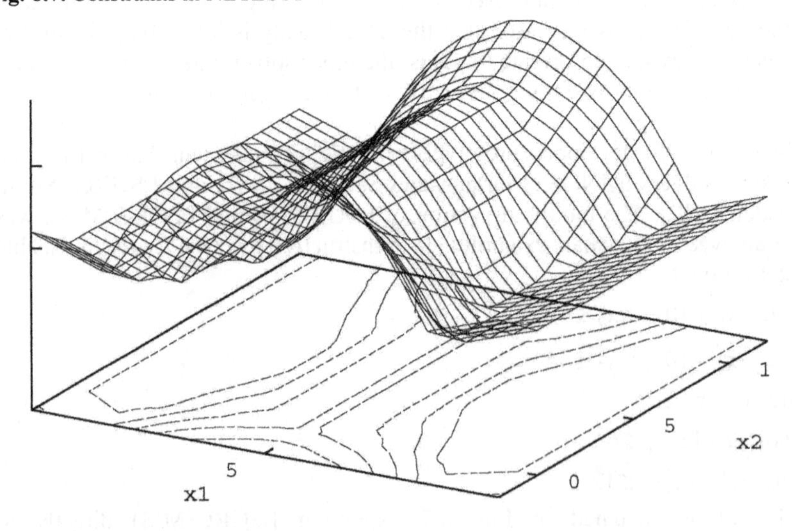

Fig. 6.8. Sum of the absolute constraint violations in NLTEST1

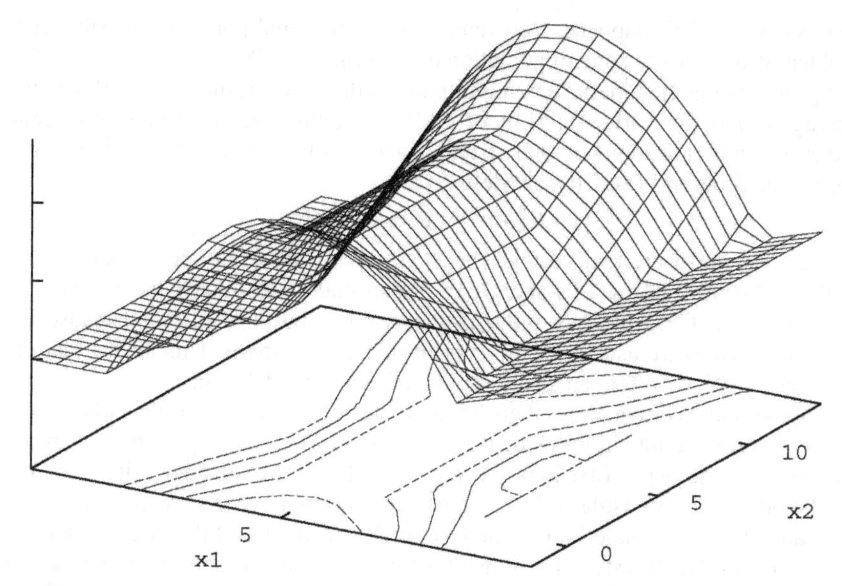

Fig. 6.9. Sum of absolute constraint violations for NLTEST1 MIS1

Using an initial point of (0,4) leads LSGRG(MIS) to isolate a different MIS2:{*row1*, *row2*, $x_1 \geq 4$} due to the symmetry of the IIS. The reasoning is similar to the above. By inspection of Fig. 6.8, it is obvious that any initial point having $x_1 < 5$ will result in the isolation of MIS2, and any initial point having $x_1 > 5$ will result in the isolation of MIS1. The symmetry of NLTEST1 places the "saddle ridges" in Figs. 6.8 and 6.9 along $x_1 = 5$, with the saddle ridge demarking the border of the two zones of attractions of the infeasible local minima created by the MISs.

A related feasible problem is created by eliminating the lower bound on x_1. Using the initial point (5.5,5.5) again leads LSGRG(MIS) to isolate MIS1 (an example of case 2), but using the initial point (0,4) leads to the correct conclusion that the model is feasible (an example of case 1). Analysis of the MIS found when (5.5,5.5) is used as the initial point may lead to a new placement of the initial point such that it satisfies all of the constraints in the MIS; this should help achieve feasibility.

This example illustrates the difficulty of correctly deciding feasibility when constraints have multiple intersections, as predicted by Debrosse and Westerberg (1973).

6.4.1.1 Speeding the Isolation by Grouping Constraints

Nonlinear solvers are not able to make use of advanced starts in the same way that LP solvers can. However, in the context of the deletion filter or the additive method, it is possible to re-use the final point from one solution as the initial point for the next solution. Unfortunately, this "point-chaining" can lead the isolation process astray. This happens, for example, when the final point of a feasible subproblem is outside the zone of attraction of the original MIS.

Speed-ups via the binary search grouping methods were studied by Atlihan and Schrage (2006) for various classes of NLPs, including QCPs, second order cone programs (generalizations of QCPs), and general NLPs. The constraints in second-order cone programs (SOCPs) are defined as

$$\|A_i x + b_i\| + c_i^T x - d_i \leq 0 \; \forall i = 1 \ldots m$$

where c_i, b_i, and x are real vectors, the A_i are real matrices of appropriate dimensions, and d_i are scalars. LPs and QCPs are special cases of SOCPs. For convex QCQPs and SOCPs, the sensitivity filter may apply but it often fails because optimal dual prices are usually in the interior of the dual space. This means that the deletion/sensitivity filter can also be applied to these model forms.

Atlihan and Schrage show that the generalized binary search algorithm gives the best overall results in terms of isolating IISs more frequently than MISs when compared to the depth-first binary search and the simple deletion filter. For the QCP models, IISs are isolated in 22 of 30 models tested, for the second order cone programs, IISs are isolated for 37 of 46 models tested. Only MISs are isolated for the 11 general NLPs tested. The sensitivity filter can also be applied in the case of QCPs and SOCPs, and this speeds the solution process as well as identifying slightly more IISs. The generalized binary search is also usually the fastest of the methods. Other grouping approaches, such as fixed sizes, or adaptive methods can also be applied. Empirical studies to identify the best approach should be done.

One obvious speed-up is the use of the much faster linear IIS isolation methods if it should happen that an infeasible set of constraints consists entirely of linear constraints.

6.4.2 IIS Isolation by the Method of Debrosse and Westerberg

The algorithm by Debrosse and Westerberg (1973) for finding a feasible initial point for a set of general nonlinear constraints is described in Alg. 5.7 of Sec. 5.8. If the algorithm is not able to find a feasible point, then it outputs an IIS, providing that it is able to correctly solve for the intersections of arbitrary subsets of the constraints. To illustrate the application of Alg. 5.7 to an infeasible set of constraints, we repeat here Example 3 from Debrosse and Westerberg (1983), which is shown in Fig. 6.10.

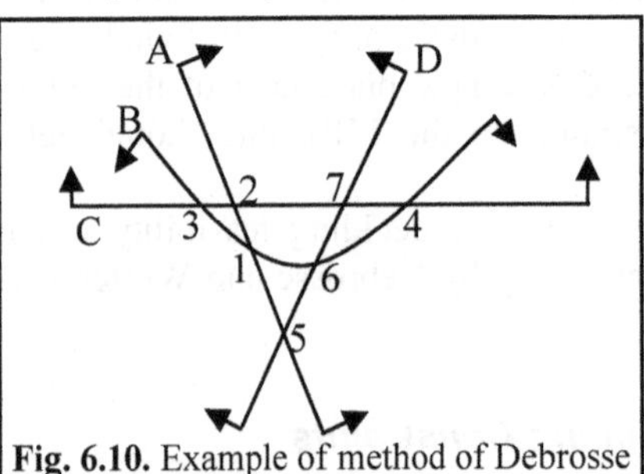

Fig. 6.10. Example of method of Debrosse and Westerberg

There are four constraints A, B, C, and D, of which constraint B is nonlinear. The numbers in Fig. 6.10 indicate the various constraint intersections that are solved in the course of the algorithm. Table 6.3 shows the steps that Alg. 5.7 carries out.

As shown in the table, several of the intersection points are found more than once. In addition, the enumeration of subsets is shown. The number of subsets enumerated grows substantially with the cardinality of ***H***.

Table 6.3. Example IIS isolation by method of Debrosse and Westerberg (1973)

steps	hypothesis set ***H***	constraints being solved	resulting point x	constraints violated at x
0,1		A,B	1	C
2	A,B,C			
9.1		A,B	1 (again)	C
9.1		A,C	2	B
9.1		B,C	3	A
2nd pt		B,C	4	D (*H* disproved)
1,2	B,C,D			
9.1		B,C	3 (again)	A (*H* disproved)
1,2	A,B,C: cycle detected. No alternative standard *H*. New *H* is A,B,C,D. (Cycle recovery routine not shown in Alg. 5.7)			
9.1		as before: (A,B), (A,C), (B,C)	1,2,3,4	C,B,A,D
9.1		A,D	5	C
9.1		B,D	6	C
9.1		C,D	7	B
14.1	Exit with {A,B,C,D} as an IIS.			

As described in Sec. 5.8, this algorithm is suited only for models that are highly structured such that each constraint involves only a few variables. By inspection of Fig. 6.10 and Table 6.3 you can see the difficulty that is caused by the inability to find all of the intersections of a set of constraints. If the solver does not find the second intersection of constraints B and C, then it will eventually wrongly identify {A,B,C} as an IIS. This difficulty is even more pronounced for very complex constraints that have multiple intersections that are difficult to enumerate.

Note that if an MIS is first isolated by some other method, it is possible to apply the method of Debrosse and Westerberg directly to it. The algorithm should run much more quickly on the small portion of the model isolated by the MIS. The difficulties of structure, equality constraints, and multiple intersections remain however.

6.4.3 Methods for Quadratic Programs

Obuchowska (1998, 1999) developed an adaptation of the deletion filter for the specific case of systems of convex quadratic inequalities. She considers the case of quadratically constrained quadratic programs (QCQP), whose constraints are defined as

$$R = \{x \in \Re^n \mid Q_i(x) := a_i^T x + \tfrac{1}{2} x^T B_i x \le b_i, i \in I = \{1,\ldots m\}\}$$

where the matrices B_i are $n \times n$, real-valued, symmetric and positive semidefinite, and the n-vectors a_i are real-valued.

Obuchowska's algorithm is a straightforward deletion filter, except that a set of candidate members of the IIS is found in advance, and these are tested and removed if their deletion does not render the set of constraints feasible. This initial set of candidate IIS members is the set of so-called *killing constraints*, defined as "a maximal subset of constraints that may have an impact on the feasibility status of the system after some perturbation of the right-hand sides". An algorithm for finding the set of killing constraints in $O(mn^2\min(m,n))$ operations is given (Obuchowska 1998).

More formally, the killing constraints are defined as follows (Obuchowska 1998): the kth inequality belongs to the set K of killing constraints if there exist values

$$b_k'' > -\infty \text{ and } b_i' > -\infty, i \in I$$

such that the system

$$Q_i(x) \le b_i', i \in I$$

is infeasible and the system

$$Q_i(x) \le b_i', i \in I \backslash \{k\}; Q_k \le b_k''$$

is feasible, or conversely.

It turns out that the set of killing constraints is the same as the set of implicit equalities, defined as follows: an inequality $a_i^T s \le 0$ in the system

$$B_i s = 0, \forall i \in I; a_i^T s \le 0, \forall i \in I$$

is an implicit equality if $a_i^T s = 0$ for all s satisfying the system. A related theorem shows that if the system has no implicit equalities, then it is feasible and there are no killing constraints. This sets up Alg. 6.18 for identifying the set E of implied equalities.

INPUT: a set of quadratic inequality constraints as defined above.
1. $k = 1; E_0 = I$.
2. Find the set E_k of all implied equalities in the system $B_i s = 0, \forall i \in E_{k-1}; a_i^T s \le 0, \forall i \in E_{k-1}$
3. IF $|E_k| = 0$ THEN the system is feasible and $K = \varnothing$; exit.
4. IF $E_k = E_{k-1}$ THEN $K = E_k$; exit.
5. $k = k+1$; go to Step 2.

OUTPUT: the set K of killing constraints (implied equalities).

Alg. 6.18. Finding the set of killing constraints

Obuchowska (1998) elucidates some properties of implied equalities, killing constraints and IISs in Thm. 6.23.

Theorem 6.23: *Implied equalities, killing constraints, and IISs* (Obuchowska 1998, Lemma 3.1). Where T is an IIS: (a) Every inequality in the system

$B_i s = 0, \forall i \in T; a_i^T s \leq 0, \forall i \in T$ is an implicit equality. (b) For every $k \in T$ the system $B_i s = 0, \forall i \in T \setminus \{k\}; a_i^T s \leq 0, \forall i \in T \setminus \{k\}$ contains inequalities that are not implicit inequalities in the system. (c) Any IIS belongs to K, that is $\bigcup_\alpha IIS_\alpha \subset K$. ■

Thm. 6.23 sets up Alg. 6.19, a variation on deletion filtering for isolating IISs in QCQP (Obuchowska 1998). The distinction from an ordinary deletion filter is that the set of constraints is first reduced by running Alg. 6.18 to identify the killing constraints.

INPUT: an infeasible set of quadratic inequality constraints as defined above.
1. Run Alg. 6.18 to find the set of killing constraints K.
2. FOR every constraint j in K:
   3.1 IF the system $Q_i(x) \leq b_i$, $i \in K \setminus \{j\}$ is infeasible THEN:
      3.1.1 $K \leftarrow K \setminus \{j\}$

OUTPUT: the set K is an IIS.

Alg. 6.19. Deletion filtering for inequality-constrained QCQP

Obuchowska has tested Alg. 6.18 on a few small problems, but does not report any test results for Alg. 6.19. Neither of the algorithms has been implemented or tested on a commercial scale.

As mentioned in Sec. 6.4.1.1, Atlihan and Schrage (2006) apply various constraint grouping algorithms for isolating IISs in Quadratically Constrained Quadratic programs using the deletion filter and a combined deletion filter and additive method. They report good results for these methods.

6.4.4 Methods for Space-Covering Global Optimizers

Dravnieks and Chinneck (1997) considered how to isolate an IIS within a global optimization system. The underlying global optimizer is a space-covering branch and bound system in which the variable space is iteratively subdivided into smaller and smaller multidimensional boxes. A given box may have one of the following statuses: (i) infeasible if it can be proven that every point in the box violates at least one constraint, (ii) feasible if it can be proven that every point in the box satisfies all of the constraints, (iii) indeterminate if the box is not known to be feasible or infeasible. Proofs of feasibility or infeasibility of a box make use of interval calculations of the maximum and minimum values of the constraint functions over the box, which are then compared to the constraint limits. Infeasibility of the entire model is proven when all boxes are infeasible.

When infeasibility of the entire model is proven, the global optimizer outputs a list of the constraints that were used to prove infeasibility of each individual box along with the original variable bounds. This list is then submitted to a deletion filter. Because the global optimizer is able to determine the feasibility status of an arbitrary set of constraints with perfect accuracy, the deletion filter will function correctly, and an IIS will be produced. However the process is very slow since it requires the solution of numerous global optimization problems, and hence is suitable only for small models. The limit applies mainly to the number of dimensions since this has the biggest impact on how many boxes will be generated.

In one small example described in the paper, an alternative objective function that is more oriented to the feasibility problem proves helpful. The inclusion of a local optimizer to search for feasible points in boxes is also recommended.

6.4.5 Software Survey

Chinneck (1995) developed academic prototype software implementing the deletion filter for NLPs which used LSGRG to carry out the NLP solutions. Two commercial solver systems claim to have infeasibility analysis systems for NLPs. The LINGO modeling system (Lindo Systems Inc. 2007) claims the ability to isolate small infeasible subsets in NLPs, mentioning quadratic systems in particular, and uses combinations of the deletion filter and the additive method. See Atlihan and Schrage (2006) for further information. Frontline Systems Inc. (2007) makes similar claims for their solver platform.

6.5 Methods Specific to Constraint Programming

Bruni (2005) considers the problem of finding a *minimally unsatisfiable subformula (MUS)* in the clauses defining a conjunctive normal form (CNF) formula. A CNF formula is conjunction of clauses C_j in which each clause is a disjunction of Boolean variables. Each variable can take a true or false value (α_j) or can be negated ($\neg\alpha_j$). For a set of $i = 1\ldots n$ variables over $j = 1\ldots m$ clauses, and with I_j the set of variables in C_j, a CNF statement has the form $\underset{j=1\ldots m}{\wedge}(\underset{i\in I_j}{\vee}[\neg]\alpha_j)$. When the CNF formula has no solution (i.e. there is no set of true/false values for the variables that will make the statement true), then it can be valuable to find an MUS. This is a subset of the clauses in the original formula that has the property of being unsatisfiable while any proper subset is satisfiable. There is a direct analogy to the concept of an IIS.

Bruni uses Thm. 6.16 (see Sec. 6.2.3) to identify an MUS in specific cases in which the structure of the polytope defined by the linear relaxation of the CNF satisfiability problem has certain properties. The conversion of the CNF to a linear system proceeds as follows. A disjunctive clause has a set of positive (i.e. not negated) variables π and a set of negative (i.e. negated) variables ν. For the clause to be true, we require:

$$\sum_{i \in \pi} x_i + \sum_{i \in \nu} (1 - x_i) \geq 1, x_i \text{ are binary} \tag{6.1}$$

Where |ν| is the number of negated literals, this can be rewritten as:

$$\sum_{i \in \nu} x_i - \sum_{i \in \pi} x_i \leq |\nu| - 1, x_i \text{ are binary} \tag{6.2}$$

Where $\boldsymbol{B}$ is the $m \times n$ {0,1,–1} matrix whose rows correspond to clauses in this form, and $\nu(\boldsymbol{B})$ is the m-vector of the negated literals, then the CNF formula can be converted to the binary linear system $\boldsymbol{B}(\boldsymbol{x}) \leq \nu(\boldsymbol{B}) - \mathbf{1}$. The continuous-variable linear relaxation of this system is then:

$$\boldsymbol{B}(\boldsymbol{x}) \leq \nu(\boldsymbol{B}) - \mathbf{1},\ 0 \leq x \leq 1, \text{ for all } x, \text{ where } x_i \text{ are continuous} \tag{6.3}$$

This is now a standard linear program that matches the first form required in Thm. 6.16, hence the alternative form can be constructed, and IISs identified from the supports of the alternative system.

Note, though, that the original system is binary, while Thm. 6.16 applies only to the continuous linear system. Hence an IIS discovered in this manner may not necessarily correspond to a MUS. This leads to the following theorem.

Theorem 6.24: *Non-equivalence of IIS and MUS via Farkas' Theorem* (Bruni 2005). Consider the two systems of linear inequalities Eqn. 6.2 and the alternative version of Eqn. 6.3 (not shown). If the alternative version of Eqn. 6.3 is feasible, then Eqn. 6.2 is infeasible, and the supports of the alternative version of Eqn. 6.3 (when restricted to clausal inequalities) identify MUSs of Eqn. 6.2. However if the alternative version of Eqn. 6.3 is infeasible, then Eqn. 6.3 is feasible, but it is unknown whether Eqn. 6.2 is feasible or not.■

Bruni further defines the *integral-point property* as a class of polyhedra which, if non-empty, contain at least one integral point. This strengthens Thm 6.24 into the new Thm. 6.25.

Theorem 6.25: *Farkas' Theorem for polytopes having the integral-point property* (Bruni 2005). If the polyhedron for Eqn. 6.3 has the integral-point property, then the following hold. If the alternative version of Eqn. 6.3 is infeasible, then Eqn. 6.2 is feasible. If the alternative version of Eqn. 6.3 is feasible, then Eqn. 6.3 is infeasible, and each IIS given by the supports of the alternative version of Eqn. 6.3 (when restricted to clausal inequalities) identifies an MUS of Eqn. 6.2.■

Several classes of propositional CNF formulae have the integral-point property, including Horn, renamable-Horn, extended Horn, Balanced and Matched, hence Thm. 6.25 is quite useful. Bruni (2005) presents computational results which show this.

De Siqueira N. and Puget (1988) introduce a prototype of the additive method (see Sec. 6.1.3) for use in logic programming. Clauses are tested in a specific order. There are three steps in their algorithm, as follows. (1) There is a clause that has always failed during the proof of infeasibility; remove this clause from the original set and add to the set P. (2) If P is infeasible, then exit with P as a minimal infeasible set. (3) Find a solution to the set of clauses in P, apply this solution to the set of clauses remaining in the original set, and go to step 1. Note that the solution found in step 3 will necessarily cause failure of the reduced original set during step 1. This is very similar to the additive method (Alg. 6.2) except for the use of the specific solution found in step 3.

Some general constraint logic programming (CLP) languages extend ordinary logic programming to include linear constraints. In this case, methods for finding IISs in linear programs can be used when the model proves infeasible. De Backer and Beringer (1991) find IISs for the purposes of *intelligent backtracking*, using a method similar to that of Gleeson and Ryan (1990); this idea is extended by Holzbaur et al. (1996). Burg et al. (1994) present a method of finding *minimal conflict sets* (i.e. IISs) which are also used for intelligent backtracking. In constraint logic programming, constraints are processed one at a time. Burg et al. maintain the current set of constraints in a special solved form achieved by Gaussian operations. The solved form appears to be similar to van Loon's form, and the "minimal conflict sets" appear to be isolated in the same manner.

Bakker et al. (1993) rediscover the deletion filter in the context of constraint satisfaction problems. They name their algorithm DOC for "*D*iagnosis of *O*ver-determined *C*onstraint satisfaction problems".

Junker (2001) introduces three variants of an algorithm for finding *minimal conflicts* in general constraint satisfaction problems, i.e. IISs, among the constraints defining the problem. This is an extension of the work by de Siqueira N. and Puget (1988) to general constraint satisfaction problems. The RePlayXplain variant is identical to the additive method (Alg. 6.2). The RobustXplain variant is equivalent to carrying out the additive method until infeasibility is detected, followed by a reverse deletion filter (see Alg. 6.7). Finally, the QuickXplain algorithm is a variant of the additive method with binary grouping (see e.g. Alg. 6.5).

Hemery et al. (2006) work towards increasing the efficiency of isolating *minimal unsatisfiable cores* (MUCs), which are equivalent to IISs. They describe a *constructive method* equivalent to the additive method, a *destructive method* equivalent to the deletion filter, and a *dichotomic method* equivalent to a binary search (see Sec. 6.1.5) for finding MUCs. Several ideas are used to improve efficiency, such as using the *dom/wdeg* heuristic (Boussemart et al. 2004) to order the variables when assigning values. Dom/wdeg selects the variable that occurred most frequently in the constraints that were most often violated during previous steps. An additional efficiency is to first reduce the original set of constraints by eliminating all constraints containing no variables whose range was reduced during the filtering steps. In other words, the initial set of constraints is condensed to those constraints that were used to eliminate some values from the domain of any variable; this is the *wcore* process. Wcore produces an infeasible set, but it is not necessarily irreducible. Still it is normally smaller than the original set of constraints and hence provides an advanced start for the process of isolating a MUC. This is similar to Thm. 6.22 introduced by Guieu and Chinneck (1999).

Isolating IISs (or *minimal conflict sets* or *minimal cores* or *minimal unsatifiable sets*) has been a preoccupation of both the mathematical programming and the constraint programming communities for some time. Until recently, however, the two communities operated largely in isolation, and so neither was aware of the velopments in the other community. Similar ideas were re-invented numerous times. Some of the pioneering papers on several subjects are listed below.

- *The additive method.* De Siqueira N. and Puget (1988) develop a prototype of the additive method for the case of conjunction of clauses. Tamiz, Mardle and Jones (1995, 1996) introduce the additive method for use in linear programming. Junker (2001) expands on the concept for general constraint programs.
- *The deletion filter.* Dravnieks (1989) introduces the deletion filter, the sensitivity filter for linear programming, and the elastic method. This work is finalized in Chinneck and Dravnieks (1991). Bakker et al. (1993) rediscover the deletion filter for constraint satisfaction problems.
- *Additive/deletion filter.* Guieu and Chinneck (1999) show how the additive method and the deletion filter can be combined into a single method. Junker (2001) introduces a QuickXplain variant on the additive/deletion algorithm.
- *Pivoting methods.* Gleeson and Ryan (1990) show how IIS pivoting can be used to isolate IISs. De Backer and Beringer (1991) develop similar methods for constraint programming.
- *Constraint grouping.* Chinneck (1995) suggests that the deletion filter and the additive method could be improved by treating constraints in groups. Guieu and Chinneck (1999) introduce several specific grouping algorithms for the deletion filter and the additive method for mathematical programs. Junker (2001) introduces binary grouping for constraint satisfaction problems. Atlihan and Schrage (2006) introduce binary grouping for mathematical programs.
- *Advanced subset of constraints.* Guieu and Chinneck (1999) introduce the concept that only the variables that have been branched on in the solution of an infeasible MIP form an infeasible set in conjunction with their bounds and integer restrictions and the complete set of linear constraints (Thm. 6.22). This can be refined if each leaf node is analyzed via a sensitivity filter. Hemery et al. (2006) introduce the wcore concept which eliminates some of the original constraints during the search for an IIS based on the fact that they have not been used to reduce the range of any variables.

7 Finding the Maximum Feasible Subset of Linear Constraints

When a linear program is infeasible the usual first tactic is to isolate an IIS via the methods described in the previous chapter. However there is a complementary approach that has analytic value: find the smallest number of constraints to remove such that the remaining constraints constitute a feasible set. The removed constraints in some sense contribute to the infeasibility most heavily. Consider two overlapped IISs: {A,B,C} and {C,D,E} in a larger model with some number m of constraints in total. To eliminate all infeasibility from the model, we can remove one constraint from each IIS, say A from the first IIS and E from the second, leaving a feasible set of size $m-2$. However, we can also remove all infeasibility by removing just the single constraint C, which destroys both IISs simultaneously and leaves a maximum cardinality feasible set of size $m-1$. The single removed constraint C contributes to infeasibility in both of the IISs in the model, and hence is a better focus for the initial diagnostic effort.

Nothing in the previous paragraph restricts this concept to sets of linear constraints. However the current state of the art is indeed limited to methods for linear systems. For this reason we restrict our attention in this chapter mainly to linear constraints. There is wide scope for extending the methods to other forms of optimization problems.

The problem of finding the maximum cardinality feasible subset in an infeasible set of linear constraints is known most commonly as the *maximum feasible subsystem* problem (MAX FS) (Amaldi et al. 1999). The problem can also be viewed as finding the minimum number of linear constraints to remove such that the retained constraints constitute a feasible system, which is known as the *minimum unsatisfied linear relation* problem (MIN ULR) (Amaldi 1994). The two problems have complementary objective functions. In addition, all infeasible systems have one or more IISs, so an infeasible set of constraints can be made feasible by deleting at least one member of every IIS it contains. Finding the smallest cardinality set of constraints to cover all IISs is known as the *minimum-cardinality IIS set-covering* problem (MIN IIS COVER) (Chinneck 1996c), which is identical to MIN ULR.

For our purposes, MAX FS, MIN ULR, and MIN IIS COVER are the same problem and the terms will be used interchangeably. Several authors have shown that these problems are NP-hard (Sankaran 1993, Chakravarti 1994, Amaldi and Kann 1995). Amaldi and Kann (1995) showed that the problem is also NP-hard for homogeneous systems of inequalities (both strict and nonstrict) and binary coefficients. Amaldi and others (Amaldi 1994, Amaldi and Kann 1995, Amaldi et al. 1999) have also extensively analyzed the approximability of MAX FS, showing that it can be approximated within a factor of 2, but that it does not have a polynomial-time approximation scheme unless P = NP. Until relatively recently there has been little development of algorithms for actually solving the MAX FS problem, but heuristic methods are now available, with more under development, spurred by several important applications in fields such as radiation therapy planning, machine learning, signal processing, etc.

Note that the solution to a MAX FS, MIN ULR, or MIN IIS COVER problem is not usually unique. Consider a system having two IISs {A,B,C} and {B,C,D}; there are two MIN IIS COVER solutions of size one: {B} and {C}, and hence two different associated MAX FS sets {A,C,D} and {A,B,D}.

A closely related problem arises when the individual constraints are assigned weights. Now the problem is to find the minimum (or maximum) weight set of feasible constraints; see Parker (1995) and Parker and Ryan (1996).

It would be helpful to know in advance the number of distinct IISs in the model, though determining this is as difficult as solving the MIN IIS COVER problem itself. However a simple lower bound on the number of IISs is readily found using the deletion/sensitivity filter given in Alg. 6.9 (Chinneck 1994). Recall that this algorithm runs a sensitivity filter if a constraint is permanently dropped during the deletion filter. The sensitivity filter will remove the members of any IISs that overlap on the constraint just dropped permanently, unless they are also part of another IIS not yet eliminated. If k constraints are permanently removed by the deletion filter part of the deletion/sensitivity filter (i.e. at Step 3.2.2.1 of Alg. 6.9), then the model contains at least $k + 1$ IISs, if not more. The cost of running the deletion/sensitivity filter is relatively small, so it is not usually expensive to determine this lower bound, especially if the total number of constraints involved in IISs is small compared to the total number of constraints in the model.

Note carefully that a *maximum* feasible subsystem is a different concept than a *maximal* feasible subsystem. If a feasible subsystem is maximal, the addition of any further constraints renders it infeasible. A maximum feasible subsystem is a maximal feasible subsystem of largest cardinality. It is easy to construct maximal feasible subsystems by a simple inversion of the deletion filter (Alg. 6.1). Start with a single constraint, and add constraints one by one. When a newly added constraint triggers infeasibility, discard it. This is the *grow* method used by Bailey and Stuckey (2005) to find maximal feasible subsystems. This also suggests a simple, though inefficient, way to find maximum feasible subsystems: run the *grow* algorithm numerous times, randomizing the order of the constraints between each run, and return the largest cardinality feasible subsystem found over all of the runs.

7.1 Exact Solutions

Though MAX FS is known to be NP-hard, it can be formulated for exact solution. Only relatively small instances can be solved this way since the exact solutions require exponential time to run.

7.1.1 An Exact Solution via MIP

An exact solution via mixed-integer linear programming has been suggested several times, e.g. by Greenberg and Murphy (1991). Here is a variation of the formulation given by Parker (1995):

Minimize $Z = \Sigma y_i$

Subject to $\boldsymbol{a}_i \boldsymbol{x} \leq b_i + My_i$ for all constraints i of type $\leq$

$\boldsymbol{a}_i \boldsymbol{x} \geq b_i - My_i$ for all constraints i of type $\geq$

$\boldsymbol{a}_i \boldsymbol{x} = b_i + My'_i - My''_i$ for all constraints i of type $=$

where the y, y' and y'' are binary variables, and M is the usual "big-M" large positive value. Further, all variable bounds are included in the set of constraints shown above, or they can optionally be included separately in the normal way, but then the solution will consider only the functional constraints. In the usual manner, if a binary y variable takes the value 1, then the corresponding constraint is effectively loosened and has no effect due to the effect of adding or subtracting M. After solving this MIP, the maximum cardinality feasible subset of constraints is indicated by the constraints whose corresponding y (or y' and y'') variable(s) are all zero. The MIN ULR or MIN IIS COVER is given by the constraints having a corresponding y, y' or y'' variable whose value is 1. The conversion to a weighted version of problem is straightforward: simply add appropriate weights in the objective function.

As Parker (1995) points out, there are several difficulties with this formulation. Incorrect selection of the value of M can lead to the incorrect conclusion that the model is still infeasible, or can cause fractional values or numerical instability. In addition, the solution can be quite slow. Amaldi et al. (2007) suggest that for models in which all of the variables are bounded as $0 \leq x_j \leq u_j$ for all j, a reasonable choice for M is $\max_{i=1\ldots m} \left\{ b_i - \sum_{j:a_{ij}<0} a_{ij} u_j \right\}$.

Bordetski and Kazarinov (1981) also describe a branch and bound solution to an integer programming formulation for finding the maximum weight feasible subsystem of constraints in an infeasible set of inequalities.

While the exact big-M MIP formulation is useful for small models, it is not effective on larger models for reasons of both speed and accuracy. Numerical difficulties with this approach are evident in the empirical studies by Amaldi et al. (2005).

7.1.2 An Exact Formulation via Equilibrium Constraints

As shown by Amaldi (2003), MAX FS can be formulated as a mathematical program with a linear objective function, bilinear constraints and real variables, which is known as a linear program with equilibrium constraints (LPEC). The formulation is as follows:

$$\max \sum_{i=1}^{m} y_i$$
$$s.t.\quad y_i \boldsymbol{a}_i \boldsymbol{x} \{\leq, \geq, =\} y_i b_i$$
$$y_i \in \{0,1\}$$

Note that $\boldsymbol{a}_i$ denotes the row of the A matrix corresponding to the ith constraint. The binary variables y_i take the value 1 if the equation is included in the feasible subset and the value 0 if the variable is excluded. The MAX FS objective is straightforward: maximize the sum of the y_i. The binary restriction on the y_i can be relaxed and replaced by $0 \leq y_i \leq 1$ without harm since any nonzero value of y_i in that range simply amounts to a scaling of the linear constraint, and the objective function will drive any y_i to 1 if it is able to take any nonzero value at all.

The continuous version of the formulation constitutes a nonlinear global optimization problem which can be tackled via global solvers, or by standard NLP local solvers (though these latter may not return a globally optimum solution). The ability to solve mathematical programs with equilibrium constraints has improved in recent years (see e.g. Ferris et al. (2005)), which has increased the viability of this approach, but it is still relatively slow.

An equivalent formulation for sets of inequalities that has better properties for solution was developed in the machine learning community. Mangasarian (1994) introduces a slack variable s_i and a nonnegative variable y_i for each inequality to arrive at the following LPEC:

$$\min \sum_{i=1}^{m} y_i$$
$$s.t.\quad \boldsymbol{s} - \boldsymbol{Ax} + \boldsymbol{b} \geq \boldsymbol{0}$$
$$\boldsymbol{y}(\boldsymbol{s} - \boldsymbol{Ax} + \boldsymbol{b}) = \boldsymbol{0}$$
$$-\boldsymbol{y} + \boldsymbol{1} \geq \boldsymbol{0}$$
$$\boldsymbol{s}(-\boldsymbol{y} + \boldsymbol{1}) = \boldsymbol{0}$$
$$\boldsymbol{x} \in \Re^n, \boldsymbol{s} \geq \boldsymbol{0}, \boldsymbol{y} \geq \boldsymbol{0}$$

This is perhaps more naturally written as follows:

$$\min \sum_{i=1}^{m} y_i$$
$$s.t.\quad \boldsymbol{Ax} \leq \boldsymbol{b} + \boldsymbol{s}$$
$$\boldsymbol{y}(\boldsymbol{Ax} - \boldsymbol{b} - \boldsymbol{s}) = \boldsymbol{0}$$
$$\boldsymbol{y} \leq \boldsymbol{1}$$
$$\boldsymbol{s}(\boldsymbol{y} - \boldsymbol{1}) = \boldsymbol{0}$$
$$\boldsymbol{x} \in \Re^n, \boldsymbol{s} \geq \boldsymbol{0}, \boldsymbol{y} \geq \boldsymbol{0}$$

The effect of this formulation is to ensure that y_i is zero only when s_i is zero, i.e. when the ith constraint $\boldsymbol{a}_i \boldsymbol{x} \le b_i$ is satisfied. Note that the x_j are unrestricted in sign because they represent the coefficients in the separating hyperplane equation (see Sec. 10.1 for details). The two bilinear relationships can be included in the objective function with penalty terms, yielding a bilinear objective function subject to linear constraints. Mangasarian (1994) solves this using a sequential linear approximation procedure.

The solution procedure is improved by Bennett and Bredensteiner (1997), who add a parametric control element, δ. Modifying their development to align with the notation above yields:

$$\min_{\boldsymbol{y},\boldsymbol{s}} \left[\boldsymbol{y}(\boldsymbol{A}\boldsymbol{x} - \boldsymbol{b} - \boldsymbol{s}) + \boldsymbol{s}(\boldsymbol{y} - \boldsymbol{1})\right]$$

$$s.t. \quad \boldsymbol{A}\boldsymbol{x} \le \boldsymbol{b} + \boldsymbol{s}$$

$$\boldsymbol{0} \le \boldsymbol{y} \le \boldsymbol{1}$$

$$\sum_{i=1}^{m} y_i \le \delta$$

$$\boldsymbol{x} \in \Re^n, \boldsymbol{s} \ge \boldsymbol{0}$$

The δ parameter specifies an upper bound on the number of violated constraints. The Frank-Wolfe algorithm for uncoupled bilinear programs is used to solve this problem. It must be solved a number of times to determine the smallest value of δ for which the objective function can reach a value of zero. A secant method is used in an outer loop to adjust the value of δ until the objective function is able to reach zero.

Mangasarian (1996) proposes an alternative formulation in which the step function associated with including or excluding a constraint is approximated:

$$\min \sum_{i=1}^{m} (1 - e^{-\alpha z_i})$$

$$s.t. \quad \boldsymbol{z} \ge \boldsymbol{A}\boldsymbol{x} - \boldsymbol{b}$$

$$\boldsymbol{z} \ge \boldsymbol{0}$$

When z_i is zero, the ith constraint $\boldsymbol{a}_i \boldsymbol{x} \le b_i$ is satisfied, and the corresponding term in the objective function likewise goes to zero. The larger the violation of constraint i, the larger z_i becomes, and the closer the corresponding term in the objective function approaches 1. The nonnegative parameter α controls the quality of the approximation. This is again solved by successive linear approximation.

Empirical results reported by Mangasarian (1994, 1997) and Bennett and Bredensteiner (1997) show that the parametric variants are more effective. Still, the inherent nonlinearity of these models makes it difficult to solve them to optimality, and hence to guarantee that MAX FS is solved optimally.

7.2 IIS Enumeration and Covering

Parker (1995) and Parker and Ryan (1996) outline a method of enumerating the IISs in an infeasible system of linear inequalities, and then solving a set-covering problem over the IISs to solve the minimum weight IIS set cover problem (equivalent to MIN IIS COVER when the weights are all equal to 1). The basic theorems that allow the enumeration of the IISs are given in Sec. 6.2.3. The number of IISs can potentially be exponential in the size of the model (Chakravarti 1994), so some attention must be paid to the efficiency of the method. Parker and Ryan (1996) address the efficiency issue by generating IISs one at a time, making sure to find new IISs that are not covered by the current MIN IIS COVER solution.

Parker and Ryan's algorithm is shown in Alg. 7.1. There is one binary variable y_i for each constraint i. If some IIS J_k consists of constraints 3, 7 and 11 for example, then the associated set cover constraint is $y_3 + y_7 + y_{11} \geq 1$ to indicate that at least one of the three constraints must be removed from the model to render it feasible. There is a similar constraint for each IIS J_k in the set of IISs J.

d_i: the ith constraint in the model, c_i: weight on ith constraint,
y_i: binary variable (one per constraint), J_k: the kth IIS in the set J of IISs.

INPUT: an infeasible set of linear inequalities P.
1. Identify an initial set of IISs J (J may be empty).
2. Solve the minimum weight IIS set covering problem:
   minimize $\sum c_i y_i$
   subject to $\sum y_i \geq 1$ for $d_i \in J_k$, for all IISs $J_k \in J$.
   Let T index the elements of the optimal cover.
3. IF $P \backslash T$ feasible THEN exit.
   ELSE find an IIS that is not covered by T and add it to J.
   Go to Step 2.

OUTPUT: T is the minimum weight cover.

Alg. 7.1. Minimum-weight IIS set covering algorithm (Parker and Ryan 1996)

In terms of efficiency, there are three crucial elements in Alg. 7.1. The first is the solution of the integer program in Step 2. Some speed-up is achieved by substituting a quick set-covering heuristic in place of the full integer programming solution during all but the final iteration. The full integer programming solution is then run only once when the heuristic method returns a cover such that $P \backslash T$ is feasible, in order to determine whether a smaller weight cover exists. Parker and Ryan's method can thus be converted to a polynomial-time heuristic by omitting the final full integer programming solution.

The second crucial step is the generation of new IISs in Step 3. The algorithm uses a column generation strategy and it is generally not necessary to enumerate all IISs in order to cover them. In fact, it is possible that one constraint appears in all IISs and is found right away, so that the algorithm does not have to generate any other IIS. In the worst case, the IISs are all disjoint, in which case the algorithm must generate them all. Two methods are used to enumerate IISs, one based on visiting the extreme points of a specially constructed polytope (Thm. 6.16), and another based on the extreme rays of a specially constructed polyhedral cone (Thm. 6.17). Heuristics are applied in an attempt to find new IISs having little overlap with IISs already in J to reduce the number of integer programming solutions needed during Step 2.

The third crucial element is the limitation of the method to linear inequalities. Equalities are handled by converting each one into a pair of oppositely-oriented inequalities. This may cause a blow-up in the number of inequalities. Hence the suitability of the method may depend on the number of equality constraints in the system. Parker and Ryan extend Thm. 6.17 to deal with this problem, yielding a new theorem, as follows.

Theorem 7.1: *Supports of a general infeasible linear system* (Parker and Ryan 1996). Given the inconsistent system $S = \{x \in Q^n \mid \mathbf{Ax} \leq \mathbf{b},\ \mathbf{Cx} = \mathbf{d},\ \mathbf{L} \leq \mathbf{x} \leq \mathbf{U}\}$, the indices of the IISs of S are exactly the supports of the vertices of the polyhedron $P = \{\, y,w,v,z \in Q^m \mid \mathbf{y}^T\mathbf{A} + \mathbf{w}^T\mathbf{C} + \mathbf{v} - \mathbf{z} = \mathbf{0},\ \mathbf{y}^T\mathbf{b} + \mathbf{w}^T\mathbf{d} + \mathbf{v}^T\mathbf{U} - \mathbf{z}^T\mathbf{L} = -1,\ \mathbf{y,z,v} \geq \mathbf{0},\ \mathbf{w}$ unrestricted$\}$.■

Further, if $\mathbf{x}$ is bounded only by nonnegativity, then Thm. 7.1 can be simplified as follows.

Theorem 7.2: *Supports on a general infeasible linear system with only nonnegativity bounds* (Parker and Ryan 1996). Given the inconsistent system $S = \{x \in Q^n \mid \mathbf{Ax} \leq \mathbf{b},\ \mathbf{Cx} = \mathbf{d},\ \mathbf{x} \geq \mathbf{0}\}$, the indices of the IISs of S are exactly the supports of the vertices of the polyhedron $P = \{\, y,w \in Q^m \mid \mathbf{y}^T\mathbf{A} + \mathbf{w}^T\mathbf{C} \geq \mathbf{0},\ \mathbf{y}^T\mathbf{b} + \mathbf{w}^T\mathbf{d} = -1,\ \mathbf{y} \geq \mathbf{0},\ \mathbf{w}$ unrestricted$\}$.■

Thm. 7.2 means that the nonnegativity bounds do not need to be handled explicitly: you need only check the slack variables in the solution of the alternative system to determine whether a nonnegativity constraint forms part of the IIS.

There are some numerical issues when the right hand side of the second constraint in the alternative system in Thms. 7.1 and 7.2 is set to –1. Parker and Ryan (1996) present a solution to this difficulty by replacing the –1 right hand side by a new value determined by various parameters in the model.

Parker and Ryan conduct an empirical evaluation of three variations of their method using the standard test set of infeasible LPs (Chinneck 1993). The times reported are reasonable. For most problems the vast majority of the solution time is spent on identifying IISs in Step 3 of Alg. 7.1, with relatively little time spent on solving the set covering problem in Step 2.

Tamiz et al. (1995) describe an algorithm that starts with a heuristic enumeration of IISs, and then follows with a frequency-based heuristic to solve the resulting set-covering problem. It uses the additive method (Sec. 6.1.3) to find the individual IISs. See Alg. 7.2.

INPUT: an infeasible set of linear constraints C.
0. *pass* = forward.
   CoverSet = ∅.
1. Use the additive method to find an IIS I and write it out.
   Delete the first member of I from C.
   IF C is infeasible THEN go to Step 1.
   IF *pass* = forward THEN
      Reinstate the original constraint set C.
      Reverse the order of the constraints in C.
      pass = reverse.
      Go to Step 1.
2. Compare IISs found in Step 1 to eliminate duplicates, forming the set D of distinct IISs.
3. Find f, the most frequent constraint in D, and add it to *CoverSet.*
   $D = D\backslash$ {IISs having f as a member}.
   IF D = ∅ THEN exit.
   Go to Step 3.

OUTPUT: *CoverSet* is a small set of constraints covering the IISs

Alg. 7.2. The constraint frequency heuristic for the IIS cover (Tamiz et al. 1995)

Algorithm 7.2 suffers from the method used to isolate distinct IISs. The forward and reverse passes of the additive algorithm are not efficient ways of generating all possible IISs compared to the method used by Parker and Ryan since some IISs may be omitted while others may be isolated twice. An expensive Step 2 must also be used to eliminate the duplicates. Finally Step 3 is simply a standard set covering heuristic, likely similar to the fast heuristic used at intermediate steps by Parker and Ryan.

Tamiz et al. (1995) present empirical results for 16 of the 29 models from the netlib set of infeasible LPs (Chinneck 1993). Their criteria for the selection of this subset are not stated. They are able to find a true minimum cardinality IIS set cover in 14 of the 16 models examined (the exceptions are *reactor* and *greenbea*). Pfetsch (2002, 2005) advances Parker and Ryan's approach by using a branch-and-cut approach to improve the solutions of the intermediate set-covering MIPs. The main idea is to add a cut after each intermediate LP-relaxation is solved that separates the rows corresponding to IISs that are covered in the current solution from the rows corresponding to IISs that are not yet covered. Since this is NPhard, various heuristics are used. Cuts are of three types: (i) inequalities derived from IISs, (ii) special inequalities due to Balas and Ng (1989), and (iii) Gomory cuts (e.g. Nemhauser and Wooley (1988)). Cuts are stored in a pool and are added to the LP relaxation if violated. In addition, Pfetsch applies a preprocessing step to find small IISs, and uses a primal heuristic for possibly decreasing the cardinality of any covers found at intermediate steps.

The preprocessing step discovers simple small-cardinality IISs. IISs of cardinality 2 (i.e. parallel but oppositely-oriented inequalities) are found when constraints have the same left-hand-side bodies but different constants, and a mismatch in their orientations. Individual constraints are also scanned for infeasibility relative to the bounds on their variables. IISs identified during this step are used to set up the first covering MIP, but the method proceeds without difficulty if no IISs are found during this step.

The primal heuristic is applied after every k branch and bound nodes and operates as follows. Suppose that we have a current cover S. Initially set $S' = S$. Sort the elements of S' in increasing order of the fractional values of the associated variables in the covering LP-relaxation. Remove each member of S' in this order and check whether the reduced version of S' is still an IIS cover. If the reduced S' is no longer an IIS cover, then return the member and continue, else drop the member permanently.

Details on the cuts are available in the original publications (Pfetsch 2002, 2005). Empirical results show good results, though solution times are quite long. Chinneck's heuristic methods (see Sec. 7.4) give better or equivalent results in most cases in much shorter times. Codato and Fischetti (2004, 2006) introduce further cuts, termed combinatorial Benders' cuts, which are useful in this context.

Finally, simple (though inefficient) heuristics can be created based on deletion filtering or the additive method. Simply choose a random ordering of the constraints, isolate an IIS, and make note of it. Repeat the process until no new IISs are discovered in several iterations. Now solve a set-covering problem based on the IISs discovered. Note that if the sensitivity filter is applied after each randomization of the constraint ordering then certain IISs may not be discovered and so the resulting IIS set cover may be incomplete (see Sec. 6.2.2).

7.3 Phase One Heuristics

Any LP Phase 1 solution for an infeasible LP results in a set of constraints that are satisfied and a set that are violated. The set of violated constraints provides an IIS set cover, though generally not a minimum cover. However this Phase 1 cover does provide an upper limit on the cardinality of the MIN IIS COVER. Chinneck (1996c) summarizes these insights in the following observation.

Observation 7.1: *Elastic program cover* (Chinneck 1996c). Upon termination of an elastic program, the set of stretched constraints is an IIS set cover, and the number of stretched constraints is an upper bound on the cardinality of the set cover.■

Most phase 1 procedures are some variation of an elastic program (see Sec. 6.1.4) in which extra nonnegative elastic or artificial variables introduce added dimensions which allow the constraints to "stretch" in the original dimensions to accommodate infeasibility. The main differences in the variations involve which subsets of constraints are elasticized and whether the minimum sum of the constraint violations is the only Phase 1 stopping condition. Common measures of the infeasibility of a solution include the *sum of the infeasibilities (SINF)*, i.e. the sum of the nonnegative artificial or elastic variables, and the *number of infeasibilities (NINF)*, i.e. the number of violated constraints.

Several versions of elastic or Phase 1 programs are in common use, some of which are listed below.

- *Standard elastic program*. All row constraints are fully elasticized, including ≤ type, ≥ type, and = type (elasticized by adding two elastic variables). The variable bounds are not elasticized. The objective is to minimize SINF.
- *Full elastic program*. Same as standard elastic program, but the variable bounds are also elasticized.
- *Simple Phase 1*. As usually described in textbooks on linear programming, artificial variables are added only to constraints of the ≥ and = types. Just a single nonnegative artificial variable with coefficient +1 is added to each equality constraint. The objective is to minimize SINF.
- *MINOS Phase 1*. Wolfe (1965) describes a Phase 1 procedure that considers both SINF and NINF in deciding whether an LP is infeasible. The MINOS LP solver (Murtagh and Saunders 1987) permits arbitrary upper and lower bounds on the variables, so the model is fully elasticized. However the procedure usually terminates when it recognizes that NINF cannot be reduced any further, so SINF may not be at its minimum value upon termination. As an implementation detail, note that MINOS keeps direct track of the constraint violations and hence does not explicitly include artificial or elastic variables.

Any of these is sufficient to produce an IIS cover. While you might expect that the chances of returning a smaller IIS cover are improved if more constraints are elasticized, this is not always the case, as shown below (see Table 7.1).

A second simple observation can be made, as follows.

Observation 7.2: *Single member cover* (Chinneck 1996c). If an elastic program reports a single stretched constraint, then that constraint constitutes a minimum cardinality IIS set cover.■

The smallest possible IIS set cover cardinality is 1, which indicates that the cover constraint occurs in all IISs in the model. Hence if the model is infeasible and the Phase 1 or related procedure returns a cover of size 1, then it must be a minimum cardinality cover. This fact was also noted by Parker and Ryan (1996).

Chinneck (1996c) carried out a series of tests on various algorithms for finding MIN IIS COVER, including the MINOS Phase 1 procedure and the standard elastic program. The test set consisted of the infeasible models in the netlib repository (Chinneck 1993). The unassisted MINOS phase 1 reported a single violated constraint for 14 of the 29 models: *bgetam, box1, ceria3d, cplex2, ex72a, ex73a, forest6, galenet, gosh, klein1, pang, pilot4i, qual, vol1*. This single violated constraint is reported as the minimum IIS cover (Observation 7.2).

The results for the 15 remaining models are presented in Table 7.1. The true minimum cardinality is also included for comparison. Results in boldface indicate cases in which the Phase 1 method found the true MIN IIS COVER. The Phase 1 methods do find the true MIN IIS COVER with reasonable frequency over this test set (18 of 29 cases, or 62%). If they do not find a minimum cardinality cover, they often provide a cover that is not very much larger.

Table 7.1. IIS cover cardinality on difficult LPs for two Phase 1 methods (Chinneck 1996c)

Model	Minimum cover cardinality (Parker and Ryan 1996)	MINOS Phase 1cover	Standard elastic program Phase 1 cover
bgdbg1	12	23	13
bgindy	1	14	**1**
bgprtr	1	2	2
chemcom	1	11	12
cplex1	1	211	212
gran	not calculated	244	473
greenbea	2	3	**2**
itest2	2	**2**	**2**
itest6	2	3	4
klein2	1	3	5
klein3	1	4	19
mondou2	3	**3**	5
reactor	1	3	2
refinery	1	3	6
woodinfe	2	**2**	**2**

7.4 Chinneck's SINF-Reduction Heuristics

Chinneck (1996c, 2001a) develops a set of heuristic methods for MIN IIS COVER based on several observations in addition to the two mentioned above. The most important observation follows.

***Observation** 7.3: Reduction in elastic objective function value* (Chinneck 1996c). The elimination of a constraint that is a member of the minimum-cardinality IIS set cover should reduce *elastic SINF* more than the elimination of a constraint that is not a member of the minimum-cardinality IIS set cover.■

The removal of a constraint that is a member of the minimum cardinality set cover normally eliminates more than one IIS, hence its removal should reduce the elastic objective function value more than the removal of a constraint that is not a member of the minimum cardinality IIS set cover, whose removal will eliminate fewer IISs. Consider Fig. 7.1 which has IISs {A,B,D} and {A,C,D} with two different minimum cardinality IIS set covers: {A} and {D}. Eliminating either one of constraint A or D will reduce the elastic objective function to zero, while eliminating either constraint B or C will reduce the elastic objective function value, but not to zero. Hence either {A} or {D} should be removed by Observation 7.3.

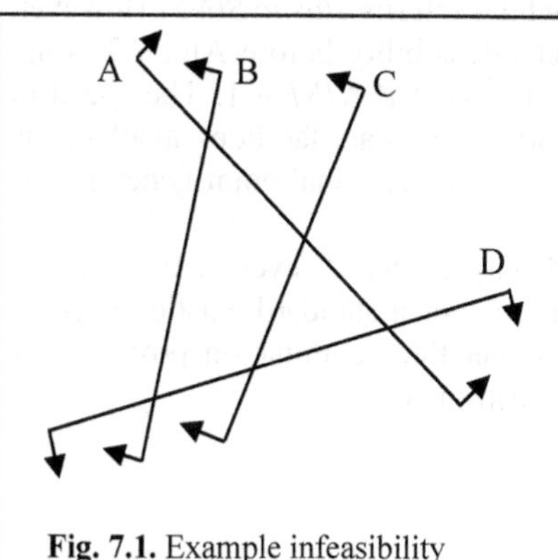

Fig. 7.1. Example infeasibility

In fact, the removal of constraint C has no effect on the value of the elastic objective function because the elastic program returns a solution in which constraint C is not stretched or tight. Minimizing the elastic objective function makes it cheaper to choose one of these three options:

- constraint A stretches to the intersection of B and D,
- constraint D stretches to the intersection of A and B,
- constraint B stretches up to C, and then A or D or both stretch.

Constraint C does not stretch because, to achieve a feasible point, B would first need to stretch to meet C and then would have to continue to stretch along with C until reaching the intersection of A and D. The cost of stretching B and C simultaneously is twice the cost of stretching a single constraint. Thus C will not be tight upon termination of the elastic program. This leads to the fourth observation.

Observation 7.4: *Elastic sensitivity* (Chinneck 1996c). Constraints to which the elastic objective function is not sensitive do not reduce the elastic objective function value when removed from the model.■

Observations 7.1–7.4 underlie Alg. 7.3, a heuristic algorithm for solving MIN IIS COVER. The algorithm takes a greedy approach: the most promising candidate is added to the cover set at each iteration of Step 2. The basic mechanism is to test each eligible constraint by temporarily removing it from the set of constraints to determine the new value of *elastic SINF*. As per Observation 7.3, the constraint whose temporary removal most reduces *elastic SINF* is added to the cover set and removed permanently. To reduce the number of constraints tested, the algorithm uses Observation 7.4 to omit testing any constraints to which the elastic objective is not sensitive.

Observation 7.2 provides an early exit from the algorithm where appropriate. The steps marked as optional in Alg. 7.3 boost the speed of the basic algorithm by noting when the cover set can be completed by a single constraint. The optional steps are easily implemented if the solver provides NINF as a matter of course, but can be ignored if NINF must be calculated by a time-consuming comparison of each constraint left hand side and right hand side.

Alg. 7.3 specifies the use of a procedure which returns *elastic SINF*. However, any phase 1 procedure can be used to detect infeasibility before Alg. 7.3 is applied, and the process can be terminated at that point if *NINF* = 1. The speed of Alg. 7.3 is greatly improved by using the advanced start facilities available in most modern LP solvers. Each elastic LP starts at the terminal point generated by the previous elastic LP.

Note that variable bounds are also tested during Step 2, even if the variable bounds are not elasticized, as in a simple phase 1 or a standard elastic program. Variable bounds are added to *CoverSet* whenever their elimination provides the lowest value of *elastic SINF*, just as for row constraints.

```
INPUT: Constraints defining an infeasible system of linear constraints.
0.  CoverSet = ∅.
    Set up elastic LP.
1.  Solve elastic LP.
    IF NINF = 1:
        Add the single violated constraint to CoverSet.
        Exit.
    HoldSet = {constraints to which elastic objective is sensitive}.
2.  MinSINF = ∞.
    CandidateSet = HoldSet.
    FOR each constraint in CandidateSet:
        Delete the constraint.
        Solve elastic LP.
        IF elastic SINF = 0 THEN:
            Add constraint to CoverSet.
            Exit.
        IF elastic SINF < MinSINF THEN:
            Winner = currently deleted constraint.
            MinSINF = elastic SINF.
            HoldSet = {all constraints to which elastic objective is sensitive}.
            IF NINF = 1, NextWinner = single violated constraint.    (optional)
            ELSE NextCand = ∅.                                       (optional)
        Reinstate the constraint.
3.  Add Winner to CoverSet.
    IF NextWinner ≠ ∅ THEN:                                          (optional)
        Add NextWinner to CoverSet.                                  (optional)
        Exit.                                                        (optional)
    Delete the Winner constraint permanently.
    Go to Step 2.
OUTPUT: CoverSet is a small set of constraints covering the IISs.
```

Alg. 7.3. Heuristic 1 for MIN IIS COVER (Chinneck 1996c)

Alg. 7.3 is easily altered to find minimum weight IIS covers by assigning weights to the elastic variables in the elastic objective function. Constraints assigned higher weights will tend to be included in *CoverSet* because their elimination gives a greater reduction in the elastic objective function value.

Empirical results (see Table 7.2) show that Alg. 7.3 is very effective in practice, but as a heuristic, it cannot guarantee to identify a minimum cardinality IIS in all cases. This is illustrated by the example in Fig. 7.2. The constraints are:

A_0: $x_1 + x_2 \geq 8$
A_1: $x_1 + x_2 \geq 9$
A_2: $x_1 + x_2 \geq 20$
B_0: $-x_1 + x_2 \geq 2$
B_1: $-x_1 + x_2 \geq 3$
C_0: $-0.25x_1 + x_2 \leq 2.75$
C_1: $-0.25x_1 + x_2 \leq 1.75$

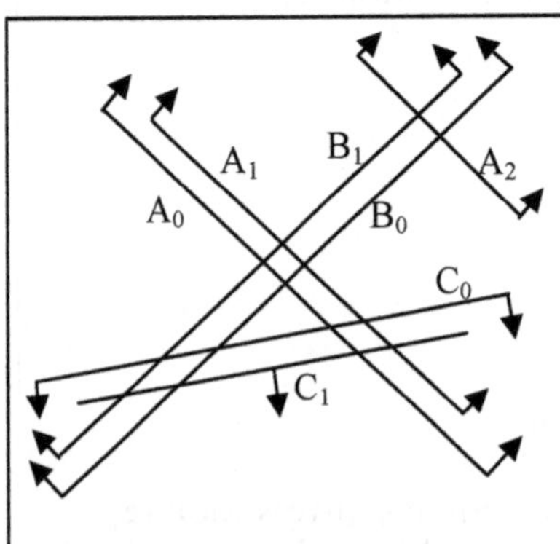

Fig. 7.2. A pathological counter-example (Chinneck 1996c)

Assuming a full elastic program, the elastic objective function is initially sensitive to the outermost constraints A_2, B_1, and C_1. The spacing between the two constraints in the B and C sets is equal to one, the space between A_1 and A_2 is 11, so *SINF* is most reduced by removing A_2. By inspection of Fig. 7.2, this leads to a cover of cardinality 3, even though smaller covers of cardinality 2 exist (either $\{B_0,B_1\}$ or $\{C_0,C_1\}$). The cover reported for this problem by a software implementation of Alg. 7.3 is $\{A_2,C_1,C_0\}$, with the members found in that order. This cover is not of minimum cardinality, and is also not minimal. In a *minimal cover*, every member of the cover set is needed in order to eliminate all of the infeasibility in the model. Similar pathological cases are unlikely in practice since it depends on there being a great deal of redundancy with a particular pattern.

Alg. 7.3 is affected by the elastic programming variant that is used because the variants will report different values of *SINF* for a given set of constraints. The different variants elasticize different subsets of the constraints, and hence Alg. 7.3 will make different decisions concerning which constraint to drop permanently in Step 3. A full elastic program is preferred for any implementation of Alg. 7.3 since this will prevent any artificial distortion of *SINF*.

While the MINOS phase 1 procedure does not guarantee to minimize *SINF*, it does work towards minimizing *NINF* until it recognizes that it cannot be reduced to zero. This behaviour can be used to identify set cover candidates by looking for the greatest drop in *NINF* when a constraint is removed, rather than the greatest drop in *SINF* as in Alg. 7.3, with the drop in *SINF* used to break ties between candidates with equivalent drops in *NINF*. Since the MINOS phase 1 permits literal constraint violations rather than using elastic variables, Step 2 of Alg. 7.3 must test violated constraints in addition to constraints to which the phase 1 objective function is sensitive. Since the MINOS phase 1 provides both *NINF* and *SINF*, the steps marked optional in Alg. 7.3 are naturally included.

Chinneck (2001a) later studied how to improve the speed of Alg. 7.3 by reducing the number of candidate constraints tested, i.e. the size of *CandidateSet*, which is normally determined by the number of constraints to which the elastic objective function is sensitive. New observations lead to new criteria for inclusion in *CandidateSet*; objective-function sensitivity is no longer the single sufficient criterion.

Alg. 7.3 is extremely effective, so any modifications should ideally delete the same constraints, and in the same order. In the best case, we would like to directly identify the correct constraint for removal at each iteration, and place only that single constraint in *CandidateSet*. While this is not possible, the two observations following below allow the assessment of each potential candidate quickly without solving an LP. This permits the addition of only a very few of the most promising candidate constraints to *CandidateSet* for testing via LP solution. Very often, the first constraint on the list is indeed the correct constraint for removal.

Observation 7.5: *Estimating SINF reduction for violated constraints* (Chinneck 2001a). For constraints that are violated in the original model (i.e. whose elastic variables are stretched in an elastic solution), a good predictor of the magnitude of the drop in *SINF* that will be obtained by deleting the constraint is given by the product (constraint violation) × |(constraint sensitivity)|.■

When converted to a full elastic program, "constraint violation" in the original model is given by the value of the elastic variable associated with a constraint. If there are two elastic variables associated with a constraint, as for equality and range constraints, then the constraint violation is the maximum value of the two elastic variables. "Constraint sensitivity" refers to the reduced cost of the variable associated with the constraint. The absolute value of the constraint sensitivity is used because the sign, determined by the constraint sense ($\leq, \geq, =$), is irrelevant since all violations are relaxations of the constraint, regardless of constraint sense.

Having a nonzero elastic variable in the elasticized model is equivalent to changing the right-hand-side value of the constraint in the original model. Thus the product in Observation 7.5, obtained from the elastic version of the model, is the same as operating on the original model to estimate the change in the objective value caused by relaxing the right hand side by the amount given by the nonzero elastic variable. As shown in elementary texts on simple sensitivity analysis, this is a perfectly accurate estimator of the change in *SINF*, provided that the basis in the original model does not change. Of course, the basis in the original model does change when an active constraint is deleted, so Observation 7.5 provides an underestimate of the change in *SINF*. Chinneck (2001a) carries out a small study of the accuracy of the estimates provided by Observation 7.5 on two difficult classification problems and concludes that the maximum-product heuristic is remarkably accurate in predicting Δ*SINF*: it is over 95% accurate in 87% of the cases examined, and over 90% accurate in 94% of the cases examined.

Observation 7.5 suggests a revision to Alg. 7.3. In Steps 1 and 2, instead of setting *HoldSet* = {constraints to which the elastic objective function is sensitive}, find *HoldSet* as follows:

1. Select the violated constraints, and arrange them in order from largest to smallest value of the product (constraint violation) × |(constraint sensitivity)|.
2. Fill *HoldSet* with the top k elements of the ordered list (or all of the elements of the list if there are fewer than k).

Chinneck (2001a) refers to this variant as Algorithm 2(k), where k refers to the length of the candidate list. Empirical results using this algorithm are presented later. Note that a list length of 1 is frequently successful. Because we wish to estimate the effect of every constraint via the product in Observation 7.5, Algorithm 2(k) requires the use of a fully elastic version of the original model (i.e. variable bounds must be elasticized as well as rows). This is straightforward in solver implementations that already permit literal constraint violations during their Phase 1 procedure (by violating the bounds on the variable associated with the constraint), but it may require the explicit addition of elastic variables in other solver implementations.

Concentrating solely on the violated constraints is often successful because the elastic objective function is itself trying to minimize *SINF*, hence it tends to violate the constraints that cause the least increase in *SINF*. However, in some models having numerous infeasibilities, it may be possible to obtain a larger drop in *SINF* by deleting a constraint that is not currently violated. Observation 7.6 describes an indicator for identifying satisfied constraints that are good candidates for deletion.

Observation 7.6: *Identifying candidate satisfied constraints* (Chinneck 2001a). For constraints that are satisfied in the original model (i.e. their associated elastic variables are zero), a good predictor of the relative magnitude of the drop in *SINF* that will be obtained by deleting the constraint is given by |(constraint sensitivity)|.■

This observation allows an ordering of the set of satisfied constraints for possible inclusion in *CandidateSet*. Observation 7.6 does not provide a direct estimate of the size of the drop in *SINF* expected when the constraint is deleted, only the relative size (i.e. a constraint with a larger |sensitivity| is expected to provide a larger drop in *SINF*).

Observations 7.5 and 7.6 can be combined to provide another variant of Alg. 7.3. In Steps 1 and 2, instead of setting *HoldSet* = {constraints to which the elastic objective function is sensitive}, find *HoldSet* as follows:

1. Select the violated constraints, and arrange them in order from largest to smallest value of the product (constraint violation) × |(constraint sensitivity)|.
2. Fill *HoldSet* with the top k elements of the ordered list (or all of the elements of the list if there are fewer than k).
3. Select the satisfied constraints to which the elastic objective function is sensitive, and arrange them in order from largest to smallest |(constraint sensitivity)|.
4. Add the top k elements of this ordered list to the bottom of *HoldSet* (or all of the elements of the list if there are fewer than k).

Chinneck (2001a) refers to this variant as Algorithm 3(k), where k refers to the length of each of the two lists. Note that a list length of k implies the solution of up to $2k$ LPs to identify the winning candidate.

We can also improve on Alg. 7.3 by taking better advantage of Observation 7.1 to provide a safety exit when the more advanced algorithms perform poorly. Because the cardinality of the IIS cover provided by the phase 1 procedure is already known, any subsequently applied algorithm can be halted when its cover cardinality exceeds the cardinality of the IIS cover already provided by the phase 1 procedure.

Further, more than one phase 1 procedure would probably be applied in a practical implementation because the solver-native phase 1 procedure is unlikely to be a full elastic SINF minimization. The implementation would normally proceed as follows:

1. Native phase 1 method detects infeasibility and records the *NINF* and IIS cover.
2. Convert to full elastic version of model.
3. Minimize *SINF* in fully elastic model (using an advanced start provided by the native phase 1 solution) and record the *NINF* and IIS cover.

The smallest *NINF* provided by the native phase 1 or the elastic phase 1 then acts as a stopping condition for any more advanced algorithm.

Alg. 7.4 combines all of these observations into a generic framework. The possible selection criteria for inclusion in *HoldSet* include (i) phase 1 objective function sensitivity (as in Alg. 7.3), (ii) high values of the product for violated constraints (i.e. Algorithm 2), or (iii) both high values of the product for violated constraints and high phase 1 objective function sensitivities (i.e. Algorithm 3).

Chinneck (1996c, 2001a) conducted a number of empirical tests of Algorithms 7.3 and 7.4. Results for 14 of the more difficult infeasible LPs in the netlib set (those for which the MINOS phase 1 does not find a single-member IIS cover) are summarized in Table 7.2. Numbers in boldface indicate solutions that return an IIS cover of true minimum cardinality, "NINF" indicates the cardinality of the IIS cover, and "LPs" indicates the number of LPs solved (excluding the initial phase 1 solution that signalled infeasibility). The results for Algorithms 2(k) and 3(k) are for simply using the selection criteria for those models directly in Alg. 7.3. The results for "Alg. 7.4 with 3(7)" are derived by combining the two phase 1 results in Table 7.1 with the results of Algorithm 3(7) to infer the outcome (Alg. 7.4 with 3(7) was not actually implemented). The *SafetySet* established during Step 0 of Alg. 7.4 is actually used for *mondou2*. If Alg. 7.4 is combined with the Algorithm 2(1), 2(7), or 3(1) selection criteria, the *SafetySet* is used even more often.

```
INPUT: Linear constraints defining an infeasible model.
0.  CoverSize = 0, CoverSet = ∅, SafetySize = 0, SafetySet = ∅.
    If native phase 1 procedure detects feasibility, then exit.
    SafetySize = (cardinality of native phase 1 cover).
    SafetySet ← {members of native phase 1 cover}.
    IF SafetySize = 1 THEN:
        CoverSize = 1.
        CoverSet ← SafetySet.
        Exit.
    Set up elastic LP.
    Solve elastic LP using advanced start from the native phase 1 solution.
    IF (elastic cover cardinality) < SafetySize THEN:
        SafetySize = (cardinality of elastic phase 1 cover).
        SafetySet ← {members of elastic phase 1 cover}.
    If SafetySize = 1 then:
        CoverSize = 1.
        CoverSet ← SafetySet.
        Exit.
    HoldSet = {constraints meeting selection criteria}.
1.  MinSINF = ∞.
    CandidateSet ← HoldSet.
    FOR each constraint in CandidateSet:
        Delete the constraint.
        Solve elastic LP.
        IF SINF = 0 THEN:
            Add constraint to CoverSet.
            CoverSize = CoverSize + 1.
            Exit.
        If SINF < MinSINF then:
            Winner = currently deleted constraint.
            MinSINF = SINF.
            HoldSet ← {constraints meeting selection criteria}.
            If NINF = 1, NextWinner = single violated constraint.
            Else NextWinner = ∅.
        Reinstate the constraint.
2.  Add Winner to CoverSet.
    CoverSize = CoverSize + 1.
    IF NextWinner ≠ ∅ THEN:
        Add NextWinner to CoverSet.
        CoverSize = CoverSize + 1.
        Exit.
    Delete the Winner constraint permanently.
    IF CoverSize ≥ (SafetySize – 1) THEN:
        CoverSet ← SafetySet.
        CoverSize = SafetySize.
        Exit.
    Go to Step 1.
OUTPUT: CoverSet is an IIS cover of cardinality CoverSize.
```

Alg. 7.4. Heuristic 2 for MIN IIS COVER (Chinneck 2001a)

The *gran* model is omitted because it causes numerical difficulties. The list length for Algorithms 2 and 3 can be set as desired. Shorter lists are faster, but longer lists are more accurate. With a sufficiently long list, Algorithm 3 is equivalent to Alg. 7.3. Experimentation with shorter lengths showed that a length of 7 is quite effective, particularly for Algorithm 3. Results with list lengths of 1 and 7 for Algorithms 2 and 3 are given in Table 7.2.

Table 7.2. Comparison of algorithms on difficult infeasible LPs (Chinneck 2001a)

	Alg. 7.3		Alg. 2(1)		Alg. 2(7)		Alg. 3(1)		Alg. 3(7)		Alg. 7.4 with 3(7)
model	NINF	LPs	NINF	LPs	NINF	LPs	NINF	LPs	NINF	LPs	NINF
bgprtr	**1**	1	**1**	0	**1**	0	**1**	0	**1**	0	**1**
itest2	**2**	7	**2**	1	**2**	2	**2**	2	**2**	6	**2**
mondou2	**3**	384	7	6	5	25	6	11	5	53	**3**
reactor	**1**	25	**1**	0	**1**	0	**1**	0	**1**	0	**1**
woodinfe	**2**	47	**2**	1	**2**	2	**2**	2	**2**	4	**2**
bgdbg1	**12**	645	**12**	11	**12**	65	**12**	22	**12**	142	**12**
bgindy	**1**	1	**1**	1	**1**	1	**1**	1	**1**	1	**1**
chemcom	**1**	2	**1**	1	**1**	1	**1**	1	**1**	1	**1**
greenbea	**2**	404	**2**	1	**2**	6	**2**	2	**2**	13	**2**
itest6	**2**	10	4	4	**2**	7	4	7	**2**	8	**2**
klein3	**1**	53	9	9	**1**	7	4	8	**1**	7	**1**
cplex1	**1**	213	211	210	211	1455	4	8	**1**	9	**1**
klein2	**1**	17	3	2	3	7	2	4	**1**	11	**1**
refinery	**1**	36	3	2	3	9	3	4	2	18	2
# min NINF	14		8		10		8		12		13
avg. NINF	2.2		18.5		17.6		3.2		2.4		
avg. LPs		131.8		17.8		113.4		5.1		19.5	

As expected, Alg. 7.3 requires the most LP solutions on average (131.8) while Algorithm 3(1) requires the fewest (5.1). These average results are skewed by *cplex1*. In 12 of the 14 models, Algorithm 2(1) requires the smallest number of LPs. Algorithms 2(1) and 3(1) are both very quick in comparison to Alg. 7.3. It is instructive to look in detail at the 4 models that require more than 100 LPs for solution by Alg. 7.3. In most cases, Algorithms 2 and 3 solve far fewer LPs than Alg. 7.3 (the exception is *cplex1*), and are reasonably accurate.

Algorithms 2 and 3 are slightly less accurate than Alg. 7.3. Algorithm 2 does poorly on *cplex1*, in terms of both accuracy and speed. Ignoring *cplex1* gives Algorithm 2(1) an average of 3.0 LPs (instead of 17.8), and Algorithm 2(7) an average of 10.2 LPs (instead of 113.4). A corollary observation is that a change of algorithms can have a dramatic impact on accuracy for a particular model. Alg. 7.3 and 3(7) are the only ones able to achieve the true MIN IIS COVER for *cplex1*.

Table 7.2 is broken into five groups. The MIN IIS COVER in models 1–5 is found by one or both of the two phase 1 procedures applied. In fact, the MINOS phase 1 is the only procedure to find a MIN IIS COVER for *mondou2*. Because of this, Alg. 7.4

finds the true MIN IIS COVER for all five of these models in conjunction with any of the Algorithm 2 and 3 candidate selection criteria. This argues for the inclusion of the MINOS-style phase 1 procedure in Alg. 7.4, as does the excellent performance of the MINOS phase 1 on the other 14 models for which it found single-member IIS covers.

The MIN IIS COVER is found by all of the algorithms for models 6–9 in Table 7.2, including the fast short-list versions. The reduction in the number of LPs solved as compared to Alg. 7.3 is dramatic for *bgdbg1* and *greenbea*. This underlines the effectiveness of the new algorithms.

Models 10 and 11 of Table 7.2 require a longer list length to find a MIN IIS COVER. Each of the four short-list algorithms requires about the same small number of LPs to arrive at a solution, but smaller cardinality solutions are returned by Algorithms 2(7) and 3(7). This argues for the longer list lengths in Algorithms 2 and 3.

Algorithm 3 is the best approach for models 12 and 13 of Table 7.2. Even with a list length of 1, a cover of cardinality 4 is found for *cplex1* using Algorithm 3 versus a cover cardinality of 211 using Algorithm 2. Algorithm 3(7) finds the true MIN IIS COVER in both cases. This argues for the use of the selection criteria of Algorithm 3 in the framework of Alg. 7.4.

Finally, a MIN IIS COVER is not found using any heuristic method for the last model in Table 7.2. However, the best result, provided by both the MINOS phase 1 and by Algorithm 3(7), is very close to the optimum at only 1 greater than the true minimum cardinality.

These empirical results indicate that an effective version of Alg. 7.4 would incorporate a MINOS-style phase 1 procedure, and would use the selection criteria of Algorithm 3 at list length 7. This provides a significant speedup for general LP problems with little loss in accuracy: it fails to find a MIN IIS COVER only for *refinery*, and the cover is too large by just 1 member in that case. This algorithm is about 7 times faster than Alg. 7.3 on average.

For maximum speed at reasonable accuracy, use Alg.7.4 with selection criteria from Algorithm 3(1). This does not give a poor result on any of the test models. On the five models for which this combination does not achieve a MIN IIS COVER, the maximum distance from optimality is 3, and the average is 1.8. This algorithm is about 25 times faster than Alg.7.3 on average, and dramatically faster on many models.

Sadegh (1999) modifies Alg. 7.3 by substituting a minimax solution for the minimization of the sum of the elastic variables. This is easily done by adding constraints requiring that every elastic variable be less than some value β, and substituting the objective function min β. Sadegh reports good results on a number of test problems.

7.5 Two-Phase Relaxation-Based Heuristic

Amaldi et al. (2007) observe that the exact solution for MAX FS using a big-M MIP formulation (Sec. 7.1.1) often results in numerical difficulties for large models, but works well when the problem is small or moderate in size. This observation leads them to develop an interesting two-phase algorithm. In the first phase, a heuristic is applied to isolate a feasible subset, which is then frozen. In the second phase, the exact big-M MIP or other methods are used to expand the initial feasible set as much as possible. This has the pleasing feature of reducing the size of the problem in the second phase sufficiently that a big-M MIP can be effective. The success of the method hinges on the pairing of the methods applied in the two phases.

Amaldi et al. develop the method for the case in which all variables are bounded, which is easily transformed into a version in which all variables are nonnegative and upper-bounded. It is also straightforward to adapt the method for general linear constraints.

The most important new first phase heuristic is a linearization of the exact nonlinear bilinear formulation of the MAX FS problem seen in Sec. 7.1.2. This uses a substitution of the new variable z_{ij} for the bilinear terms $y_i x_j$, with further restrictions added so that z_{ij} is a closer approximation of $y_i x_j$. The resulting linearization is:

$$\max \sum_{i=1}^{m} y_i$$

$$subject\ to:$$

$$\sum_{j:a_{ij}<0} a_{ij} z_{ij} + \sum_{j:a_{ij}\geq 0} a_{ij} x_j \geq y_i b_i,\ i = 1...m$$

$$z_{ij} \leq u_j y_i,\ i = 1...m, j = 1...n,\ s.t.\ a_{ij} < 0$$

$$z_{ij} \leq x_j,\ i = 1...m, j = 1...n,\ s.t. a_{ij} < 0$$

$$x_j - u_j(1 - y_i) \leq z_{ij},\ i = 1...m, j = 1...n,\ s.t.\ a_{ij} < 0$$

$$l_j \leq x_j \leq u_j,\ j = 1...n$$

$$0 \leq y_i \leq 1,\ i = 1...m$$

$$z_{ij} \geq 0,\ i = 1...m, j = 1...n,\ s.t.\ a_{ij} < 0$$

An optimum solution to this model does not guarantee that the y_i variables, which are binary in the original LPEC formulation, will all have binary values. However it does provide a reasonable heuristic for identifying a large feasible subsystem, i.e. all of those constraints for which $y_i = 1$ are satisfied at the resulting point $\mathbf{x}$.

The important observation made by Amaldi et al. (2007) is that for the linear relaxation described above, as well as for a linear relaxation of the big-M MIP formulation, the inequalities for which $y_i < 1$ in the solution are not always inconsistent with those for which $y_i = 1$. This means that the feasible subsystem returned by the relaxation can possibly be augmented by further constraints, and this is where the second phase of the two-phase method comes into play. Note, though, that it is equally true that some of the constraints for which $y_i = 1$ in the relaxation solution may not belong to any maximum feasible subsystem, i.e. are wrongly included in the first phase feasible subsystem when the goal is to find a maximum cardinality set. Since the feasible subsystem returned by the first phase solution is frozen, there is no way to remove these constraints later, even if it would allow a larger feasible subsystem to be constructed.

The overall logic of the two-phase algorithm is summarized in Alg. 7.5. The output feasible subset consists of those constraints for which $y_i = 1$ in the first phase solution plus those constraints identified during the second phase procedure.

INPUT: an infeasible set of linear inequalities.

First Phase:

1. Solve a relaxation of MAX FS to obtain a solution y.
2. $I_1 \leftarrow \{i: y_i = 1, i = 1 \ldots m\}$

Second Phase:

3. Solve an exact formulation of MAX FS in which $y_i = 1$ is fixed for all $i \in I_1$.

OUTPUT: a feasible subset.

Alg. 7.5. Overall logic of the two-phase relaxation-based heuristic (Amaldi et al. 2007)

The authors apply their two-phase method, written in the AMPL language (Fourer et al. 2003) over several combinations of methods. While the exact big-*M* formulation is always used for the reduced-size second phase problem, various methods are used for the first phase:

- A linearization of the big-*M* formulation.
- The linearization of the bilinear LPEC formulation described above.
- An ordinary LP phase 1 procedure (see Sec. 7.3).

The two-phase variations are compared with several complete methods:

- A re-implementation of Chinneck's first SINF-reducing algorithm (Alg. 7.3) in AMPL.
- A branch-and-cut algorithm (Pfetsch 2002).
- A combinatorial Bender's cut algorithm (Codato and Fischetti 2004).

These methods are compared over a variety of random models as well as models derived from linear classification and machine learning instances and problems arising in digital video broadcasting. Different methods dominate in the different test sets, and various tradeoffs between solution speed and accuracy are seen.

For the complete methods, a straightforward application of the big-*M* MIP exact formulation to the entire problem is not able to reach optimality for very large models, as expected. However it frequently reaches a very good incumbent solution within the imposed time limits, which can then be used as a heuristic solution. The branch-and-cut method provides excellent results on the random test set, the only set to which it is applied. The combinatorial bender's cut algorithm performs very well on the single test set to which it is applied, though it fails completely on several of the larger instances. The *SINF*-reducing algorithm (Alg. 7.3) performs very well throughout most of the tests, though it times out on several of the extremely large digital video broadcasting problems whereas the two-phase methods and the exact big-*M* method are able to reach heuristic solutions at an incumbent before timing out.

The size of the feasible system produced during the first phase varies according to the method applied. The big-*M* relaxation gives the largest feasible subsystems for the random instances and some of the machine learning instances, but is dominated by the other two methods in some of the other test sets. It's an open question as to whether it is better to find a larger rather than a smaller feasible subsystem during the first phase. A small first-phase feasible subsystem gives the exact algorithm more room to maneuver during the second phase, but by the same token may make the second phase problem too large for exact solution. The true test is in the result returned after the completion of the second phase algorithm.

Generally speaking, the two-phase method with a bilinear relaxation first phase provides the best results over all of the data sets, though the *SINF*-reducing algorithm (Alg. 7.3) outperforms on the small classification dataset. The two-phase method with big-*M* relaxation first phase provides results that are almost as good as those produced by the bilinear first phase method, but is also much faster. The two-phase method with ordinary LP phase 1 first phase is dominated.

7.6 Randomized Thermal Relaxation Algorithms

Amaldi et al. (2005) consider the problem of solving MAX FS for extremely large systems of linear inequalities. More specifically, because the systems that they consider are so large (up to tens of millions of inequalities) they are content with solutions that are simply large feasible subsystems when it is not possible to obtain the true MAX FS. The *randomized thermal relaxation* (RTR) algorithm described in this section can be considered as a heuristic phase 1 procedure that tries to maximize the number of satisfied inequalities.

The approach taken by Amaldi et al. (2005) for these very large models is to use randomized variants of projection algorithms (see Sec. 2.8) to iteratively attempt to satisfy as many of the inequalities as possible. When the algorithm halts, the set of constraints currently satisfied constitutes the approximate solution to MAX FS. The projection variant they use is based on the thermal perceptron heuristic (Frean 1992) which iteratively relaxes all violated inequalities except for one, while moving orthogonally to the selected constraint to reduce its violation. This is similar to a sequential projection algorithm (Censor et al. 2001), but with randomization in the selection of the constraint to consider and the decision to accept an update. Details follow.

Constraints are of the form $\boldsymbol{a}_i\boldsymbol{x} \geq b_i$, $i = 1\ldots m$. At iteration i, constraint k_i is chosen randomly, and the current point x_i is updated as follows: $\boldsymbol{x}_{i+1} = \boldsymbol{x}_i + \eta_i \boldsymbol{a}_{k_i}$ with probability $p_i > 0$ if constraint k_i is violated, or $\boldsymbol{x}_{i+1} = \boldsymbol{x}_i$ otherwise. The randomized acceptance of an update is similar to the basic method in simulated annealing (Kirkpatrick et al 1983), and along with the random selection of a constraint for update helps avoid roundoff difficulties. The thermal variant gradually shifts attention from selecting constraints with large violations towards selecting constraints with small violations as the updating process proceeds, with the idea that as much feasibility as possible should be retained near the end of the process, rather than continually concentrating on very large updates which can seriously affect the number of violated constraints. Here again we see the interaction between reducing the sum of the infeasibilities (i.e. *SINF*) and the number of violations (*NINF*).

The shift in attention from large to small constraint violations is controlled by a temperature schedule, another feature borrowed from simulated annealing. The temperature t_i is a positive number that starts at a large value and gradually reduces, e.g. $t_0 = 1000$, and $t_{i+1} = t_i \times 0.0001$. The update step length η_i is determined by the violation of the selected constraint $v_i = \max\{0, b_{k_i} - \boldsymbol{a}_{k_i}\boldsymbol{x}_i\}$ and the current temperature as follows: $\eta_i = \frac{t_i}{t_0} e^{-v_i / t_i}$. Using this relationship, when t_i is large near the start of the process, large violations yield large updates, but when t_i is small near the end of the process, only small violations yield significant updates. The best solution $\boldsymbol{x}_{\text{best}}$ seen so far, in terms of the maximum number of satisfied constraints, is retained. After a preset number of iterations, $\boldsymbol{x}_{\text{best}}$ is returned as the approximate solution to MAX FS. Variations of these randomized thermal relaxation (RTR) methods are obtained by (i) changing the rule by which an inequality is selected for update (randomly, with or without replacement), (ii) changing the rule by which η_i is determined, and (iii) changing the rule by which p_i is set.

Amaldi et al. (2005) provide probabilistic termination guarantees that $\boldsymbol{x}_{\text{best}}$ almost surely solves MAX FS optimally after a finite number of iterations. The termination proofs use very long update sequences and very slow temperature decrease schedules, however empirical results are much better than might be expected from the proofs. Some modifications are implemented in practice:

- After each cycle of m randomly chosen inequalities, t_0 is reset as follows: $t_0 = \frac{1}{3}t_0 + \frac{2}{3}\sum_{k=1}^{m} v_k$.
- The maximum number of iterations is preset and is used to reduce the temperature as follows: $t_i = \left(1 - \frac{i}{MaxIterations}\right)t_0$.
- When an update yields a new point in which some of the variables fall outside of their bounds, those variables are projected back onto the bounds before proceeding.
- Good sub-optimal solutions are found by using a block-iterative update in which the update direction is given by a convex combination of the $\boldsymbol{a}_k$ from the violated inequalities in the block. The block size is decreased as the iterations proceed.
- A search is conducted along the line segment between x_i and x_{i+1} to find the point that satisfies the most constraints.

- When there has been no improvement for a predetermined number of iterations, a local search is conducted by altering the values of individual variables to satisfy more constraints. This can also be done in a grouped manner.
- A preprocessing step similar to an LP presolve is applied to identify constraints that cannot be satisfied within the current variable bounds, constraints that are always satisfied within the current variable bounds, and variables that can be fixed to their upper or lower bounds (possibly with suitable changes to b).

Amaldi et al. (2005) report on experiments in digital video broadcasting network planning, protein folding potentials, and discriminant analysis. They compare their RTR method to a big-M based MIP solution (see Sec. 7.1.1) using Cplex 8.1 (Ilog 2006). The value of M is easily determined from the problem, but is very large, which negatively impacts the MIP approach due to numerical difficulties. A two hour time limit is imposed on the MIP solutions. Despite the generous time allowed for the MIP solutions, Cplex is able to solve only relatively small instances, and RTR performs about as well as Cplex over these small instances. Over the larger instances, Cplex is generally not able to complete at all within two hours, though it occasionally returns a first solution within that time. RTR, on the other hand generally returns high quality solutions very quickly, and occasionally improves these somewhat if given more computation time.

In an interesting experiment, Amaldi et al. also apply their RTR algorithm to a set of six large feasible instances having 200,176 to 401,115 inequalities and 301 variables. RTR is able to satisfy almost all constraints within a relatively short time (less than 75 seconds in all cases, usually closer to 30 seconds on a 2.8 GHz PC). The number of unsatisfied constraints is very small in most cases, ranging between 0 and 6, with an average of 2.5 unsatisfied inequalities over the 6 instances.

In a third experiment, Amaldi et al. apply a hybrid method in which RTR is run for 5000 iterations, at which point a MIP is formulated in which only the currently unsatisfied constraints have the possibility of relaxation via the inclusion of a big-M. This MIP is then solved to optimality. This is a two-phase algorithm (Sec. 7.5) in which the first phase is solved by RTR. The results are again very promising, with the hybrid method producing near-optimal results in just a few seconds in most cases. The results are compared to both Cplex and a new Combinatorial Bender's Cuts method (Codato and Fischetti 2004). The latter method proves slightly better overall, but uses the entire two-hour time limit on several of the instances. Instance sizes are relatively small in these experiments: 169 to 1066 inequalities.

The randomized thermal relaxation methods are an excellent approach for heuristically solving MAX FS for very large sets of linear inequalities.

7.7 An Interior-Point Heuristic

Meller et al. (2002) develop an approximate solution for MAX FS for extremely large sets of linear inequalities that makes use of the properties of interior-point LP solvers. Specifically, for a null objective function, interior point algorithms find a point near the analytic centre of a set of linear inequalities, i.e. the point that minimizes $\sum_{i=1}^{m} \ln(b_i - \boldsymbol{a}_i\boldsymbol{x})$ where there are m inequality constraints of the form $\boldsymbol{a}_i\boldsymbol{x} \le b_i$. This is a form of barrier function that tends to push the current iterate away from the satisfied constraints and generally towards the centre of a polytope. Note that the final point reached is affected by the extra push from redundant constraints, so it may not be near the geometric centre of a polytope.

The *maximum feasibility guideline* algorithm developed by Meller et al. (2002) operates as follows. First obtain an initial solution $\boldsymbol{x}_0$, which will satisfy some subset $P(\boldsymbol{x}_0)$ of the inequality constraints in this infeasible system. For the protein folding application of interest to Meller et al. $\boldsymbol{x}_0$ is conveniently provided by statistical potentials, but may be more difficult to obtain in other contexts. The success of the method depends greatly on a good choice of $\boldsymbol{x}_0$. Next, find the analytic centre of the subset of constraints $P(\boldsymbol{x}_0)$ and denote this by $\boldsymbol{x}_1$. The analytic centre will necessarily satisfy all of the constraints in $P(\boldsymbol{x}_0)$, and may satisfy a number of additional constraints; this possibly larger set of constraints satisfied at $\boldsymbol{x}_1$ is denoted by $P(\boldsymbol{x}_1)$. Now iterate the process, finding a new analytic centre for $P(\boldsymbol{x}_k)$ and denoting this as $\boldsymbol{x}_{k+1}$, halting the iterations when $P(\boldsymbol{x}_k) = P(\boldsymbol{x}_{k+1})$.

Meller et al. (2002) test their method on several very large infeasible LPs resulting from protein folding problems. In one problem with 627,567 inequalities, the method progresses from an initial guess that violates 57,211 constraints to violating 6,800 constraints after the first analytic centre is found, and finally to 1,928 violated constraints when the method converges. Two different methods of generating the initial solution are used, based on knowledge of the application. One method gives better results than the other, but in both cases there is a significant increase in the size of the feasible subset as the algorithm iterates. The authors note that it can require up to 15 analytic centre solutions before the method converges, each solution requiring several minutes of workstation time. In their implementation, each analytic centre solution is independent, without the benefit of a warm start for the interior point solution.

This approach is reminiscent of the bootstrapping method for achieving feasibility for sets of nonlinear constraints as outlined in Sec. 5.3.

7.8 Working with IIS Covers

An IIS cover is directly useful as a tool to focus the analytic effort, but it also has other applications. It can be used to quantify the importance of constraints relative to the infeasibility, and it can also be used as a basis for finding individual IISs. Details follow below.

7.8.1 Single Member IIS Covers

Small cardinality IIS covers are generally the most useful in focussing the analytic effort on the constraints that cause the greatest difficulties. However, an IIS cover having only a single member may not be especially helpful for two reasons. If the model contains just a single IIS, then any member of the IIS forms a single-member set cover, so the single-member IIS cover focuses inappropriately on a random member of the IIS. The same problem of possibly misleading focus applies when there are IISs overlapped on a common subset of constraints, each of which is a candidate for an IIS cover of cardinality 1.

What needs to be determined is the complete set of single member IIS covers. If given a complete IIS in addition, it is then easy to determine whether there is only a single IIS, or whether there are overlapped constraints. Alg. 7.6 outlines a procedure first used in LINDO (Schrage 1991). The members of the IIS are labelled either "necessary" (i.e. necessary to the IIS) or "sufficient" (i.e. sufficient to remove all infeasibility in the model). "Sufficient" constraints are single member IIS set covers.

INPUT: Constraints defining an infeasible model.
1. Find an IIS.
2. FOR each member of the IIS:
   - Temporarily remove the current member from the model.
   - Test feasibility of the reduced model via solution of phase 1 LP.
   - IF the reduced model is feasible, label the current member "sufficient".
   - ELSE (reduced model infeasible), label the current member "necessary".
   - Return the current member to the model.

OUTPUT: An IIS with all members labelled "necessary" or "sufficient".

Alg. 7.6. The IIS member labelling scheme

Alg. 7.6 has some useful properties, as described in the following theorems.

Theorem 7.3: *Single IIS* (Chinneck 1997a). All of the IIS members are labelled "sufficient" by Alg. 7.6 if and only if there is only a single IIS in the model.

Proof: If there is another IIS in the model that does not overlap with the original IIS, then none of the members of the original IIS will be labelled "sufficient" since none of them can eliminate all of the infeasibility in the model. If there are other IISs in the model which overlap the original IIS, then some of the members of the original IIS will be labelled "necessary" rather than "sufficient" (recall that it is impossible to have a subset of an IIS that is itself infeasible since IISs are irreducible). Hence all members of the IIS are labelled "sufficient" if and only if there is only a single IIS in the model. ■

Corollary 7.4: *Overlapped IISs* (Chinneck 1997a). Alg. 7.6 labels some of the IIS members "sufficient" and some "necessary" if and only if there is a single cluster of IISs with some elements common to all IISs in the cluster.

Proof: As in Theorem 7.3, if there is another IIS in the model that does not overlap with the original IIS, then none of the members of the original IIS will be labelled "sufficient" since none of them can eliminate all of the infeasibility in the model. Hence there is only a single cluster of IISs that overlaps on the "sufficient" members. And since some members are labelled "necessary" (because their removal does not eliminate all of the infeasibility) there must be other IISs in the model. ■

If some constraints are labelled "sufficient" and some are labelled "necessary", then additional information is obtained because the "sufficient" members are more probably incorrect than the "necessary" members.

Algorithm 7.6 can be used when the set cover cardinality is greater than one by first applying Alg. 7.7 (described below) to create subsets of the model in which the set cover cardinality is one. Thm. 7.3 and Corollary 7.4 can then be used to extract additional information about the infeasibility.

If Alg. 7.6 is applied to the first IIS isolated, without knowledge of the set cover cardinality, it can give some idea of the set cover cardinality by using Thm.7.3 and Corollary 7.4.

Theorem 7.5: *Set cover cardinality of 1* (Chinneck 1997a). The minimum IIS set cover cardinality is one if and only if any member of the IIS is labelled "sufficient" by Alg. 7.6.

Proof: If the removal of any constraint in the model eliminates all of the infeasibility, then it is a set cover, by definition. Further, since it has cardinality one, it must be of minimum cardinality (see Observation 7.2). ■

Corollary 7.6: *Set cover cardinality >1* (Chinneck 1997a). No members of the IIS are labelled "sufficient" if and only if the minimum IIS set cover cardinality is greater than one.

Proof: If no members of the IIS are labelled "sufficient", then no single constraint is able to eliminate all of the infeasibility in the model. Hence the minimum set cover cardinality must be greater than one. ■

7.8.2 Finding Specific IISs Based on IIS Covers

In repairing an infeasible LP, it is very helpful if one IIS covered by each member of the cover set is found. This is simple to do if the covering algorithm is operating on a list of IISs, even if the list is incomplete, as in Parker and Ryan's method or the constraint frequency heuristic. Otherwise, a simple algorithm due to Chinneck (1996c) can be used when the set cover is provided by a method which does not first list IISs. See Alg. 7.7.

```
INPUT: (i) Constraints defining original infeasible model, (ii) IIS cover.
FOR each member of IIS cover:
        Eliminate all members of IIS cover except the current member.
        Test feasibility via solution of phase 1 LP.
        IF the reduced model is feasible THEN:
                Issue message and remove current member from IIS cover.
        ELSE (reduced model is infeasible):
                Isolate and report an IIS having few rows.
        Reinstate all members of IIS cover.
OUTPUT: One IIS for each member of (possibly reduced) IIS cover.
```

Alg. 7.7. Finding one IIS for each member of the IIS set cover

Alg. 7.7 will sometimes identify constraints which have been added to the IIS cover in error. When all but one member of the cover are removed from the model and it becomes feasible, then it is obviously not necessary to remove the single remaining member as well in order to eliminate all infeasibility, hence the current member is not part of a minimal cover.

Note that when all but one member of the cover are eliminated in Alg. 7.7, subsets of the model are created in which the retained set cover member is a single member IIS set cover.

The satisfiability community has taken a different approach to finding IISs based on IIS covers. Given the availability of efficient solvers for the maximum satisfiability (MAXSAT) problem (see Sec. 4.1), Liffiton and Sakallah (2005) first generate the complete set of IIS covers, and then use this set to generate the complete set of IISs. To generate the complete set of IIS covers, they allow the maximum satisfiability solver to eliminate at most k constraints while seeking a feasible solution for all constraints that remain. Of course, each maximum satisfiability solution yields an associated IIS cover. All covers at some value k are found by adding constraints that block out solutions already found. k is incremented from an initial value of 1 until no more IIS covers can be found.

Now a second algorithm is applied to generate a single IIS from the complete set of IIS covers, as shown in Alg. 7.8. It operates on the principle that every member of every IIS must be in some IIS cover in the complete set. At each iteration the algorithm chooses a particular cover and a constraint from that cover is added to the IIS that is being constructed. It then removes the constraints from the set of covers to make sure that in later steps we will only find IISs that contain the chosen constraint. The process repeats until a single IIS has been constructed, which is signalled by the emptying of the list of IIS covers. The main idea is that a minimal cover of the cover sets is an IIS.

INPUT: *COVERS*, the complete set of IIS covers in the model.
0. *IIS*=∅
1. WHILE *COVERS* ≠ ∅:
   1.1 *CurrentCover* ← select a cover in *COVERS*.
   1.2 *Constraint* ← select a constraint in *CurrentCover*.
   1.3 *IIS* ← *IIS* ∪ *constraint*.
   1.4 Remove all constraints in *CurrentCover**Constraint* from all covers in *COVERS*.
   1.5 Remove all covers in *COVERS* that contain *Constraint*.
2. Return *IIS*.

OUTPUT: a single IIS.

Alg. 7.8. Finding a single IIS given the complete set of IIS covers (Liffiton and Sakallah 2005)

An example follows. Suppose the model contains the IISs {A,B,C}, {C,D,E} and {F,G,H}, then the complete set of IIS covers is {A,D,F}, {A,D,G}, {A,D,H}, {A,E,F}, {A,E,G}, {A,E,H}, {B,D,F}, {B,D,G}, {B,D,H}, {B,E,F}, {B,E,G}, {B,E,H}, {C,F}, {C,G}, {C,H}. The algorithm proceeds as follows:

- Step 1.1: Select cover {A,D,F}.
- Step 1.2: Select constraint A.
- Step 1.3: *IIS* ←{A}.
- Step 1.4: Eliminate constraints D and F from all covers. *COVERS* ←{ {A}, {A,G}, {A,H}, {A,E}, {A,E,G}, {A,E,H}, {B}, {B,G}, {B,H}, {B,E}, {B,E,G}, {B,E,H}, {C}, {C,G}, {C,H} }.
- Step 1.5: Eliminate all covers containing constraint A. *COVERS* ←{ {B}, {B,G}, {B,H}, {B,E}, {B,E,G}, {B,E,H}, {C}, {C,G}, {C,H} }.
- Step 1.1: Select cover {B,E,G}.
- Step 1.2: Select constraint B.
- Step 1.3: *IIS* ←{A,B}.
- Step 1.4: Eliminate constraints E and G from all covers. *COVERS* ←{ {B}, {B}, {B,H}, {B}, {B}, {B,H}, {C}, {C}, {C,H} }.
- Step 1.5: Eliminate all covers containing constraint B. *COVERS*←{ {C}, {C}, {C,H} }.
- Step 1.1: Select cover {C}.
- Step 1.2: Select constraint C
- Step 1.3: *IIS* ←{A,B,C}.
- Step 1.4: *COVERS* ←{ {C}, {C}, {C,H} }.
- Step 1.5: Eliminate all covers containing constraint C. *COVERS* ← ∅.
- Step 2: Return {A,B,C}.

There is also an effort to use Alg. 7.8 to generate all IISs in the model, basically by branching to reorder the choices made in Steps 1.1 and 1.2 of Alg. 7.8. Given that the number of IISs is potentially exponential, this is not an efficient approach.

Bailey and Stuckey (2005) also find IISs by operating on the set of IIS covers, but they do not assume that the complete set of IIS covers is provided in advance. Their *dualize and advance* heuristic instead finds IIS covers one by one using the *grow* algorithm mentioned at the beginning of this chapter. Set covers of the partial set of IIS covers are found as the algorithm proceeds. Again, this algorithm is only practical in where a fast method for finding maximum feasible subsystems is available, as in the MAXSAT context.

7.9 The Minimum Number of Feasible Partitions Problem

Amaldi and Mattavelli (2002) propose a generalization of the MAX FS problem, which they designate the *minimum number of feasible partitions problem* (MIN PFS): given a possibly infeasible system of linear constraints, find a partition of this system into a minimum number of feasible subsystems. In the previous part of this chapter we have considered the separation of an infeasible system of linear constraints into two partitions: the MAX FS set and the MIN IIS COVER (equivalently, MIN ULR) set. Note that it is possible that the MIN IIS COVER set is itself infeasible, hence further partitioning may be needed to solve the MIN PFS problem. MIN GRAPH COLOURING is a special case of MIN PFS.

William Pulleyblank and others showed that any set of linear inequalities $\boldsymbol{Ax} \geq \boldsymbol{b}$ can be partitioned into two sets that are both feasible. The proof is provided by Greenberg (1996a) in the following theorem.

***Theorem** 7.7: MIN PFS cardinality for linear inequalities* (Greenberg 1996a, Theorem 18). Suppose a set S of linear inequalities is inconsistent. There exists a partition of S, say $S' \cup S''$ such that S' and S'' are each consistent and S' is a maximal consistent subsystem (in which case $X(S') \cap X(S'') = \emptyset$).

Proof: Construct a line that intersects each hyperplane, $H_i = \{\boldsymbol{x} | \boldsymbol{a}_i \boldsymbol{x} = b_i\}$ where $\boldsymbol{a}_i \neq \boldsymbol{0}$ for each i. Totally order the points along the line; rename and reorder so that $\boldsymbol{x}^i$ is the point on H_i. Now initialize $S' = \{\boldsymbol{a}_1 \boldsymbol{x} \geq b_1\}$ and continue to add $\boldsymbol{a}_i \boldsymbol{x} \geq b_i$ to S' as long as $\boldsymbol{a}_i \boldsymbol{x}^k \geq b_i$ for all $k < i$. The first time this fails, initialize $S'' = \{\boldsymbol{a}_i \boldsymbol{x} \geq b_i\}$. For each $i > k$, the halfspace $X(\{\boldsymbol{a}_i \boldsymbol{x} \geq b_i\})$ intersects either $X(S')$ or $X(S'')$, so the inequality can be added to S' or S'' respectively. Test first if $S' \cup \{\boldsymbol{a}_i \boldsymbol{x} \geq b_i\}$ is consistent and if so add this inequality to S'. It then follows that all inequalities not in S' are precisely those whose augmentation renders inconsistency. This means that S' is a maximal consistent subsystem (and that $X(S') \cap X(S'') = \emptyset$). ■

Note that Thm. 7.7 applies only when all of the constraints in the model are linear inequalities. It does not apply when equalities are included. Consider, for example, a set of three or more parallel and separated hyperplane equality constraints. Now the cardinality of the MIN PFS solution is equal to the number of hyperplanes. For similar reasons, Thm. 7.7 also does not apply in the case of *hyperslabs*, pairs of complementary inequalities that define a slab in hyperspace, when the pairs must be handled together (i.e. both satisfied in the same partition). This is important in the sequel.

Amaldi and Mattavelli (2002) raise MIN PFS in the context of a particular application in estimating piecewise linear models (see Sec. 11.8). The goal is to model a set of noisy data points with a small number of linear pieces. Each known point $\boldsymbol{d}_i$ in j dimensions can be transformed into an equation of the form $\Sigma_j d_{ij} w_j = w_0$, where $\boldsymbol{w}$ is the vector of variables $w_1, w_2 \ldots w_n$ and $\boldsymbol{w}^+$ is $\boldsymbol{w}$ augmented with w_0. It is then possible to solve a MIN PFS problem to find the smallest number of piecewise linear sections to model the data, but the noise in the data means that there will be an unnecessarily large number of linear sections. To handle the noise in the data points, each point is instead rendered as a pair of *complementary inequalities* defining a slab of limited thickness in n-space: $\boldsymbol{d}_i \boldsymbol{w} \leq w_0 + \varepsilon$ and $\boldsymbol{d}_i \boldsymbol{w} \geq w_0 - \varepsilon$. This defines a slab of width 2ε, where ε may not be identical for all pairs of complementary inequalities. After the heuristic MIN PFS solution, each partition is feasible, so each partition allows a solution for the variables in $\boldsymbol{w}^+$. This defines a slab that contains all of the data points whose corresponding pairs of complementary inequalities are satisfied in the partition.

Amaldi and Mattavelli (2002) propose a greedy heuristic for the MIN PFS solution over the pairs of complementary inequalities. It first finds a close-to-maximum feasible subsystem (feasible subsystem containing a close-to-maximum number of pairs of complementary inequalities), and the complementary close-to-minimum cardinality set of removed pairs of complementary inequalities, for the original set of complementary inequalities. The process is then repeated on the set of removed complementary inequalities. This cycle continues until the final set of removed complementary inequalities is itself feasible. In this way the system is subdivided into a small number of feasible partitions. Empirical results for the piecewise linear modelling application are very good where the MAX FS subproblems are solved by a randomized thermal relaxation algorithm (see Sec. 7.6).

The same general greedy approach can be applied in the more general case that does not feature slabs defined by pairs of inequalities. Simply recursively solve MAX FS until the MIN IIS COVER partition is itself feasible. This approach is clearly not guaranteed to solve MIN PFS exactly, as Amaldi and Mattavelli (2002) show in this example: A: $x_1 + x_2 = 0$, B: $x_1 - x_2 = 0$, C: $x_2 = 1$, D: $x_2 = 2$. The cardinality of MAX FS for this system is two. There are five feasible systems: A and B are satisfied at (0,0), B and C are satisfied at (1,1), B and D are satisfied at (2,2), A and C are satisfied at (−1,1), and A and D are satisfied at (−2, 2). If the greedy algorithm finds {A,B} as the first MAX FS solution, then it will return a cardinality 3 MIN PFS solution, even though a solution of cardinality 2 exists, e.g. {B,D} and {A,C}. Of course the MAX FS subproblems can be solved by any suitable method.

Murty et al. (2000) outline the same minimum number of feasible partitions problem and the same greedy approach, though they approach it by looking separately at systems composed entirely of equalities, and systems composed entirely of inequalities. They show that the problem is NP-hard for systems composed entirely of equalities.

Bemporad et al. (2005) improve on the greedy algorithm in the context of solving a problem of identifying piecewise affine models of discrete-time nonlinear and hybrid systems from input-output data. They alter the greedy algorithm by allowing a degree of backtracking to see whether solutions developed for later partitions might provide better solutions at earlier stages of the process.

The main algorithm developed by Bemporad et al. (2005) is summarized in Alg. 7.9. U is the entire set of complementary inequalities in the original infeasible system; U_k is the set of complementary inequalities addressed by the kth MAX FS heuristic solution. S_k is the set of complementary inequalities satisfied by the slab solution $\boldsymbol{w}^+_k$ for the kth feasible partition. Hence $U_k = U \setminus \{S_1 \cup S_2 \cup \ldots \cup S_{k-1}\}$. Note that the sets S_k and U_k may be changed as the algorithm backtracks.

The main idea in Alg. 7.9 is that the slab solution developed at some stage k may actually provide a better solution at some earlier stage d than the solution originally returned for stage d. In this case (Step 7.3) the algorithm backtracks to stage d and replaces its solution with the solution for stage k, then resumes from stage d. This helps mitigate the greedy aspect of the original algorithm by Amaldi and Mattavelli.

Bemporad et al. use a variation of the Randomized Thermal Relaxation algorithm (Sec. 7.6) to solve the individual MAX FS problems. Their variation is specifically designed to improve performance in solving the MIN PFS problem. The main idea is that if the number of cycles in the RTR algorithm becomes too high, then the current best solution $\boldsymbol{w}^+_{\text{best}}$ (i.e. the solution seen so far that satisfies the most pairs of complementary inequalities) is replaced by an improved version over the same set of satisfied constraints. The improved solution is obtained by solving for the l_∞ projection norm over the set D of points contained within the $\boldsymbol{w}^+_{\text{best}}$ slab. The l_∞ projection norm is defined as $\arg\min_{\boldsymbol{w}, w_0} \max_{D} | w_0 - \boldsymbol{d}_i \boldsymbol{w} |$, which has the effect of setting $\boldsymbol{w}^+$ so that the distance from the farthest data point to the centerline of the slab is minimized. This is found by solving a linear programming problem. The l_∞ projection norm is used because it has favourable properties in a later refinement procedure that works with data points that satisfy more than one of the linear models to try to assign them to a single model. However Bemporad et al. mention that different measures could be used to return an improved model for the set, including least squares.

Where the maximum number of RTR cycles is C, recalculation of $\boldsymbol{w}^+_{\text{best}}$ is carried out for all cycles above $0.7C$ or $0.8C$. This value was determined experimentally.

Bemporad et al. (2005) carry out a set of experiments to compare their modified algorithm (both the RTR modifications and the backtracking heuristic in Alg. 7.9) with the original RTR (Sec. 7.6). The experiments use randomized data generated in such a way that the minimum number of feasible partitions is known a priori to be 4. Over repeated trials, the original RTR algorithm generates an average of 18 feasible partitions, with a range of 12 to 22, and a high variance. Bemporad et al.'s modified algorithm generates an average of 5 feasible partitions, with a range of 4 to 7, and low variance. It is also interesting that there are very few points in any partitions beyond the 4th one.

INPUT: an infeasible set of complementary inequalities U.
0. $k \leftarrow 0$; $S_1, S_2, S_3 \ldots \leftarrow \varnothing$
1. $k \leftarrow k+1$
2. $U_k \leftarrow U \backslash \{S_1 \cup S_2 \cup \ldots \cup S_{k-1}\}$.
3. IF $U_k = \varnothing$ THEN:
   3.1 $k \leftarrow k-1$
   3.2 Exit.
4. $\mathbf{w}^+_k \leftarrow$ slab equation returned by heuristic MAX FS solution for U_k.
5. $S_k \leftarrow$ {all constraints in U_k satisfied by slab $\mathbf{w}^+_k$}.
6. $d \leftarrow 1$
7. WHILE $d < k$ DO:
   7.1 $U_d \leftarrow U \backslash \{S_1 \cup S_2 \cup \ldots \cup S_{d-1}\}$
   7.2 $N_{kd} \leftarrow$ number of constraints in U_d satisfied by slab $\mathbf{w}^+_k$.
   7.3 IF $N_{kd} > |S_d|$ THEN:
      7.3.1 $\mathbf{w}^+_d \leftarrow \mathbf{w}^+_k$
      7.3.2 $S_d \leftarrow$ {all constraints in U_d satisfied by slab $\mathbf{w}^+_d$}
      7.3.3 $k \leftarrow d$
      7.3.4 Go to Step 1.
   7.4 $d \leftarrow d+1$
8. Go to Step 1.

OUTPUT: k feasible partitions with slabs $\mathbf{w}^+_1 \ldots \mathbf{w}^+_k$.

Alg. 7.9. Backtracking greedy algorithm for MIN PFS (Bemporad et al. 2005)

Once a heuristic solution for MIN PFS has been found, Bemporad et al. (2005) add a final refinement stage. This stage allows partitions to be merged, discarded, and updated. It also handles *undecidable* points that are contained within more than one partition slab. The main steps are summarized in Alg. 7.10. The algorithm begins (Step 0) by calculating the best slab equation for each partition using the l_∞ projection norm. In Step 1, partitions whose slab equations are too similar (as measured by the ratio of matrix norms) are merged, and replaced by a new slab equation. α is a user-specified control parameter. Partitions may also be discarded in Step 3 if their slab contains too few data points. β is a user-specified control parameter.

Step 4 deals with the undecidable points. The user-specified control parameter c sets the number of nearest-neighbour feasible points to use in the decision. The undecidable point is assigned to the partition whose slab contains the most nearest-neighbour feasible points, provided it is contained in the slab associated with that partition. It is possible that an undecidable point will retain its undeciable status after this process.

It is also possible that the process will terminate with some points still in the infeasible category. If an infeasible point is far from any feasible slab, then Bemporad et al categorize it as an outlier that should be ignored.

Note that there is a considerable literature on other approaches to solving the piecewise linear model estimation problem, and it is certainly possible that some of those other approaches can be adapted to solving the MIN PFS problem. Note particularly the literature on PieceWise affine AutoRegressive eXogenous (PWARX) models. See Juloski et al. (2005) for a comparison of four procedures for this problem, including the method by Bemporad et al.

7.10 Partial Constraint Satisfaction in Constraint Programming

In an overconstrained constraint satisfaction problem, not all of the constraints can be satisfied simultaneously. This is addressed in 3 different ways in constraint programming: the MAXSAT problem, partial constraint satisfaction, and soft constraints.

As described in Sec. 4.1, the MAXSAT problem consists of finding a set of true/false values for the Boolean literals such that the largest number of clauses is satisfied. This is directly analogous to the problem of finding the maximum feasible subset of constraints that is the subject of this chapter. The Boolean literals are equivalent to binary variables, and the clauses are equivalent to constraints. Consider the example Boolean satisfaction problem given in Sec. 4.1: $(A \vee B) \wedge (\neg A \vee C \vee D) \wedge (\neg B \vee \neg D) \wedge (\neg C)$. The conversion of a Boolean satisfiability problem to a binary integer programming problem is well-known and direct:

clause in Boolean variables	*constraint in binary variables*
$A \vee B$	$A + B \geq 1$
$\neg A \vee C \vee D$	$(1-A) + C + D \geq 1 \rightarrow -A + C + D \geq 0$
$\neg B \vee \neg D$	$(1-B)+(1-D) \geq 1 \rightarrow -B-D \geq -1 \rightarrow B+D \leq 1$
$\neg C$	$(1-C) \geq 1 \rightarrow -C \geq 0 \rightarrow C \leq 0$

This implies that the maximum feasible subset problem can be attacked for certain binary integer programming problems by converting them to Boolean satisfiability problems and applying MAXSAT solution algorithms. This can also go the other way: de Givry et al. (2003) examine a number of options for solving the weighted MAXSAT problem, including converting it for solution by a standard MIP solver.

INPUT: a heuristic solution for MIN PFS consisting of s partitions.

0. Calculate the best slab model w^+_i for each partition using the l_∞ projection norm.
1. Merge partitions:

   1.1 Find partitions i and j such that $\mu(w_i^+, w_j^+) = \dfrac{\|w_i^+ - w_j^+\|}{\min\{\|w_i^+\|, \|w_j^+\|\}}$ is minimized.

   1.2 IF $\mu(w_i^+, w_j^+) \le \alpha$, THEN merge the data points associated with each partition into a new partition.

   1.3 Calculate the new slab for this partition using the l_∞ projection norm.

   1.4 $s \leftarrow s - 1$.
2. Data point reassignment:

   2.1 FOR each data point $\boldsymbol{d}_k$:

   2.2 Select case:

   a. IF $\boldsymbol{d}_k$ is contained within the slab associated with exactly one partition, THEN assign it to that partition and mark as *feasible*.

   b. IF $\boldsymbol{d}_k$ is contained within the slab associated with more than one partition, THEN mark as *undecidable*.

   c. ELSE mark $\boldsymbol{d}_k$ as *infeasible*.
3. Discard partitions:

   3.1 Find partition i whose associated slab contains the smallest number of points.

   3.2 IF (data points contained in slab for partition i)/(total number of data points) $\le \beta$ THEN discard partition i.

   3.3 $s \leftarrow s - 1$.

   3.4 Go to Step 2.
4. Assign undecidable data points:

   4.1 For each undecidable point $\boldsymbol{d}_k$:

   4.1.1 Find the c closest feasible points to $\boldsymbol{d}_k$.

   4.1.2 Identify the partition i to which the greatest number of the c closest feasible points belong.

   4.1.3 IF $\boldsymbol{d}_k$ is within the slab associated with partition i THEN assign $\boldsymbol{d}_k$ to partition i and mark it as *feasible*.
5. Update parameters:

   5.1 Calculate $(w^+_i)_{\text{best}}$ for each partition using the l_∞ projection norm.
6. Termination:

   6.1 IF $\|(w^+_i)_{\text{best}} - w^+_i\| \le \gamma\|w^+_i\|$ for all $i = 1\ldots s$, THEN exit.

   6.2 $w^+_i \leftarrow (w^+_i)_{\text{best}}$ for all $i = 1\ldots s$.

   6.3 Go to Step 1.

OUTPUT: s partitions with associated slabs $(w^+_i)_{\text{best}}$ for $i = 1\ldots s$.

Alg. 7.10. Partition refinement (Bemporad et al. 2005)

Freuder and Wallace (1992) raise the idea of *partial constraint satisfaction.* Their main goal is to find values for a subset of the variables such that a subset of the constraints is satisfied. One obvious goal is to find variable values such that a maximum number of constraints are satisfied. Mechanisms used include branch and bound, and more efficient use of the search tree via backjumping, backmarking, arc-consistency, forward checking, and other techniques. Constraints can also be assigned an order, with constraints added to the set in the given order, stopping when infeasibility is reached. As well, constraints can be assigned a weight or strength, which implies an order. Again, constraints are added to the set in decreasing order of strength, with ties among constraints of identical strength being broken in various ways. The idea of partial constraint satisfaction gave rise later to the concept of *soft constraints* that are not necessarily satisfied at the solution, unlike hard constraints that must be satisfied; the same idea under the same name is also seen in the context of multi-objective programming in optimization (see Sec. 9.3).

Meseguer et al. (2003) summarize the state of the art in solving over-constrained constraint satisfaction problems. The techniques are similar to those found in the optimization literature for finding the best solution for an infeasible model or for a multi-objective program, including fuzzy and probabilistic approaches (see Sec. 8.1.5), lexicographical ordering (see Sec. 9.3), weighting methods, and hierarchical approaches. Meseguer et al. recognize that solving an overconstrained constraint satisfaction problem in the "best" way amounts to an optimization problem, and hence various optimization techniques are also employed; we will return to this theme in Chap. 8. In addition, various constraint programming techniques have been modified to deal with soft constraints. See also Petit et al. (2000). Research in constraint programming techniques to deal with soft constraints is ongoing.

Freuder and Wallace (1992) raise the idea of *partial constraint satisfaction*. Their main goal is to find values for a subset of the variables such that a subset of the constraints is satisfied. One common goal is to find variable values such that a maximum number of constraints is satisfied. Mechanisms used include branch and bound and more efficient variants of the search such as backjumping, backmarking, arc consistency, forward checking, and other techniques. Constraints can also be satisfied in order, with constraints added to the set in the given order, stopping when no solution is reached. As well, constraints can be assigned a weight or strength, which implies an ordering: constraints are added to the set in decreasing order of strength, with ties among constraints of identical strength broken in various ways. Hierarchical constraint satisfaction gave rise to the concept of soft constraints, which may be violated if they cannot be satisfied at the solution, unlike hard constraints that must be satisfied; the same idea underlies the concept of goals in multi-objective programming in general (see Sec. 9.1).

Meseguer et al. (2003) summarize the state of the art in solving over-constrained constraint satisfaction problems. The techniques are similar to those found in the optimization literature for turning a hard infeasible model into a feasible model or for multi-objective programs, including fuzzy and probabilistic approaches (see Sec. 8.1.5), lexicographical ordering (see Sec. 9.3), weighting methods, and iterational approaches. Meseguer et al. recognize that solving an over-constrained constraint satisfaction problem is the direct analogue to an infeasibility problem and hence various relaxation methods are also analyzed, as with relation to this theme in Chapter 8. A number of constraint programming techniques have been modified to deal with soft constraints; see also Rolf et al. (2006). Research in constraint programming continues to deal with soft constraints in general.

8 Altering Constraints to Achieve Feasibility

Chapters 6 and 7 present two different approaches for analyzing infeasible systems of constraints. Both try to discover useful information about the system (an IIS or a maximum feasible subsystem) that the modeler can use to correct the model. However it is possible to approach the analysis in a completely different way by asking this question: what is the smallest adjustment to the constraints in the model that will render if feasible?

Most available methods in this category address only infeasible linear systems, though at least one method can be applied to nonlinear systems. The methods divide into two broad classes: those that only consider shifting the constraints via a change to the right hand side constant (i.e. a parallel translation of the constraint), and those that consider the much harder problem of finding the minimum change to all of the constraint coefficients, including both the constraint bodies and the right hand side constants. Oddly, none of the research addresses the issue of incorrect relationship directions. For example, the model might be rendered feasible if a ≥ relationship is changed to a ≤ relationship, or if an = is changed to a ≥, etc. The general unaddressed question is this: what is the smallest number of constraint relationships to change such that the model is made feasible?

As we will see in this chapter, there are several ways to define the "smallest adjustment" that will render a model feasible, each having different solution complexities and yielding different results. All are based on minimizing some kind of matrix norm that expresses the difference between the "corrected" version of the model and the original version.

8.1 Shifting Constraints

The most straightforward approach to altering constraints to attain feasibility is to simply shift (or "translate") them in space by adjusting the value of the constant, also known as the "right hand side" (RHS). Many researchers have developed methods for finding the "best" set of constraint shifts according to some criteria (e.g. minimum total distance moved, etc.). The most attention has been paid to infeasible linear programs, but some more recent methods can also be applied to nonlinear models.

In the case of infeasible linear systems, Murty et al. (2000) distinguish four different ways to measure the "best" adjustment of the right hand sides needed to attain feasibility:

- *The smallest number of shifted constraints* (called the "smallest changes model" by Murty et al.). This is equivalent to the maximum feasible subset problem of Chap. 7, or more exactly, it is equivalent to the MIN ULR problem.
- *The smallest total penalty for fixed penalties.* The penalty cost of changing a RHS is fixed for each individual constraint (regardless of the size of the shift), and the goal is to choose the constraints to shift such that the total penalty is minimized. This model may apply when there is a fixed cost to break a contract, for example. This is the same as the previous model if the penalties are identical across all constraints.
- *The smallest total penalty for variable penalties.* The penalty cost of shifting a constraint is a linear function of the size of the shift, and the goal is to minimize the total penalty. This is equivalent to a weighted elastic programming model (see Sec. 6.1.4 and Sec. 8.1.1).
- *The smallest total penalty for variable penalties with bounds.* This is identical to the previous model, except that some or all of the elastic variables are bounded, indicating that the associated constraint can be shifted only a limited distance. This is again equivalent to elastic programming with simple bounds on the elastic variables. Note that if the bounds on the elastic variables are too tight, then even the elastic formulation may be infeasible.

Which way to measure the "best" adjustment of the constraint RHSs needed to achieve feasibility depends on the application. Other measures, including nonlinear penalties, can also be imagined.

Note that row scaling can significantly affect the results for many of the measures described below. It is best to apply these methods to a fully scaled "original" version of the model.

8.1.1 Using the Phase 1 Result

The phase 1 formulation for a linear program includes artificial variables which effectively allow the constraints to shift so that an initial feasible solution is readily available to start the simplex method (see Sec. 7.3). If the LP is infeasible, then at least one of these artificial variables cannot be forced to zero at the phase 1 optimum. However, if the original LP is modified by adjusting the right hand side constants as indicated by the nonzero artificial variables, then a feasible solution is obtained. If the phase 1 objective function seeks to minimize the sum of the artificial variables, then this is a "minimum" adjustment of the constraints in that sense.

Note that the usual phase 1 formulation does not add artificial variables to all constraints, nor does it allow equality constraints to shift in both directions. To obtain a true minimum adjustment of the constraints, a fully elastic program should be used. This allows all constraints to shift, and allows equality constraints to shift in either direction (see Sec. 6.1.4). This is equivalent to the smallest total penalty for a variable penalties model in which the penalty rates are all identical and equal to one.

8.1.2 Minimizing the l_1 Norm

The l_1 matrix norm for some matrix $\boldsymbol{D}$ is $\|\boldsymbol{D}\|_{l_1} = \sum_{ij} |d_{ij}|$. For a system of inequality constraints of the form $\boldsymbol{Ax} \geq \boldsymbol{b}$, a full elastic program is identical to minimizing an l_1 matrix norm objective function subject to the elastic constraints, i.e. $\min \|(\boldsymbol{b} - \boldsymbol{Ax})^+\|_1$ where $(\bullet)^+$ indicates component-wise application of the operator $\max\{0,\bullet\}$. Minimizing the l_1 matrix norm to obtain the best feasible correction to an infeasible set of linear relations was first suggested by Charnes and Cooper (1961) in the context of *goal programming*. Without using the term elastic programming, which arrived much later, they added elastic variables to some of the constraints in an infeasible model and minimized their sum, i.e. minimized the l_1 matrix norm. In the goal programming application, some goals, expressed as constraints, are known to be incompatible, and it is only these constraints that are elasticized.

Minimizing the l_1 matrix norm can of course be accomplished by standard linear programming applied to the fully elasticized model, but Dax (2006) describes a more efficient affine-scaling method. Once the solution point $\boldsymbol{x}^*$ for the l_1 matrix norm minimization is known, the adjustments for the constraint right hand sides can be seen directly from the values of any nonzero elastic variables. If a constraint is violated, then its right hand side should be relaxed by an amount equal to the magnitude of the elastic variable, with appropriate sign.

The Cplex 10.0 (Ilog 2006) LP solver incorporates a *FeasOpt* option which allows the user to specify preferences that result in a weighted elastic solution for the infeasible model. The weighted penalty function takes the form $\sum_i v_i / p_i$ where v_i is the constraint violation and p_i is the user-specified preference value. Values of p_i that are zero or negative indicate that constraint i is not to be modified, and higher values of p_i indicate a greater preference for constraint i to be modified, if necessary. In addition, upper and lower bounds can be specified for the extent of adjustment of the right hand side constant for each constraint. This arrangement lets the user specify a fully or partially elasticized version of the model with preference weights on the elastic variables and limits on their adjustment. The bound changes determined by the weighted elastic solution are returned, along with the solution point and objective function value now permitted by the relaxed model.

8.1.3 Least-Squares Methods

One of the earliest approaches to finding the best correction for an infeasible set of linear inequalities is to find the point that has the smallest sum of squared constraint violations. As before, once the solution point is known, the necessary constraint shifts are easily found by substituting the point into each constraint.

The least-squares problem for a system of linear inequalities $\boldsymbol{Ax} \geq \boldsymbol{b}$ is

$$\min \sum_{i=1}^{m} [(b_i - \boldsymbol{a}_i \boldsymbol{x})^+]^2$$

where $\boldsymbol{a}_i$ is the ith row of the A matrix and is violated. Solution methods for this problem have been considered by Han (1980), Censor and Elfving (1982), and De Pierro and Iusem (1985). Byrne and Censor (2001) show that simultaneous projection methods converge for infeasible sets of convex constraints to a point that minimizes a Bregman distance function that in some sense measures the total violation. In the usual case this amounts to a weighted sum of the squared Euclidean distances from the solution point to feasible points on all violated constraints.

The least-squares optimization is considered less robust than minimizing the l_1 norm (Sec. 8.1.2) because a single outlier can dominate the solution.

8.1.4 Roodman's Bounds on Minimum Constraint Adjustments

Roodman (1979) was the first to develop methods for finding the minimum adjustment of the constraints in an infeasible system, in the case of linear systems. He first described a very simple approach that is effective if there is only a single IIS in the model, and then developed more advanced methods that apply when the infeasibility is more complex. The main idea is to find the smallest adjustments of the constraint right hand sides that are needed to provide a feasible solution. Roodman's motivations in developing his method are interesting. He began with a feasible investment problem, and then pushed this to become infeasible. He next looked for parametric adjustments to make the problem feasible again, with the idea of seeing how this changed the original solution.

When the adjustment of a single constraint is sufficient to render the model feasible, then simple methods can be applied. This is the case when there is a single IIS, or when the IIS set cover has a cardinality of one. Methods to determine this a priori were not available in Roodman's time, but it is easy to identify some of the constraints that are involved in the infeasibility, e.g. constraints that are violated. Given a constraint that is probably involved in the infeasibility, convert the constraint to an objective function with a sense (maximize or minimize) that tries to tighten the constraint as much as possible. This is equivalent to elasticizing the constraint appropriately and then minimizing the elastic variable. If the final solution for this modified problem is now feasible for the original model (with the exception of the constraint that was converted to an objective function), then we know the minimum adjustment of the constraint right hand side that will render the model feasible. This process can be repeated over all of the likely constraints, and the smallest adjustment of any violated constraint can be recommended as the minimum adjustment of the constraints to render the model feasible.

This approach has several drawbacks. As described by Roodman, the method has no way of making sure that all of the relevant constraints are selected for testing (applying the sensitivity filter would take care of this, but the method was not available in Rooman's time). More importantly, this method cannot deal with more complex infeasibilities that require the adjustment of more than one constraint in order to achieve feasibility. For this reason, Roodman develops more advanced methods that try to find lower bounds on the minimum adjustment of the right hand side for every linear constraint that is needed to achieve feasibility.

Define $A(b_i)$ as the optimum value of the objective function associated with some right hand side b_i of constraint i in a feasible model. $A(b_i)$ is a piecewise convex linear function of b_i. Each linear piece is given by a different basic feasible solution and has a slope equal to the dual price of constraint i. This same principle applies to a phase 1 solution, which is artificially "feasible". Let b_i^* be the original right hand side value in constraint i, and let Π_i^* be the dual price of constraint i in the phase 1 solution. Finally, let b_i^f denote the value of b_i such that $A(b_i^f)=0$ in the phase 1 solution, with all other parameters unchanged.

Given the convexity of $A(b_i)$, Roodman calculates bounds on the value of b_i such that feasibility is just achieved using the relationship:

$$A(b_i^*) + \Pi_i^*(b_i^f - b_i^*) \leq A(b_i^f) = 0.$$

This relationship yields bounds on the minimum adjustments of the constraint right hand sides, as covered by suitable algebra in the following three cases:

- If $\Pi_i^* < 0$ then $(b_i^f - b_i^*) \geq -A(b_i^f)/\Pi_i^*$. b_i must be increased by at least $-A(b_i^f)/\Pi_i^*$ to reach feasibility.
- If $\Pi_i^* > 0$ then $(b_i^f - b_i^*) \leq -A(b_i^f)/\Pi_i^*$. b_i must be decreased by at least $A(b_i^f)/\Pi_i^*$ to reach feasibility.
- If $\Pi_i^* = 0$ then no change in b_i will permit a feasible solution. This is equivalent to the sensitivity filter.

Roodman defines $s_i = 1$ if $\Pi_i^* < 0$ and $s_i = -1$ if $\Pi_i^* > 0$. Hence a lower bound on the size of the appropriate relaxation of the right hand side for constraint i is $R_i = -s_i(A(b_i^*)/\Pi_i^*)$, and the actual change is given by $s_i R_i$ to take into account the sense of the constraint. The feasible solution $\boldsymbol{x}$ is not determined by this method, which only provides lower bounds on the RHS adjustments that are needed.

Roodman also develops heuristics for finding phase 1 solutions for infeasible systems that have better properties for subsequent analysis. These methods are based on the observation that the most violated constraints provide the tightest lower bounds on the right hand side adjustment needed to reach feasibility. Accordingly, it is worthwhile to find a phase 1 solution in which all of the constraints are violated by approximately the same amount so that the bounds determined by the analysis above are more consistent. The main idea is to assign weights to the artificial variables to try to achieve this outcome. Roodman provides some suggestions on how to assign the weights on the artificial variables:

- Assign a weight that is inversely proportional to the units used in the constraint.
- Assign the smallest positive weights to the constraints for which it is most important to find tight bounds. This low weight means that the constraint is more likely to be violated, and by a larger amount, and hence its adjustment bound will be the tightest.

The bounds can be further tightened after the weighted phase 1 solution is obtained. Define l_j and u_j as the lower and upper bounds on w_j (the weight associated with the jth artificial variable) such that for $l_j \leq w_j \leq u_j$ the final basis is unchanged. Let $R_i(w_j)$ define R_i as a function of w_j. $R_i(w_j)$ is monotone so that the two end-points $R_i(l_j)$ or $R_i(u_j)$ will both provide lower bounds on the adjustment to the

RHS, hence choose the tighter of the two. Given G basic artificial variables in the phase 1 solution, then the tightest lower bounds are given by

$$R_i^* = \max_{j=1...G} \{\max[R_i(l_j), R_i(u_j)]\}\ .$$

Roodman (1979) reports that some of the lower bounds are usually made exact by this computationally inexpensive process.

Roodman goes on to develop a very similar approach for the case in which the dual simplex method is used to solve the phase 1 problem. In addition he briefly outlines an approach based on parametric programming for modifying multiple constraint RHSs simultaneously.

8.1.5 A Fuzzy Approach to Constraint Shifting

León and Liern (2001) use a fuzzy sets approach to shifting constraints to repair infeasibility in sets of linear inequalities (easily extendable to equalities). The main idea is that the fuzzy membership function expresses the degree to which a particular point satisfies a given constraint. The membership function makes use of Roodman's limits: the membership value is 0 if the point violates the original constraint to an amount greater than Roodman's limit; varies linearly between 0 and 1 if the point violates the original constraint to an amount between Roodman's limit and satisfying the constraint; and is equal to 1 if the point satisfies the original constraint. They then solve the problem of finding a point that maximizes the minimum membership function value for any constraint.

Given a set of inequalities defined as $\boldsymbol{A}_1\boldsymbol{x} \geq \boldsymbol{b}^1$ and $\boldsymbol{A}_2\boldsymbol{x} \leq \boldsymbol{b}^2$ with $\boldsymbol{x} \geq \boldsymbol{0}$, this amounts to finding a value λ to transform the original RHS vector $\boldsymbol{B} = (\boldsymbol{b}^1, \boldsymbol{b}^2)^{\mathrm{T}}$ into $\boldsymbol{B}(\lambda) = (\boldsymbol{b}^1 - \lambda\boldsymbol{R}, \boldsymbol{b}^2 + \lambda\boldsymbol{S})^{\mathrm{T}}$, where $\boldsymbol{R}$ and $\boldsymbol{S}$ are vectors of Roodman's limits with appropriate signs. The value of λ will be at least equal to $1/k$ where k is the number of nonzero dual prices. A smaller λ represents a smaller perturbation of the model relative to the original, and is in some sense the "best" adjustment of the constraints.

Gupta et al. (2004) take a similar fuzzy approach to finding a best approximate solution to an infeasible generalized linear complementarity problem.

8.1.6 A Goal Programming Approach to Constraint Shifting

Yang (2006) points out that any method based on a weighted sum of elastic variables to determine the constraint shifts can arrive at only a limited set of possible solutions: those that appear at the cornerpoints of the solution space. A goal programming approach, on the other hand, allows solutions that arrive at any point on the efficient frontier, giving the modeller a vastly larger set of possible constraint shifts that provide a feasible solution.

For an infeasible continuous optimization problem (linear or nonlinear) having inequality constraints of the form $f_i(x) \leq 0, i = 1...m,$ Yang formulates the constraint shifting problem as a multi-objective program as follows: minimize $(y_1, y_2,...,y_m)$

subject to $f_i(x) - y_i \leq 0, y_i \geq 0, i = 1...m$ where the y_i are nonnegative elastic variables. A common solution approach for a multi-objective program is to minimize the weighted sum of the objective functions, in this case: $\min z = \sum_{i=1}^{m} w_i y_i$ subject to $f_i(x) - y_i \leq 0$ for $i = 1...m$ where $y_i \geq 0$ for $i = 1...m$, $\sum_{i=1}^{m} w_i = 1$ and $w_i \geq 0$ for all $i = 1...m$. This amounts to a weighted version of an elastic program whose solution will always yield one of a limited number of basic feasible solutions.

The multi-objective program can instead be handled by a goal programming formulation. The simplest such formulation is $\min \|y - \bar{y}\|$ where $\bar{y} = (\bar{y}_1, ..., \bar{y}_m)$ is a reference point and $\|\bullet\|$ is a norm operator, for which several different choices are available. The usual reference point is $\bar{\mathbf{y}} = \mathbf{0}$. Given $\mathbf{0}$ as the reference point, the commonly-used l_1 norm objective function is $\min z = \sum_{i=1}^{m} |w_i y_i|$ where $\sum_{i=1}^{m} w_i = 1$ and $w_i \geq 0$ for all $i = 1...m$. Given nonnegative $\boldsymbol{w}$ and $\boldsymbol{y}$, this is again equivalent to a weighted elastic program, and has the same shortcoming of producing only a limited number of basic feasible solutions.

The l_∞ norm, on the other hand, has different properties. The objective function based on the l_∞ norm is $\min z = \max_{i=1...m} |w_i y_i|$ where $\sum_{i=1}^{m} w_i = 1$ and $w_i \geq 0$ for all $i = 1...m$. This can be implemented as min z subject to $w_i y_i \leq z$, $i = 1...m$, $f_i(x) - y_i \leq 0, y_i \geq 0, i = 1...m$ and the usual $\sum_{i=1}^{m} w_i = 1$ and $w_i \geq 0$ for all $i = 1...m$. The absolute value is again not needed because $\boldsymbol{w}$ and $\boldsymbol{y}$ are nonnegative. Yang (2006) shows that an optimal solution to the l_∞ norm optimization $(\boldsymbol{x}^*, \boldsymbol{y}^*)$ is guaranteed to be a weakly efficient solution to the multi-objective program. A second theorem shows that an optimal solution to the l_∞ norm optimization that is unique in $\boldsymbol{y}^*$ is an efficient solution to the multi-objective program. Yang presents a way to improve any solution returned by the l_∞ norm optimization to make it efficient by reducing the members of $\boldsymbol{y}$ as much as possible. This is done by solving a second optimization based on the $\boldsymbol{y}^*$ returned by the original l_∞ norm optimization: $\min z = \sum_{i=1}^{m} |w_i y_i|$ subject to $y_i \leq y_i^*$ and $f_i(x) - y_i \leq 0, y_i \geq 0, i = 1...m$ where $\sum_{i=1}^{m} w_i = 1$ and $w_i \geq 0$ for all $i = 1...m$. A third theorem (Yang 2006) shows that the l_∞ norm optimization for the set of constraints constituting an IIS is necessarily an efficient solution for the multi-objective problem. Yang demonstrates that this approach is also effective for nonlinear problems, providing of course that the various nonlinear optimizations can be solved correctly.

The l_∞ norm optimization with subsequent tightening allows any solution on the efficient frontier to be reached by adjusting the $\boldsymbol{w}$ weights. This provides the modeler with a vastly increased set of possible ways to shift the constraints to achieve feasibility. The choice of the weights is left to the modeler, but this may be obvious by context. In practice the modeler may omit elastic variables entirely for constraints expressing basic physical laws, and may assign large weights to constraints that he prefers not to violate.

8.1.7 Constraint Shifting in Sequential Quadratic Programming

Sequential Quadratic Programming (SQP) is a technique for solving nonlinear programs in which the original model is approximated by a sequence of quadratic programs that are much easier to solve. Each quadratic subproblem has linear constraints that approximate the original nonlinear constraints at the current trial point, and a quadratic objective function that approximates the original nonlinear objective function at the current trial point (or more accurately approximates the Lagrangian of the original objective function). The solution point for the current quadratic subproblem becomes the trial point at which the next set of constraint and objective function approximations are created. This cycle of approximation and solution continues until certain stopping conditions are met, typically that the current point and the last point are sufficiently close.

As pointed out by Boman (1999), a few unusual situations arise in solving SQPs:

1. The original NLP model is feasible, but some of the quadratic subproblems are infeasible.
2. The original NLP model is infeasible, but some of the quadratic subproblems are feasible.
3. The quadratic subproblem is feasible, but we don't want to move to the resulting optimum point.

In the first case, Boman suggests that some form of approximate solution for the subproblem that is in some sense "closest to feasibility" is needed to permit the algorithm to proceed. In the second case, the main algorithm should return a solution that is in some sense "closest to feasibility" for the complete original model. In the third case, some measure of the closeness of the subproblem optimum point to feasibility in the original NLP helps to determine whether the algorithm wishes to accept the updated point. In all three cases, the issue of determining the closest feasible and infeasible points arises. Not surprisingly, minimization of l_1 and l_∞ norms emerges as a major theme of Boman's work.

Boman's l_1 norm minimization is carried out by elasticizing the constraints of the quadratic program in a manner similar to that used for linear programs (see Sec. 6.1.4), but using an objective function that combines the original objective and a minimization of the sum of the elastic variables. For an NLP of the form min $F(\boldsymbol{x})$ subject to $\boldsymbol{c}(\boldsymbol{x}) \geq \mathbf{0}$, the elastic version is:

$$\min_{x,t} F(\boldsymbol{x}) + \gamma \boldsymbol{e}^T \boldsymbol{t}$$

subject to $\boldsymbol{c}(\boldsymbol{x}) + \boldsymbol{t} \geq \mathbf{0}$, $\boldsymbol{t} \geq \mathbf{0}$, where $\boldsymbol{t}$ are the elastic variables, γ is a penalty weight, and $\boldsymbol{e}$ is a column vector of 1s. Boman considers differing weights for individual constraints, bounded elastic variables, elastic bounds, switching between elastic and non-elastic modes, updating the penalty parameters and various other alternatives in developing an improved SQP solution algorithm. He compares his method with the long history of usage of elastic models in solvers such as SNOPT (Gill et al. 2005). The l_∞ norm minimization is more space efficient in that only a single elastic variable is needed:

$$\min_{x,\tau} F(\boldsymbol{x}) + \gamma\tau,$$

subject to $\boldsymbol{c}(\boldsymbol{x}) + \tau\boldsymbol{e} \geq \mathbf{0}$, $\tau \geq \mathbf{0}$, as formulated by Boman.

8.1.8 Violating a Limited Number of Constraints by a Limited Amount

In the context of a radiation therapy planning problem (Sec. 11.1) Censor et al. (2006) propose a method for finding a feasible solution by allowing the violation of a limited number of constraints by a limited amount. Individual linear inequalities can be violated by up to an amount β. Specifically, where a particular linear inequality i has the form $\boldsymbol{a}_i\boldsymbol{x} \leq b_i$, it may be relaxed up to $\boldsymbol{a}_i\boldsymbol{x} \leq (1+\beta)b_i$, where $0 \leq \beta \leq \beta_{\max}$. The fraction of the constraints that can be so violated is limited by $\alpha_{\max}$ where $0 \leq \alpha \leq \alpha_{\max} \leq 1$.

The problem is formulated as follows. Replace each inequality i of the form $\boldsymbol{a}_i\boldsymbol{x} \leq b_i$ by $\boldsymbol{a}_i\boldsymbol{x} \leq t_i b_i$ where $0 \leq t_i \leq (1+\beta)$. A *check inequality* can then be formulated to help insure that the number of violated inequalities does not exceed $\alpha_{\max}m$ where there are m inequalities in total:

$$\sum_{i=1}^{m} t_i \leq \alpha m(1+\beta) + (1-\alpha)m = m(1+\alpha\beta).$$

The objective function is to minimize Σt_i. Of course, the LP solution of this problem does not guarantee that no more than $\alpha_{\max}m$ constraints are violated; a MIP formulation is needed to guarantee this. However, for the radiation planning problems studied by Censor et al. (2006), the solution of this *approximating LP* frequently provides a solution which respects the limit on the number of constraints violated. This is partly due to the effect of the objective function which tries to keep the t_i small, thereby reducing the number of unnecessarily violated constraints.

It is difficult to know in advance what values of α and β should be chosen, so Censor et al. define a heuristic iterative procedure to determine small values of α and β that solve the problem, as shown in Alg. 8.1. $\Delta\alpha$ and $\Delta\beta$ are small increments for the two main parameters and are adjusted in an outer loop to find a solution that is in some sense close to feasible.

Censor et al. report good results using the approximating LP and the heuristic adjustment to find small values of α and β when the procedure is applied to radiation treatment planning problems.

INPUT: α_{max}, β_{max}, $\Delta\alpha$, $\Delta\beta$.
0. $k = \alpha_0 = \beta_0 = 0$.
1. Solve the approximating LP using α_k and β_k. IF feasible THEN:
   1.1 IF $k = 0$ THEN exit with the solution $\boldsymbol{x}$.
   1.2 IF the α and β conditions are satisfied at $\boldsymbol{x}$ THEN exit with the solution $\boldsymbol{x}$.
2. $\rho = \beta_k + \Delta\beta$.
3. IF $\rho \leq \beta_{max}$ THEN $\beta_{k+1} = \rho$ and $\alpha_{k+1} = \alpha_k$.
   ELSE
   3.1 $\sigma = \alpha_k + \Delta\alpha$.
   3.2 IF $\sigma \leq \alpha_{max}$ THEN $\alpha_{k+1} = \sigma$ and $\beta_{k+1} = \beta_k$.
   ELSE exit unsuccessfully.
4. $k = k + 1$; go to Step 1.

OUTPUT: a solution $\boldsymbol{x}$ that satisfies the α and β conditions or a failure message.

Alg. 8.1. Adjusting the α and β conditions (Censor et al. 2006)

8.2 Adjusting the Constraint Matrix

It is much more difficult to determine the best way to achieve feasibility for an infeasible model by adjusting the constraint coefficients in addition to adjusting the right hand side constants. However a number of researchers have addressed this problem for infeasible LPs. The motivation is generally not to provide an automatic fix for the model, but instead to provide some insight into the infeasibility by showing the "nearest" feasible model.

There are various ways to measure how closely the adjusted feasible model approximates the original infeasible version. Amaral et al. (2006) describe the general problem for a set of linear inequality constraints as *min* $\varphi(\boldsymbol{H},\boldsymbol{p})$ subject to $(\boldsymbol{A}+\boldsymbol{H})\boldsymbol{x} \leq \boldsymbol{b}+\boldsymbol{p}$ where $\boldsymbol{x}$ is a convex set (usually limited by upper and lower bounds). $\boldsymbol{H}$ is a set of corrections to the constraint body matrix $\boldsymbol{A}$ and $\boldsymbol{p}$ is a set of corrections to the right hand side vector $\boldsymbol{b}$ such that feasibility is achieved for the infeasible original set of inequalities $\boldsymbol{Ax} \leq \boldsymbol{b}$. The objective function $\varphi(\boldsymbol{H},\boldsymbol{p})$ measures the closeness of the adjusted system to the original system and is taken as a suitable matrix norm. Some common choices for the matrix norm include the l_1 norm, the l_∞ norm $\|\boldsymbol{D}\|_{l_\infty} = \max_{ij}|d_{ij}|$ for some matrix $\boldsymbol{D}$, and the ∞-norm $\|\boldsymbol{D}\|_\infty = \max_i \sum_j |d_{ij}|$.

Vatolin (1992) developed an LP-based solution for the problem of minimizing the l_∞ norm for a system of linear inequalities. He defines the original system as

$\boldsymbol{Ax} \leq \boldsymbol{b}$, $\boldsymbol{x} \geq \boldsymbol{0}$, whose rows are subdivided into the sets M_0 (not to be corrected) and M_1 (available for correction). The columns are also subdivided into the sets J_0 (not to be corrected) and J_1 (available for correction). Where $\boldsymbol{a}_i$ represents row i of $\boldsymbol{A}$ and the corrections vector for row i is $\boldsymbol{h}_i = [\boldsymbol{h}_i', h_{i,n+1}]$ of dimension $1\times(n+1)$ (which includes corrections to both the constraint body and the right hand side constant), the corrected system is $\boldsymbol{a}_i\boldsymbol{x} \leq b_i$, $i \in M_0$; $(\boldsymbol{a}_i + \boldsymbol{h}_i')\boldsymbol{x} \leq (b_i - h_{i,n+1})$, $i \in M_1$; $\boldsymbol{x} \geq \boldsymbol{0}$. Note that $h_{ij} = 0$ for $i \in M_1$ and $j \in J_0$.

This formulation is nonlinear due to the bilinear $\boldsymbol{h}_i'\boldsymbol{x}$ terms, but can be rendered linear by a change of variables. Let $\boldsymbol{h}_i = t_i\boldsymbol{c}$ for $i \in M_1$ where the definition of $\boldsymbol{c}$ depends on the matrix norm used in the optimization; for the l_∞ norm, $c_j = 0$ for $j \in J_0$ and $c_j = -1$ for $j \in J_1$. Further, $\boldsymbol{ch}_0 = -1$ where $\boldsymbol{h}_0 > \boldsymbol{0}$. The change of variables is then

$$\boldsymbol{x} = \frac{(h_{0,1}, h_{0,2}, \ldots h_{0,n})^T}{h_{0,n+1}},$$

resulting in the following LP for minimization of the l_∞ norm:

$$\begin{aligned} &\min \theta \\ \text{subject to:}\quad &[\boldsymbol{a}_i, -b_i]\boldsymbol{h}_0 \leq 0 \text{ for } i \in M_0 \\ &[\boldsymbol{a}_i, -b_i]\boldsymbol{h}_0 \leq t_i \text{ for } i \in M_1 \\ &0 \leq t_i \leq \theta \text{ for } i \in M_1 \\ &\sum\nolimits_{j \in J_1} h_{0,j} = 1 \\ &h_{0,j} \geq 0 \text{ for } j = 1 \ldots n+1 \end{aligned}$$

The solution of this LP yields values for $\boldsymbol{h}_0$ and $\boldsymbol{t}$, from which the individual corrections can be found using $\boldsymbol{h}_i = t_i\boldsymbol{c}$, i.e. for $i \in M_1$, $h_{ij} = 0$ for $j \in J_0$ and $h_{ij} = -t_i$ for $j \in J_1$. The magnitude of the largest individual correction is given by θ. If desired, the best correction point $\boldsymbol{x}$ can also be recovered by reversing the change of variables. If the l_1 norm is to be minimized using Vatolin's approach, then $|J_1|$ LPs must be solved, each LP correcting a single column of $[\boldsymbol{A},\boldsymbol{b}]$. Popescu (2001) considers the use of interior-point LP algorithms for the solution of the Vatolin formulations.

Amaral (2001) extends this idea to minimization of the ∞-norm and derives limits on the number of LPs needed for this optimization. However she observes that this approach involves only a single column of $[\boldsymbol{A},\boldsymbol{b}]$ at a time and results in particular patterns of the corrections. To provide greater freedom in the resulting corrections patterns, Amaral and colleagues (Amaral and Barahona 2005, 2005a, Amaral et al. 2006) study the use of the Frobenius norm

$$\|\boldsymbol{D}\|_F^2 = \sum\nolimits_{i=1}^{m}\sum\nolimits_{j=1}^{n}|d_{ij}|^2$$

for the objective function $min\ \varphi(\boldsymbol{H},\boldsymbol{p})$, yielding the problem (P) $\min\|[\boldsymbol{H},\boldsymbol{p}]\|_F^2$ subject to $(\boldsymbol{A}+\boldsymbol{H})\boldsymbol{x} \leq \boldsymbol{b} + \boldsymbol{p}$, where $\boldsymbol{l} \leq \boldsymbol{x} \leq \boldsymbol{u}$ and $\boldsymbol{l}$ and $\boldsymbol{u}$ are vectors of the lower and upper bounds on $\boldsymbol{x}$, respectively. The Frobenius norm has several attractive properties compared to the l_1, l_∞, and ∞ norms. Imposing upper and lower bounds on the variables improves the solution properties, guaranteeing that (P) has a global minimum, and is not restrictive in practice.

Minimization of the Frobenius norm correction is a nonconvex and nonlinear global optimization problem that can be solved by any suitable global optimization technique. Amaral et al. (2006) use a branch and bound approach incorporating the Reformulation-Linearization-Convexification Technique due to Sherali and Tuncbilek (1992) to solve the global optimization. The tree is constructed by subdividing the range of a selected variable to create two child nodes for each parent node, as in the standard branch and bound algorithm for MIPs. An approximation to (P) is solved at each node to provide a lower bound on the value of the Frobenius norm. The approximation is constructed to be a convex NLP that is solved efficiently due to its special structure. See Amaral et al. (2006) for details of the optimization approach.

As an example of the kind of information provided by adjusting both the constraint matrix and the right hand side, consider the small demonstration example in Amaral et al. (2006). The original infeasible system consists of the three constraints $-x_1 - x_2 \leq -7, x_2 \leq 3, 2x_1 - x_2 \leq -2$, with bounds $\mathbf{1} \leq \boldsymbol{x} \leq \mathbf{5}$. Applying the Reformulation -Linearization-Convexification Technique results in a 17-node branch and bound tree with the resulting optimum corrected system being $-1.0398x_1 - 1.1193x_2 \leq -6.9749$, $-0.1067x_1 + 0.6805x_2 \leq 3.0672$, $1.9745x_1 - 1.0764x_2 \leq -1.9839$. This does show that a feasible solution is available with a relatively similar system of inequalities, though it is difficult to know exactly how to use the information. For example, the corrected system introduces x into the second inequality though it seems unlikely that a change of this type has real physical meaning.

Amaral et al. (2006) apply the method to a selection of small problems. The results are promising in that solutions are returned for about half of the problems in a very small number of nodes, indicating that the lower bound approximation is quite tight. Computation times for all models are small, though this is partly due to artificially tight bounds being imposed on the variables. The authors conclude that their technique will handle infeasible LPs with fewer than 100 rows and 100 columns. This is very restrictive compared to the size of linear programs encountered in practice, but the technique can be applied to subsets of the constraints. In particular, in may be useful to first isolate an IIS via the techniques of Chap. 6 and to then apply the techniques of Amaral et al. (2006) to find an optimal correction.

8.3 Related Research

There are several research directions that are related to the topic of this chapter, but that are sufficiently different that they are given only a cursory treatment here. References are provided so that readers can follow up on these topics.

A number of authors have studied condition measures for mathematical programs, an important aspect of which is the distance to ill-posedness (including infeasibility) of feasible models. Here again various matrix norms are used to measure this distance. See e.g. Vera (1998) or Ordonez and Freund (2003).

Renegar (1994) shows that LPs have large or sensitive optimal values only if they are nearly primal or dual infeasible.

Another problem on feasible models is the *best approximation problem* in which the best feasible point relative to a given infeasible point is sought. This arises when a solution may have been obtained by other means, prior to the addition of various constraints. Now a feasible solution is found that is as similar as possible to the original, but now infeasible, solution. Projection algorithms of various types are often used to solve this problem, see e.g. Censor (2006).

PART III: APPLICATIONS

Part I of this book addressed methods for reaching a feasible solution quickly in optimization models. Part II addressed the analysis of infeasible models. As you might imagine, there are countless direct applications for algorithms in these classes: reducing the time to reach a feasible solution reduces the overall solution time; good tools for analysis of infeasible instances reduces the overall time for the complete modeling and solution cycle. However, there are a surprising number of applications beyond these straightforward ones. These "spin-off" applications are the subject of Part III.

Closest to home is the application of methods for the analysis of infeasibility to the analysis of other model forms. Unboundedness in primal linear programs is directly related to infeasibility in the dual (Sec. 9.1). Network models can be plagued by the inability to carry flow in some of the arcs, a condition known as *nonviability* that can be analyzed by a simple transformation to an infeasibility problem (Sec. 9.2). The interaction of the objectives in multiple objective linear programs can also be transformed into an infeasibility problem and analyzed with the assistance of the tools developed in Part II (Sec. 9.3).

Other important and seemingly unrelated problems can also be addressed by simple transformations to infeasibility analysis problems. Many of these are well addressed by methods for the solution of the maximum feasibility problem (see Chap. 7). A standard problem in classification and data mining is the placement of hyperplanes to separate data points of one type from data points of other types in the training set. This is easily transformed into a MAX FS problem, and hence solutions are returned that tend to minimize the number of misclassified points, whereas more traditional approaches may have minimized other measures such as the sum of the squared misclassification distances (Sec. 10.1). This is the same as providing the initial training for a neural network. A related problem is determining the *data depth* of a particular point in a multidimensional cloud of data points, defined as the minimum number of data points on one side of a hyperplane through the point in question (Sec 10.2). This again is easily transformed into a MAX FS problem and addressed via the methods of Chap. 7. Massive data sets, such as census data, are routinely screened for errors using linear relationships. The data validation rules can be analyzed for internal inconsistencies using the methods of Chaps. 6 and 7 (Sec. 10.3).

Several specific applications are well addressed as instances of the MAX FS problem. Radiation treatment planning results in a large set of linear inequalities that express the fact that diseased tissue must receive more than some minimum amount of radiation while nearby healthy tissue and important organs should receive less than some maximum dose. Since these requirements often conflict, this can be addressed as an instance of a MAX FS problem (Sec. 11.1). The problem in protein folding is to find the natural folding shape that minimizes the energy, which will be smaller than for other folded shapes. Given the energy inequalities associated with similar "decoy" shapes, information about the natural folded shape is gleaned by solving a MAX FS problem that is typically extremely large (Sec. 11.2). The digital video broadcasting problem is another MAX FS problem of very large scale (Sec. 11.3). Here the broadcast coverage area is subdivided into small regions, each of which should receive a signal with a minimum amount of power from a set of transmitters. This is again a MAX FS problem.

The best approximation methods of Chap. 8 are needed in automated test assembly (Sec. 11.4) in which the idea is to meet the requirements imposed by the test assembly rules as well as possible. IIS isolation is used to analyze problems of buffer overrun in computer programs (Sec. 11.5) when linear constraints describing the growth in the size of the buffer generate infeasibilities. User preferences used to rank the value of internet pages can be expressed as linear inequalities (Sec. 11.6), but these can result in infeasibilities, which are analyzed using either a best correction approach (Chap. 8) or a MAX FS strategy (Chap. 7).

IIS analysis is a common feature in tree-structured search, such as in branch and bound solution of MIPs or modern constraint programming systems, and is used to direct the backtracking process more efficiently (Sec. 11.7). Piecewise linear models are often used to approximate various physical phenomena such as signals. Estimation of such models (Sec. 11.8) can be represented as an instance of the MIN PFS problem of Sec. 7.9. Finding sparse solutions for systems of linear equations amounts to a MAX FS problem (Sec. 11.9). Various NP-hard problems can also be converted to the MAX FS problem (Sec 11.10), though some of these are in binary rather than continuous variables, for which we do not as yet have good solution heuristics.

9 Other Model Analyses

The various algorithms for analyzing infeasibility turn out to be useful in analyzing other types of models and modeling problems. This chapter describes three such examples.

9.1 Analyzing Unbounded Linear Programs

It is well known that if the primal form of a linear program is unbounded, then the dual form is infeasible (see e.g. Winston and Venkataramanan (2003)). For this reason, if an LP is found to be unbounded then an infeasibility analysis of the dual provides insight into the reason for the unboundedness. This is exactly the approach taken in the LINDO software (Schrage 1997) to return a *minimal unbounded set* of variables. At least one of the variables in this set must be finitely bounded to eliminate the unboundedness in the model. The thought process for the analyst is similar to the process for analyzing infeasibility: there may be other minimal unbounded sets of variables that must be found and fixed one by one in order to eliminate all unboundedness in the model. It is also possible to find a minimal set of variables to restrict so that all unboundedness is removed (similar to an IIS set cover). Since unboundedness difficulties are usually caused by missing constraints or bounds, the usefulness of this approach lies in providing clues as to where constraints have been omitted.

9.2 Analyzing the Viability of Network Models

Network models are among the largest constrained optimization problems regularly solved. Because of their scale and complexity, automated methods for "debugging" formulation errors are especially welcome. Networks are susceptible to a special kind of modeling error called *nonviability* (Chinneck 1990a, 1990b, 1992) that results in some of the arcs being unable to carry flow. Note that zero flow is still a feasible solution for an arc, but the fact that it is the only solution for an arc indicates a modeling error. This is a special case of a *forcing substructure* (Greenberg 1996a), i.e. a structural problem that forces a variable to take on a particular fixed value.

There are a number of variations of network models:

- *Pure networks* contain only regular nodes which conserve flow.
- *Generalized networks* have one or more arcs in which a gain factor multiplies the flow into the arc to determine the flow out of the arc.

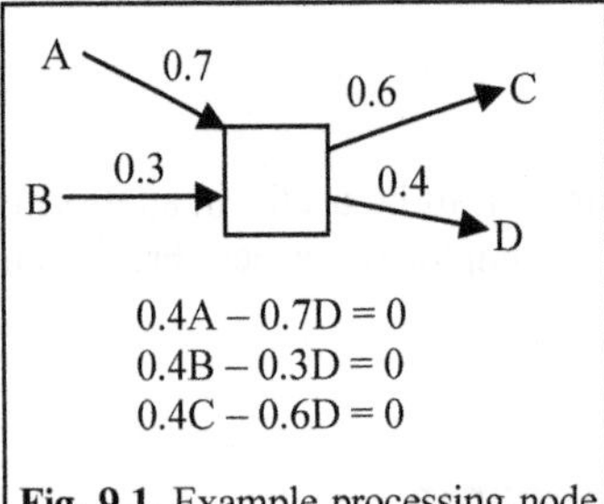

Fig. 9.1. Example processing node with ratio equations

- *Pure processing networks* have regular nodes and one or more *processing nodes* in which the flows in the incident arcs are constrained to fixed proportions. Processing nodes are usually depicted as squares with labels on the incident arcs showing the proportions of flow. Flow conservation holds at all nodes. An example appears in Fig. 9.1.
- *Nonconserving processing networks* are processing networks in which at least one of the processing nodes does not conserve flow.

Further information on processing networks is available from Koene (1982) or Chinneck (1990a, 1992). Chinneck (1992) shows how generalized networks are easily transformed into nonconserving processing networks. All network forms can then be considered as various special cases of nonconserving processing networks.

Nonviability is a property of the network structure (i.e. the pattern of interconnection of the arcs and nodes). It is particularly common in processing networks. The *structural relationships* of a network model include:

- for regular nodes: the flow conservation equation,
- for a processing node having t terminals: a set of $t-1$ independent ratio equations that specify the proportions of flow in each incident arc,
- for arcs: the nonnegativity constraints on the arc flow.

Note that other bounds on the arc flows are not considered to be part of the network structure, nor are the arc costs per unit of flow, the objective function, or any extra side constraints. Network viability and nonviability are defined as follows:

Definition 9.1: *Network viability and nonviability (Chinneck 1997a).* If the complete set of structural relationships in a network model does not provide a unique solution for any of the arc flow variables, then the network is *viable.* In a *nonviable* network, the set of structural relationships provides a unique solution for one or more of the arc flow variables. ■

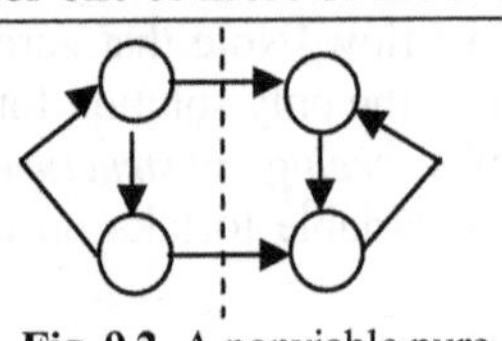

Fig. 9.2. A nonviable pure network

Because the structural relationships are all homogeneous equations or nonnegativity constraints, nonviability forces arc flows to zero. A simple example is shown in Figure 9.2 in which both of the arcs cut by the dashed line are nonviable.

Nonviability is generally an undesirable property of a network model since it indicates that a portion of the model can never be used, even before nonstructural arc bounds or an objective function are considered. Modellers should be alerted to nonviability in the same way that they are alerted to infeasibility so that corrections can be made if warranted. However, unlike infeasibility, nonviability is "silent": it is not reported automatically. Explicit measures must be taken to screen networks for this property.

Chinneck (1990a) presents a procedure for detecting, localizing, and analyzing nonviabilities in pure processing networks and provides a taxonomy of structures causing nonviability. A prototype of the procedure was implemented in software (Chinneck 1990b). A second paper (Chinneck 1992) extends the method to non-conserving processing networks, thereby covering all network forms, including generalized networks. However, a simpler approach using standard IIS isolation techniques can be adapted to the isolation of nonviabilities. The method is based on the following theorem.

Theorem 9.1: *Nonviability and infeasibility* (Chinneck 1990a). A network is viable if and only if it is feasible for all variables to be positive simultaneously.

Proof: The structure of the network consists of the homogeneous structural equations and the arc flow nonnegativity bounds. These define a convex polyhedral cone feasible region. There is either a unique solution to such a system with $\mathbf{x}=\mathbf{0}$ or there are multiple semipositive solutions (Murty 1983). A unique solution for a variable is possible only at the intersection of the homogeneous equations and the nonnegativity constraints (i.e. a nonviable solution of zero). Each variable must have more than one possible solution value for viability, which implies that each variable must be positive in at least one feasible solution. Because the feasible region is a convex polyhedral cone, if a variable can be positive in some solution, then there must exist feasible solutions in which all of the variables are positive simultaneously. If one such solution exists, then there are many (Murty 1983). Therefore if we can find any feasible solution in which all variables are positive simultaneously, then the network must be viable. The converse follows easily. ■

Thm. 9.1 makes it easy to test for viability by placing arbitrary *positivity* bounds on the arc flow variables, e.g. $x_j \geq 1$, and then testing the *feasibility* of the resulting system using a standard LP phase 1. If the modified system is feasible, then the original network model is viable. If the modified system is infeasible, then the original network model is nonviable, and IIS isolation techniques can be applied to the infeasible modified model to isolate the nonviability in the original network model. If desired, the isolated portion of the model can be tested using the original algorithm (Chinneck 1990a) to arrive at a classification of the cause of the novabiity. Stripping the isolated IIS of its added posibility bounds yields the minimal nonviabiityabiity.

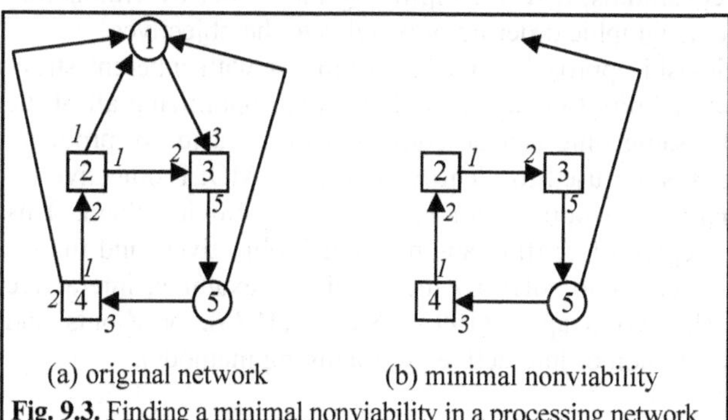

Fig. 9.3. Finding a minimal nonviability in a processing network model (Chinneck 1996b)

A simple example is shown in Fig. 9.3. The processing node proportions are shown in italics. The minimal nonviability found by assigning positivity bounds to all arcs and isolating an IIS is shown in Fig. 9.3 (b). It is easy to explain. Assume, for simplicity, that the arc connecting node 5 to node 4 is carrying 3 units of flow. Following the flow ratios imposed by the processing nodes, this becomes 1 unit of flow between nodes 4 and 2, then 0.5 units of flow between nodes 2 and 3, and finally 1.25 units of flow between nodes 3 and 5. Hence we have 1.25 units of flow entering node 5, but expect 3 units (or more given the outflow from node 5 to node 1) to leave node 5. This is clearly not possible and results in the minimal nonviability shown in Fig. 9.3 (b). Note that the arc connecting nodes 5 and 1 is part of the minimal nonviability because it is part of the simple flow balance of regular node 5 and its nonnegativity bound prevents backward flow. While structurally nonviable, the model in Fig. 9.3 (a) is feasible since it admits the solution in which all arc flows are set to zero. It only becomes infeasible when the positivity constraints on the arc flows are added.

The IIS isolation approach to nonviability analysis is used in a method for analyzing petri net models for various classes of modelling errors (You 1993). Petri nets are commonly used in the design of software systems.

9.3 Analyzing Multiple-Objective Linear Programs

A multiple-objective linear program (MOLP) consists of a set of linear constraints and a set of more than one linear objective functions. Formulating a large multiple objective linear program can be difficult due to questions about whether relationships should be cast as objectives or constraints, and about how the different objectives interact and interfere with each other.

Chinneck and Michalowski (1996) show how techniques for the analysis of infeasible LPs can assist in the formulation of MOLPs, both by analyzing infeasible constraint sets and by illuminating the interactions among objectives and constraints. IIS analysis can help in deciding whether relationships should be represented as objectives or constraints, and in simplifying the model by eliminating objectives or assigning lexicographic ordering or weights to the objectives.

Chinneck and Michalowski approach MOLPs as models with inherent structural inconsistencies. The inability to reach a single solution optimizing all of the objective functions at the same time can be converted to a feasibility problem. Techniques for isolating IISs are used to identify sources of MOLP objective incompatibility and to analyze the options for reducing or removing them. This analysis may reduce the degree of conflict among MOLP objectives, and the resulting problem should be easier to solve by means of, for example, interactive methods (see Michalowski and Szapiro (1992), Steuer (1986), or Zionts and Wallenius (1983) for a discussion of interactive programming methods).

Classifying the mathematical relationships in a MOLP as constraints or objectives can be difficult. For example, should a cost expression be an objective ("minimize cost"), or a constraint ("cost must not exceed $100,000")? Chinneck and Michalowski define four classes of mathematical relationships below; a similar distinction has been used by Ignizio and Cavalier (1994):

- *Hard constraint:* a mathematical relationship that is definitely classed as a constraint. These are often basic physical relationships such as conservation of flow in a network.
- *Soft constraint:* a mathematical relationship that is currently classed as a constraint, but which could be considered an objective by dropping the right hand side and adding an objective sense (maximize or minimize).
- *Hard objective:* a mathematical relationship that is definitely classed as an objective (maximize or minimize).
- *Soft objective:* a mathematical relationship that is currently classed as an objective, but which could be considered as a constraint if an appropriate right hand side and relationship sense were added.

Constraint to objective conversions are straightforward: drop the right hand side and the relationship symbol, and add the appropriate optimization sense: $\geq$ constraints become maximizations, $\leq$ constraints become minimizations, $=$ constraints can become either (though $=$ constraints are unlikely candidates for conversion to objectives). Objective to constraint conversions are more complicated. First the constraint sense must be determined. Generally maximization objectives become $\geq$ constraints, and minimization objectives become $\leq$ constraints, but either could be converted to an $=$ constraint and used as a goal constraint with the addition of appropriate deviational variables. Second, the right hand side value must be set; this requires some knowledge of the application. The value assigned to the right hand side in a soft constraint is called the *aspiration level* of the constraint. Similarly, when a soft objective is converted to a constraint, an aspiration level must be assigned.

Current practice assumes that the modeller can make an initial assignment of every relationship to one of the four classes described above. Given this initial classification of the relationships, the modeller faces two main MOLP formulation problems, in addition to the usual LP formulation problems:

1. *Final Classification:* arriving at a final classification of the soft constraints and objectives. Should a soft constraint be converted to an objective? Should a soft objective be converted to a constraint, and if so, what should the aspiration value be? This produces the final form of the MOLP, which can then be solved.
2. *Simplification:* elimination of constraints and objectives, rewriting of constraints, resetting of aspiration values etc. to arrive at a simpler or clearer formulation.

Chinneck and Michalowski approach these two formulation problems by providing algorithmic tools for *Interaction Analysis*, the process of analyzing the intercations and interferences between objectives and between objectives and constrains. Interaction Analysis may also assist in the solution of the MOLP.

9.3.1 Interaction Analysis of the Constraints

The MOLP constraint set is analyzed in two stages: (i) test the feasibility of the complete set of hard constraints, and then (ii) test the entire set of constraints including both the hard and soft constraints. Infeasibility during step (i) indicates ordinary LP infeasibility that can be analyzed by the techniques outlined in Chaps. 6–8. Infeasibility during step (ii) shows that the aspiration levels of the soft constraints are unrealistic. The IIS isolated by the infeasibility analysis in step (ii) is especially revealing because it shows the *set* of constraints which interact to cause infeasibility. There are two cases to consider.

- *Case 1: only one soft constraint and a set of hard constraints in the IIS.* This shows that the aspiration level of the soft constraint is unrealistic and conflicts with basic hard constraints, such as physical limitations of equipment. This implies that either (i) the aspiration level of the soft constraint must be relaxed to a feasible level, or (ii) the soft constraint should be converted to an objective.
- *Case 2: more than one soft constraint in the IIS.* In this case, the aspiration levels of the soft constraints interact with each other, and any hard constraints in the IIS, to create the infeasibility. The changes needed to repair the model are similar to the first case: (i) change the aspiration level of one or more of the soft constraints, or (ii) convert one or more of the soft constraints to objectives. In addition, we know that if more than one of the soft constraints is converted to an objective, then these converted objectives will interfere with each other. The kind of analysis needed at this point is then similar to the methods described below.

9.3.2 Interaction Analysis of the Objectives

Infeasibility analysis deals only with constraints, so to use IIS isolation to analyze the interaction of the objectives in the MOLP, all objectives (both soft and hard) must first be converted to constraints. This is easily accomplished via a two-step process which produces the *converted MOLP*.

1. *Find the extreme feasible aspiration level of each objective.* Create and solve one LP for each objective, in which only that objective appears while all of the other objectives are temporarily removed. Note the optimum terminating point(s) (this is the extreme feasible aspiration point, $\boldsymbol{x}_A^{\text{opt}}$, for some objective A) and the extreme feasible aspiration level ($y_A = f_A(\boldsymbol{x}_A^{\text{opt}})$ for some objective A) for each LP.
2. *Convert each objective to a constraint.* This is done by rewriting each objective as a constraint, with the relationship sense determined appropriately (maximization objectives become $\geq$ constraints, minimization objectives become $\leq$ constraints), and the right hand side equal to the extreme feasible

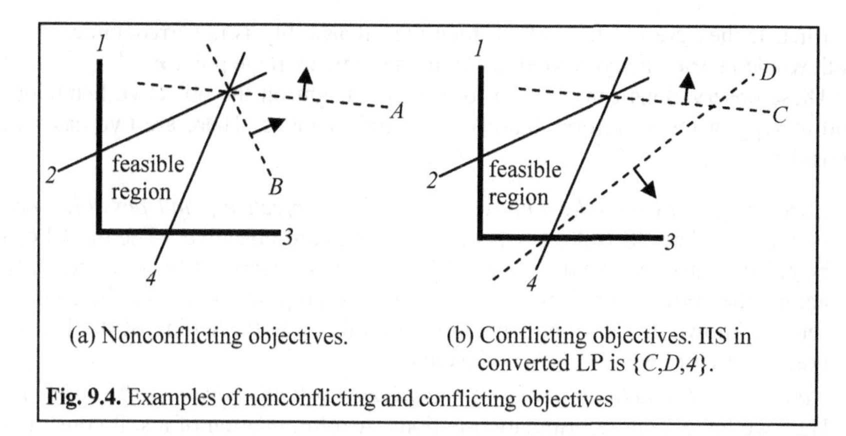

(a) Nonconflicting objectives. (b) Conflicting objectives. IIS in converted LP is {*C*,*D*,*4*}.

Fig. 9.4. Examples of nonconflicting and conflicting objectives

aspiration level determined in Step 1. For numerical reasons, it is better to relax the right hand side from the extreme feasible aspiration level by a small epsilon amount.

Observation 9.1: *Identifying objectives that are not in conflict* (Chinneck and Michalowski 1996). Any objectives whose Step 1 LPs terminate at the same extreme point in Step 1 (or who have the same extreme points as alternative optima) are not in conflict. Consider Fig. 9.4 for example.■

Observation 9.2: *Identifying models that are not true MOLPs* (Chinneck and Michalowski 1996). If all of the Step 1 LPs terminate at the same extreme point (or have the same extreme points as alternative optima), then none of the objectives are in conflict, and the model is not a *true MOLP*.■

If the condition described in Obs. 9.2 holds, then the converted MOLP is feasible because the common extreme point exists, and it satisfies all of the original (hard and soft) constraints, and all of the converted objectives. This leads directly to the following theorem.

Theorem 9.2: *Converted MOLP infeasibility* (Chinneck and Michalowski 1996). If the original constraint set (hard and soft) is feasible, then the converted MOLP is infeasible if and only if the original model is a true MOLP.

Proof: The constraint set is feasible, so infeasibility can happen only due to an interaction involving one or more of the converted objectives. If the converted objectives are not in conflict, then a feasible point exists, as discussed above. Thus an infeasible converted LP can be constructed only if one or more of the converted objectives are in conflict. This is the definition of a true MOLP.■

Observation 9.3: *Objectives in converted MOLP infeasibility* (Chinneck and Michalowski 1996). If the original constraint set is feasible and the model is a true MOLP, then any IIS in the converted MOLP will involve more than one converted objective. The extreme feasible aspiration levels are defined such that each converted objective is feasible with respect to the constraint set. Infeasibility then requires at least two converted objectives.■

Observation 9.4: *Conflicting objectives in converted MOLP IIS* (Chinneck and Michalowski 1996). Any converted objectives appearing together in an IIS are in conflict. If they are not in conflict, then the infeasibility is not irreducible, since both would restrict the converted model to the same extreme point.■

These observations can be used to provide insight on the objective behaviour and to suggest model reformulations and simplifications. There are two cases to consider:

- *Case 1: The IIS includes only converted hard objectives, and possibly hard constraints.* The IIS isolates sets of conflicting hard objectives. The model can be reformulated by abandoning an objective that is clearly of lesser importance among the converted objectives in the IIS, if appropriate. Or, once the conflict sets are known, the information can be used to guide the setting of the lexicographic order or weights on the objectives.
- *Case 2: The IIS includes at least one converted soft objective or soft constraint.* Here we have more reformulation options. A reformulation of a soft constraint may permit the objectives to achieve their extreme feasible aspiration levels. For example, assume that constraint *4* in Fig. 9.4 (b) is a soft constraint. Objectives *C* and *D* can simultaneously achieve their extreme feasible aspiration levels if constraint *4* is relaxed back to the intersection of *C* and *D*.

Constraint *4* in Fig. 9.4 (b) can be relaxed in two ways: (i) it can be converted to an objective instead of a constraint, or (ii) its aspiration level can be adjusted. Further, we can discover exactly how much to relax constraint *4* by constructing and solving a small elastic program (see Sec. 6.1.4) from the IIS: elasticize just constraint *4* and solve the LP consisting of the IIS constraints only. The value of the elastic objective gives the relaxation in constraint *4* that is needed to allow the other objectives to achieve their extreme feasible aspiration levels.

If the IIS includes a soft objective, then similar considerations apply: if it is important that the other objectives achieve their extreme aspiration levels, then ignore the soft objective, or convert it to a soft constraint whose aspiration level is set appropriately.

9.3.2.1 Generating Different Interacting Sets of Objectives

You normally deal with a single IIS at a time when considering only constraints: each IIS found is repaired before analysis proceeds. When considering objectives as well, the IISs in the converted MOLP simply provide information about interacting sets of relationships, so we wish to be able to shift our focus from one interacting set to another (i.e. from one IIS to another) as the analysis proceeds, while all IISs remain intact in the converted MOLP. Because IIS analysis algorithms isolate only a single IIS, this causes some difficulty, but three techniques can be applied, as described below.

1. Eliminate a converted objective from the current IIS. One method of generating a different IIS, and hence a different interacting set of relationships, is to eliminate one of the converted objectives from the current IIS. When this is done, restarting the IIS isolation algorithm will isolate a different IIS if one exists. Converted objectives can be eliminated by either (i) actual elimination from the model, or (ii) elasticization.

To examine how much the aspiration level of a particular objective must be adjusted to accommodate the other objectives in a particular IIS, the following procedure can be used: (1) choose the constraint set consisting only of the constraints and converted objectives in the IIS, (2) change the converted objective in question back to an objective, (3) solve the now-feasible LP to optimality, (4) calculate the difference between the original aspiration level and the optimum value found in step (3). The difference found in step (4) shows whether the conflicts between the objectives in the IIS are serious. A small difference indicates that the conflict is not serious, perhaps resolvable by converting one objective to a constraint with a slightly relaxed aspiration level, or by eliminating one objective entirely.

2. *Apply the IIS search guide codes.* Different IISs are often found by setting the IIS guide codes so as to encourage or discourage the inclusion of specific constraints or converted objectives. One approach is to discourage the inclusion of all of the members of the current IIS. A particularly useful technique when examining the objective interactions is to run the IIS analysis once for each objective with that objective encouraged to stay in the IIS, and the other objectives encouraged to drop out of the IIS. If a certain relationship appears in all or most of the IISs generated this way, then that relationship is particularly conflictive and is a good candidate for change in some manner (convert relationship type if soft, adjust right hand side if constraint, etc.).

3. *Use an IIS-enumerating algorithm.* The IIS-enumerating algorithm described by Gleeson and Ryan (1990) could, in principle, be used to find all of the IISs in the converted model (see Sec. 6.2.3).

Option 2 is used in the complete method described later.

9.3.2.2 Which Objectives Conflict With a Particular Objective?

One common MOLP formulation question is to find all of the other objectives which conflict with a particular objective. This is easily determined by examining the converted MOLP: all objectives which do not share the same extreme aspiration point $\boldsymbol{x}_k^{\text{opt}}$ (or an alternative extreme aspiration point) with the objective in question are in conflict with it.

9.3.2.3 Evaluating the Relative Amount of Objective Interference

The degree of interference between some objective A and some objective B is determined by (i) substituting $\boldsymbol{x}_A^{\text{opt}}$ into objective B and $\boldsymbol{x}_B^{\text{opt}}$ into objective A, then (ii) determining how much each objective moves away from its extreme aspiration level at the new point. The absolute difference found in step (ii) can be used directly, or it can be normalized by dividing by the extreme aspiration level if appropriate.

We say that objective A *interferes strongly* with objective B if the value of objective B at $\boldsymbol{x}_A^{\text{opt}}$ is greatly different (absolute or normalized difference as appropriate) from the extreme aspiration level of objective B; otherwise objective A *interferes weakly*. Objective interference is *relative* because it is possible that objective A interferes strongly with objective B while objective B interferes weakly with objective A.

This analysis may suggest model simplifications. For example, if objective B interferes weakly with objective A, then the model can perhaps be simplified by considering objective B superior to objective A and, for example, using lexicographic ordering of the objectives.

Where the tradeoffs among several objectives must be analyzed simultaneously, an *objective interference table* can be constructed. First create a table having columns for the objectives and rows for the extreme aspiration points for the objectives, e.g. $\boldsymbol{x}_A^{\text{opt}}$, $\boldsymbol{x}_B^{\text{opt}}$, etc. The element of the table for the row $\boldsymbol{x}_A^{\text{opt}}$ and the column for objective B is then $[y^{B\text{opt}} - f^B(\boldsymbol{x}_A^{\text{opt}})]$ or $[y^{B\text{opt}} - f^B(\boldsymbol{x}_A^{\text{opt}})]/y^{B\text{opt}}$ if normalized. The normalized table shows the fractional loss in the objective function value relative to the extreme aspiration level when the objective function is evaluated at the extreme aspiration point for a different objective. Examples of how it can be used are given in Section 9.3.4.

9.3.3 Summary of the Method

The steps in the method are summarized in Alg. 9.1. Note that each test is reapplied until passed because there may be more than one IIS in the model, perhaps unrelated.

Steps 1–5 of Alg. 9.1 are straightforward. Step 6 allows a great deal of flexibility. The modeller could, for example, choose to generate and analyze a single IIS in Step 6. Decisions on how to simply the model (e.g. convert a soft constraint to an objective or just modify its right hand side) require domain knowledge which cannot be incorporated into the algorithm. Similarly, only the modeller, applying domain knowledge, can determine when the model is sufficiently well formulated to proceed to the solution stage.

Analyze the constraint interactions:

1. Apply a phase 1 feasibility test to the set of hard constraints. If feasible, go to Step 2, else (infeasible) identify an IIS and repair the basic LP formulation error. Go to Step 1.
2. Select the entire set of constraints (hard and soft) and apply a phase 1 feasibility test. If feasible, go to Step 3, else (infeasible) identify an IIS and proceed as follows:

   Case 1: *only one soft constraint in the IIS*. Either (i) relax aspiration level of the soft constraint, or (ii) convert the soft constraint to an objective.

   Case 2: *more than one soft constraint in the IIS*. Either (i) relax the aspiration level(s) of one or more of the soft constraints, or (ii) convert one or more of the soft constraints to objectives.

   Go to Step 2.

Create the converted MOLP:

3. Find the extreme feasible aspiration level and extreme feasible aspiration point (and any alternative extreme feasible aspiration points) for each objective by selecting the entire set of constraints and only one objective at a time and solving to optimality.
4. Group objectives having the same extreme feasible aspiration points (or alternative extreme feasible aspiration points): members of each group are non-conflicting objectives. If there is only one group of objectives, then the model is not a true MOLP, so exit.
5. Convert each objective to a constraint by appending the extreme feasible aspiration level as the right hand side and using the constraint sense appropriate to the objective ($\geq$ for maximize and $\leq$ for minimize).

Analyze the objective interactions:

6. Identify a set of IISs as follows. For each objective, set the guide codes to encourage the inclusion of that objective in the IIS and to encourage the exclusion of the other objectives. Identify frequently occurring elements in the IISs and proceed as follows.

   Case 1: *frequent element is a hard objective or constraint*. Either (i) abandon objective(s) in the set that are of lesser importance, or (ii) use the objective interference table to set the lexicographic order or weights on the objectives appearing in the IIS.

   Case 2: *frequent element is a converted soft objective or soft constraint*. Either (i) soft constraint: adjust the aspiration level or convert to an objective, or (ii) soft objective: use the objective interference table to decide whether to ignore or convert to a soft constraint with appropriate aspiration level.
7. If analysis complete then construct objective interference table for final setting of lexicographic order or objective weights and exit. Else (analysis not complete) go to Step 3.

Alg. 9.1. Analyzing MOLPs using IIS isolation algorithms

9.3.4 Example

The example is an adapted and simplified version of a land-use problem deveoped by Steuer and Schuler (1981). See Chinneck and Michalowski (1996) for details of the relationships. The seven objectives relate to, in order, pasturage, dispersed recreation, timber production, and populations of deer, rabbits, squirrels, and quail. The first three objectives are soft (i.e. they could be considered as constraints), while the last four are hard (they are definitely objectives), since this is basically a forestry problem. Soft constraint *s1* is a budget limitation. The steps of the analysis method are applied below.

Steps 1 and 2. The complete set of constraints is feasible, so the tests in Steps 1 and 2 are passed.

Step 3. The extreme feasible aspiration levels of the various objectives are summarized below:

objective	extreme feasible aspiration level
zs1	1577.59
zs2	164.33
zs3	7437.00
zh4	28.96
zh5	81.07
zh6	40.28
zh7	100.57

Step 4. Every extreme aspiration point is different, so no groupings can be made. All objectives are in conflict and the original problem is a true MOLP.

Step 5. All objectives are converted to constraints. For example, maximize *zs1* is converted to $zs1 \geq 1577.59 - \in$, where $\in$ is about 0.01.

Step 6. The set of IISs identified by using the guide codes follows. Notice how each objective appears in its respective IIS.

{*zs1, zs2, s1, h2, h3*}
{*zs1, zs2, s1, h2, h3*}
{*zs1, zs3, s1, h2, h6, h8, h12, h16*}
{*zs1, zh4, s1*}
{*zs1, zh5, s1*}
{*zs1, zh6, s1*}
{*zs1, zh7, s1, h2, h3, h6, h8, h12*}

Frequently occurring elements in the IISs are *zs1* and *s1* which are members of all seven IISs. We elect to remove the budget constraint *s1* entirely, which is equivalent to adjusting the right hand side to a high value.

Step 7. Analysis is not complete, so proceed to Step 3.

Step 3. The new extreme feasible aspiration levels are shown below. Notice that all of the extreme feasible aspiration levels have increased (except *zs3*) with the removal of *s1*.

objective	extreme feasible aspiration level
zs1	1853.88
zs2	185.65
zs3	7437.00
zh4	31.87
zh5	95.58
zh6	44.01
zh7	134.88

Step 4. Every extreme aspiration point is different, so no groupings can be made. All objectives are in conflict and we are still dealing with a true MOLP.

Step 5. All objectives are converted to constraints. For example, maximize *zs1* is converted to $zs1 \geq 1853.88 - \in$.

Step 6. The set of IISs identified by using the guide codes follows:

{*zs1, zs2, s2–s7, h1, h3–h5, h8–h9, h12–h13, h16*}
{*zs1, zs2, s2–s7, h1, h3–h5, h8–h9, h12–h13, h16*}
{*zs1, zs2, s2–s7, h1, h3–h5, h8–h9, h12–h13, h16*}
{*zs1, zh4, s2–s7, h1–h5, h8–h9, h13*}
{*zs2, zh5, s2–s7, h1, h3–h5, h8–h9, h12–h13, h16*}
{*zs2, zh6, s2–s7, h1, h3–h5, h9, h13*}
{*zs1, zh7, s2–s7, h1, h3–h5, h9, h13*}

The results are less conclusive this time. The IISs contain many more elements, *zs3* does not appear in any of the IISs generated, there are several common soft constraints (*s2–s7*), and all of the IISs contain a lengthy list of hard constraints. Because there is no clear-cut conclusion to be drawn, we construct a normalized objective interference table, as shown in Table 9.1. Rows in the table are for the extreme aspiration points for the named objective, columns are for the objectives, and the elements are the fractional decreases in the objectives at the points (i.e. $[y^{Bopt} - f^{B}(x_A^{opt})]/y^{Bopt}$ for table element *AB*).

Table 9.1. Normalized objective interference table

	zs1	*zs2*	*zs3*	*zh4*	*zh5*	*zh6*	*zh7*
zs1	0.000	0.426	1.000	0.550	0.490	0.435	0.154
zs2	0.103	0.000	0.000	0.063	0.014	0.219	0.016
zs3	0.827	0.816	0.000	0.670	0.795	0.583	0.753
zh4	0.250	0.132	0.000	0.000	0.046	0.258	0.056
zh5	0.000	0.006	0.000	0.092	0.000	0.201	0.023
zh6	0.000	0.357	0.000	0.267	0.423	0.000	0.021
zh7	0.183	0.449	0.000	0.207	0.430	0.053	0.000

Note that the diagonal elements of the table are zero, because each objective reaches its extreme aspiration level at its extreme aspiration point.

Since no particular relationship is clearly identified as a candidate for change, the final formulation phase activity is to make recommendations about lexicographic ordering or weighting of the objectives. We choose to use the objective interference table to examine how best to set the lexicographic order.

One way to set the order is to look at which extreme aspiration points have the least negative impact on the other objectives. This can be done in a many ways: average deterioration in objective value, smallest maximum decrease in objective value, etc. For illustrative purposes, we choose to look at the average decline in objective value by calculating the average of the off-diagonal elements in each row of the table, with results as follows:

zs1	*zs2*	*zs3*	*zh4*	*zh5*	*zh6*	*zh7*
.509	.069	.741	.125	.054	.178	.220

With the idea of ordering the objectives so that a greater degree of satisfaction of one objective early has the least negative impact on the later objectives, we can order the objectives based on increasing values of the average decline in objective values: *zh5, zs2, zh4, zh6, zh7, zs1, zs3*. The reverse of this ordering shows the objectives having the most to the least impact on the others. It is not surprising to see that timber production (*zs3*) and pasturage (*zs1*), both land-intensive activities, have the most impact on the other recreation and wildlife objectives. This sort of insight can be used to prepare the final model formulation for solution.

Step 7. Finished formulating, so exit and go to solver. Choosing a lexicographic ordering approach, our final model is thus to maximize the objectives in the chosen order (*zh5,zs2,zh4,zh6,zh7,zs1,zs3*) subject to the constraints excluding constraint *s1*.

10 Data Analysis

Techniques originally devised for analyzing infeasible linear programs turn out to have many interesting and useful applications in data analysis. The problem of placing a hyperplane in an n-dimensional space to separate two categories as well as possible can be directly transformed into an instance of the MAX FS problem, so the algorithms of Chap. 7 can be applied. This is also identical to the problem of providing the initial training for a neural network. A related problem in statistics is determining the *data depth* of a point in an n-dimensional cloud of data points, defined as the smallest number of data points on one side of hyperplane through the point in question. This is again transformable to the MAX FS problem. Finally, arithmetic constraints are often used to check massive data sets such as census results. The rules themselves may be contradictory, and this can be checked via the methods of Chaps. 6 and 7.

10.1 Classification and Neural Networks

A standard problem in machine learning and data classification is the placement of a separating surface (normally a hyperplane) in an n-dimensional space of item features in such a way that it completely separates instances in one set (e.g. of Type 0) from instances in another set (e.g. of Type 1). This is not normally possible with real data, so the objective is usually to place the separating surface so as to minimize the number of incorrectly classified instances. See the example in Fig. 10.1. This minimum misclassification cardinality problem is easily transformed into the MIN IIS COVER problem (Amaldi 1994, Parker 1995, Chinneck 1998, 2001a) as shown below. The conversion results in a MIN IIS COVER problem that has no variable bounds or equality constraints, and hence is somewhat simpler in structure.

The data instances exist as points in an n-dimensional space of attributes or features, each having an associated type (e.g. Type 0 or Type 1). This constitutes the *training set* of instances. Finding the minimum-misclassification hyperplane in the training set is an essential step in building up a decision tree that can be used to classify new instances as they are encountered. The problem of placing a hyperplane to misclassify as few of the training set instances as possible can also be viewed as the problem of determining the smallest number of points to remove from the training set such that all of the remaining points can be correctly classified by a single hyperplane. This is translated into the MIN IIS COVER (or MAX FS) problem as follows:

Given a training set of I data points ($i = 1\ldots I$) in J dimensions ($j = 1\ldots J$), in which the value of attribute j for point i is denoted by d_{ij}, and the class of each point is known (either Type 0 or Type 1), define a set of linear constraints as follows (one constraint for each data point): $\Sigma_j d_{ij} w_j \leq w_0 - \epsilon$ for points of Type 0, and $\Sigma_j d_{ij} w_j \geq w_0 + \epsilon$ for points of Type 1, where ϵ is a small positive constant. Note that the variables are the unrestricted w_j, where $j = 0\ldots J$, while the d_{ij} are known constants. A similar conversion is given by Parker (1995).

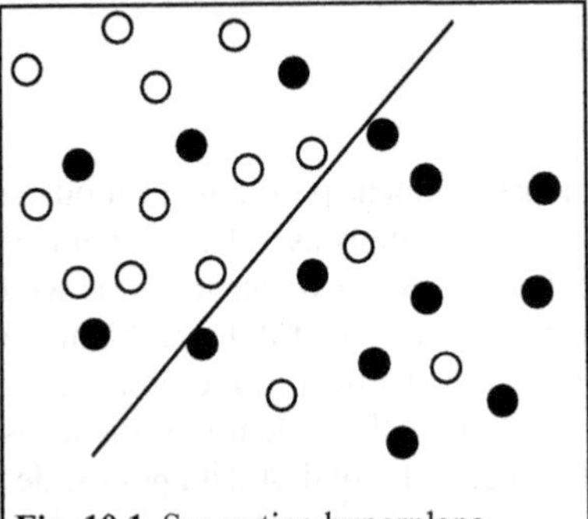

Fig. 10.1. Separating hyperplane

If the data are completely separable by a single hyperplane, then any solution to the LP resulting from the conversion will yield a set of values for the w_j, which then defines the separating hyperplane $\boldsymbol{wx} = w_0$. If the data cannot be completely separated by a single hyperplane, then the LP resulting from the conversion will necessarily be infeasible. Finding a solution to the MIN IIS COVER problem in this infeasible LP then also solves the classification problem of finding the smallest number of points to remove such that the remaining points are completely separable by a single hyperplane. The constraints in the IIS cover correspond to misclassified points in the classification problem. Because the points removed will be incorrectly classified by the resulting hyperplane, this constitutes a method of finding a hyperplane that misclassifies the smallest number of points, an important goal in classification research.

Table 10.1 provides information about nine frequently analyzed binary classification problems taken from the publicly available UCI Repository of Machine Learning Databases (Newman et al. 1998), a common source of classification test data. "Net points" is the number of data instances remaining after incomplete tuples are removed.

Table 10.1. Classification data sets

data set	net points	number of features
breast cancer	683	9
bupa	345	6
glass (type 2 vs. others)	214	9
ionosphere	351	34
iris (versicolor vs. others)	150	4
iris (virginica vs. others)	150	4
new thyroid (normal vs. others)	215	5
pima	768	8
wpbc	194	32

Table 10.2 compares the results obtained when three different algorithms for placing separating hyperplanes are applied to these data sets:

- Alg. 7.3 as implemented in CLIIS, a MINOS-based prototype for placing classifier hyperplanes (Chinneck 2001a).
- Algorithm 2(1) of Sec. 7.4 (i.e. choosing only the violated constraint having the maximum product as the single candidate each time) as implemented in CLIIS (Chinneck 2001a).
- A parametric exact LPEC formulation (see Sec. 7.1.2) solved by successive linear programming, and implemented in the MISMIN code (Bennett and Bredensteiner 1997).

All algorithms are applied to the entire data set (i.e. there is no separation into training and testing sets). The best results in terms of both accuracy (*% acc.*) and time (*secs*) are shown in boldface.

Because MISMIN is among the best of the available programs for minimizing the number of classification errors in classification problems, it is a good standard for comparison. Bennett and Bredensteiner (1997) show that MISMIN performs favorably against such other well-known programs as OC1 (Murthy et al. 1994) and CSADT (Heath et al. 1993).

Table 10.2 shows that Alg. 7.3 is the most accurate, but also the slowest, while MISMIN is the fastest. Algorithm 2(1) provides a major speed-up over Alg. 7.3 (several orders of magnitude in some cases), yielding times comparable to those for MISMIN (and sometimes faster). More significant, however, is that it does this with very little loss of accuracy.

Table 10.2. Three algorithms for classification (Chinneck 2001a)

	Alg. 7.3		Alg. 2(1)		MISMIN	
	% acc.	secs	% acc.	secs	% acc.	secs
breast cancer	**98.4**	17	**98.4**	4.3	98.2	**0.7**
bupa	75.1	159	**75.9**	1.3	73.9	**0.6**
glass (type 2 vs. others)	**81.8**	38	78.5	**0.6**	76.6	**0.6**
ionosphere	**98.3**	44	**98.3**	5.4	**98.3**	**2.6**
iris (versicolor vs. others)	**83.3**	5	**83.3**	**0.2**	82.0	0.3
iris (virginica vs. others)	**99.3**	0.4	**99.3**	**0.1**	**99.3**	0.3
new thyroid (normal vs. others)	**94.9**	3	**94.9**	**0.3**	93.5	**0.3**
pima	**80.6**	1434	80.2	7.2	80.5	**1.5**
wpbc	**96.9**	17	**96.9**	**1.2**	91.2	1.5
average:	89.8	216.2	89.5	2.3	88.2	0.9

An important difference between the approaches taken in Alg. 7.3 and in Algorithm 2(1) as opposed to many other methods is that they remove points from the data sets one at a time instead of all at once as in other methods. This raises the possibility of "guiding" the removal process as it is underway. For example, if the classification accuracy of Type 0 points is lower than that of Type 1 points at some intermediate point in the hyperplane placement process, then the point removal procedure could be coerced to prevent the removal (and hence misclassification) of any more Type 0 points until the classification accuracies are balanced.

Amaldi et al. (2007) ran various MAX FS algorithms on the same data sets as reported in Table 10.2. Algorithms include:

- A branch-and-cut implementation due to Pfetsch (2002), see Sec. 7.2.
- An exact Big-*M* branch and bound using the Cplex 8.1 solver, see Sec. 7.1.1.
- The two-phase algorithm using the linearized bilinear model for the first phase, and the exact Big-*M* branch and bound using Cplex 8.1 for the second phase, see Sec. 7.5.
- The two-phase algorithm using a linearization of the Big-*M* method for the first phase, and the exact Big-*M* branch and bound using Cplex 8.1 for the second phase, see Sec. 7.5.
- The two-phase algorithm using a simple LP phase 1 for the first phase, and the exact Big-*M* branch and bound using Cplex 8.1 for the second phase, see Sec. 7.5.
- A reimplementation of Alg. 7.3 using the AMPL scripting language and with Cplex 8.1 as the LP solver.

Results are summarized in Table 10.3. All solutions were limited to 10,000 seconds of CPU time. Alg. 7.3 is the only method that is not branch-and-bound based, and is also the only method that did not time out on any of the data sets. The accuracy shown for timed-out solutions is for the incumbent solution available at time-out. There are some differences in the accuracies and times for Alg. 7.3 between Tables 10.2 and 10.3 which are likely due to differences in coding and in the LP solvers used.

Table 10.3. More MAX FS algorithms for classification (Amaldi et al. 2007)

	branch & cut		Big-*M*		2Ph-bilin		2Ph-Big*M*		2Ph-LP		Alg. 7.3	
	%acc	sec	%acc	sec	%acc	sec	%acc	sec	%acc	sec	%acc	sec
breast cancer	**98.4**	43	98.2	*t*	**98.4**	371	**98.4**	1	**98.4**	8	**98.4**	15
bupa	65.8	*t*	71.9	*t*	**75.7**	*t*	75.4	*t*	75.1	*t*	75.1	331
glass (type 2 vs. others)	**83.2**	*t*	82.2	*t*	78.5	133	80.4	375	78.0	*t*	80.4	143
ionosphere	**98.3**	2215	**98.3**	*t*	**98.3**	2268	**98.3**	4	**98.3**	36	**98.3**	36
iris (versicolor vs. others)	**83.3**	1735	**83.3**	7630	82.7	4	82.7	1	82.7	5	**83.3**	30
iris (virginica vs. others)	**99.3**	0.1	**99.3**	0.4	**99.3**	1	**99.3**	0.2	**99.3**	0.2	**99.3**	0.1
new thyroid (nrm vs others)	**94.9**	17	**94.9**	46	**94.9**	9	**94.9**	0.2	**94.9**	0.4	**94.9**	10
pima	79.9	*t*	76.6	*t*	80.3	*t*	80.1	*t*	80.1	*t*	**80.6**	1977
wpbc	**93.3**	*t*	90.2	*t*	91.8	*t*	91.8	1863	91.8	*t*	91.8	91
avg	88.5		88.3		88.9		89.0		88.7		**89.1**	

t: algorithm timed out. Boldface indicates best accuracy in the row.

Adem and Gochet (2006) extend the mathematical programming based methods for finding separating hyperplanes for the two-class problem to the multiclass problem. One approach they use is to modify Alg. 7.3 for use in solving multiclass problems. The main idea is that if there are C classes, then C constraints are formulated for each data point. As for the two-class version of Alg. 7.3, one data point is removed at each iteration, but this is done by removing C-1 constraints (instead of just one constraint in the original two-class version). Various rules based on Chinneck (2001a) are used to identify the data point for removal at each iteration. Empirical results with the modified heuristic are very good.

As pointed out by Mangasarian (1993), the training of neural networks is equivalent to finding separating hyperplanes. Hence the conversion of the classification problem into the MIN IIS COVER problem allows the spectrum of solutions for MIN IIS COVER to be applied to the initial training of neural networks.

Solution of the MIN IIS COVER problem to find separating hyperplanes also appears to underlie the statistical technique known as *optimal data analysis* (Yarnold and Soltysik 2004).

10.2 Data Depth

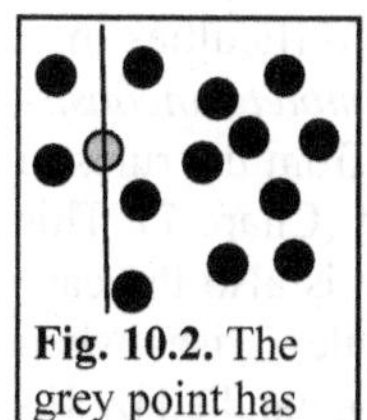

Fig. 10.2. The grey point has data depth 2

The *data depth* of a point in a cloud of points in a multi-dimensional space is a statistical concept that is related to the idea of the median of a set of points in a one-dimensional space. It is also known as the *halfspace depth*, the *location depth*, and the *Tukey depth*. The data depth of some point $\boldsymbol{p}$ in a set of points S is defined as the smallest number of points in S in any closed halfspace with boundary through $\boldsymbol{p}$. The point with the largest data depth is called the *Tukey median*.

Fukuda and Rosta (2005) provide an introduction to basic concepts and algorithms for calculating the data depth and also show that finding the data depth of a given point is equivalent to solving the MAX FS problem (or equivalently the MIN IIS COVER problem). Given the use of algorithms for solving MAX FS in classification (Sec. 10.1), this is not surprising. Here the goal is to separate a given point from as few other points as possible via a single hyperplane, as compared to separating Type 0 points from Type 1 points in data classification. The data depth problem also is identical to the closed (open) hemisphere problem,

The problem of finding the data depth for some point $\boldsymbol{p}$ is converted to a linear program as follows. Let there be I other points in the set S, each represented as a J-tuple in the J-dimensional space. Set up one constraint for every point $\boldsymbol{x}_i$ as follows: $a_1(x_{i1} - p_1) + a_2(x_{i2} - p_2) + \ldots + a_J(x_{iJ} - p_J) \geq \epsilon$, where ϵ is a small positive constant. Note that the x_{ij} and p_j are known constants while the a_{ij} are the unknown variables. If $\boldsymbol{p}$ is on the convex hull of the cloud of data points, i.e. has a data depth of 0, then the resulting LP is feasible, and the hyperplane that separates $\boldsymbol{p}$ from the rest of the points is given by $\boldsymbol{ax} \geq \epsilon$.

In the more general case where $\boldsymbol{p}$ has a data depth of greater than 0, then the resulting LP will be infeasible, and the various algorithms for solving MAX FS can be applied (see Chap. 7). More exactly, this problem is best solved by the MIN IIS

COVER or MIN ULR variants of MAX FS. The cardinality of the MIN IIS COVER or MIN ULR is the data depth of the point.

Chen (2007) reports on experiments using various branch-and-bound and branch-and-cut algorithms to calculate data depth. Bremner et al. (2006) report on experiments using so-called primal-dual methods that calculate both upper and lower bounds on the data depth and terminate when the bounds are the same. Unreported experiments by the author using Alg. 7.3 show promise.

10.3 Errors in Massive Data Sets

Mathematical and logical rules are commonly used to check the data in massive data sets, such as census or survey data (Bruni 2005a). Very simple mathematical rules may check the range of a data value, e.g. that age is between 0 and 110 years. More complex rules tie conditions together, for example ($Age - YearsMarried \geq 16$) to express the fact that the minimum age for marriage is 16. Bruni (2005a) further demonstrates how logical conditions can readily be converted to linear inequality constraints.

Given a set of rules for checking a massive data set, various difficulties may arise, such as inconsistency among the rules. In the case of *complete inconsistency*, it is straightforward to check the linear inequalities arising from the rules to identify an IIS (Chap. 6) or to find the maximum feasible subset (Chap. 7). This allows the analyst to alter the rules appropriately. However there is also the case of *partial inconsistency* in which an inconsistency among the rules arises when particular values are chosen for some of the fields. This can also be handled by IIS isolation and maximum feasible subset identification by analyzing the infeasibility that arises in that case. Depending on the format of the data, methods for the analysis of infeasibility in LPs or in MIPs may be required. This process is referred to as *validating* the data check rules.

Given a set of validated rules, it is then simple to check each data record by verifying that it satisfies all of the linear constraints that express the rules. If a record does not satisfy all of the rules, then it is possible to automatically correct the record. The goal in this case is to make the smallest possible adjustment to the record such that it satisfies all of the rule inequalities, while simultaneously disturbing the original frequency distribution of the data as little as possible. This is usually done by identifying a *donor record* that is as similar as possible to the erroneous record, and copying low-cost fields from the donor record to the erroneous record so that the erroneous record now satisfies all of the rules.

Bruni (2005a) reports very encouraging results using the methods described above for test sets based on census data for Italy. See also Bruni et al. (2001). Riera-Ledesma and Salazar-Gonzalez (2007) describe a branch-and-cut approach for the correction of records that violate some of the rule inequalities; they find the minimum number of fields to change (or more generally, the minimum weight set of fields to change).

Wu and Barbará (2002) discuss a variety of methods for imputing missing data values based on data available in summary constraints (e.g. determining missing individual values when the column total is known). The value of the summary constraint and the individual data values may conflict, and in this case techniques of infeasibility analysis can be of assistance.

11 Miscellaneous Applications

Many applications depend on representations that consist of sets of constraints. The constraints might represent limits on the amount of radiation that healthy and diseased tissue must receive when planning a course of radiation treatment (Sec. 11.1), the minimum levels of signal strength needed for digital video broadcasting (Sec. 11.3), restrictions on the questions drawn from a test bank (Sec. 11.4), etc. When the constraints conflict, common questions concern the cause of the conflict (Chap. 6), the maximum set of constraints that can be satisfied simultaneously (Chap. 7), or the smallest fix for the constraints (Chap. 8). This chapter briefly surveys some of the many specific applications that have arisen in recent years as effective techniques for the analysis of infeasibility have become available. Even more applications are sure to be discovered soon as more researchers become aware of these tools: why not add your own applications to the list?

11.1 Radiation Treatment Planning

Radiation treatment for diseases such as cancer involve careful planning such that the diseased target tissue receives a large enough dose to kill the problem cells, while non-diseased tissue receives doses that are not large enough to cause significant damage. Various kinds of optimization problems and solution methods arise in the attempt to solve the radiation treatment planning problem. See Censor (2003), Holder (2004) and Lim et al. (2006) for an overview of radiation planning models and solution techniques.

Intensity-modulated radiation therapy (IMRT) machines are designed such that small individual beams of radiation of adjustable intensity are directed through the body at various angles and intensities. The body is modeled as a collection of "volume pixels" known as *voxels*, and the resulting set of constraints, often linearized, is of the form $\boldsymbol{A}_1\boldsymbol{x} \leq \boldsymbol{u}$ for voxels in non-diseased tissue that should not receive more than a specified dose of radiation, and $\boldsymbol{A}_2\boldsymbol{x} \geq \boldsymbol{l}$ for diseased tissue that should receive a sufficiently lethal dose. The nonnegative variables $\boldsymbol{x}$ represent the intensity setting that should be assigned to each individual radiation beam.

The first IMRT problem is quickly finding a feasible solution for this set of constraints. There is a lengthy literature on this process. Standard LP and MIP solution methods can be used, depending on the problem formulation. Various forms of projection algorithms have been developed extensively for the solution of this problem. See e.g. Censor and Zenios (1997).

The second IMRT problem arises when the resulting voxel dosage inequalities form an infeasible set. A solution must be found so that the radiation therapy can proceed for the patient, but the solution must be as close as possible (in some sense) to satisfying the constraints. The measures and approaches of Sec. 8.1 can be applied to determine the minimum violation of the constraints that will result in a feasible solution. Censor et al. (2006) developed the method of Sec. 8.1.8 specifically for the solution of this problem.

Sadegh (1999) approaches the infeasible radiation therapy planning problem as an instance of MAX FS, and develops a variation of Chinneck's heuristics (see Sec. 7.4) to solve it. A MIP solution to the MAX FS problem is also described by Lee et al. (1999), in this case for a brachytherapy procedure, which involves the insertion of radioactive "seeds" in the patient. However the underlying model is similar to the IMRT models in that the body is discretized into voxels and there are lower bounds on the radiation to be delivered to diseased voxels and upper bounds on the radiation to be delivered to the healthy voxels.

11.2 Protein Folding

The prediction of the three-dimensional protein folding pattern from its amino acid sequence is currently a subject of great interest in computational biology. Knowing the folding pattern gives clues to the function of proteins. Linear constraints are often used to formulate protein energy models, resulting in LPs with extremely large numbers of constraints (on the order of tens to hundreds of millions) in a few hundred variables.

Protein folding prediction can be cast as a problem of comparing the energy of a misfolded shape to the energy of the native shape of a protein sequence. Native shapes always have a lower energy than a misfolded shape. Meller et al. (2002) express this via the relationship $\Delta E_{\text{mis,nat}} = E_{\text{misfolded}} - E_{\text{native}} \geq \varepsilon$ where $E_{\text{native}} = E(\boldsymbol{s}_{\text{nat}}, \boldsymbol{x})$ is the energy of the native structure $\boldsymbol{s}_{\text{nat}}$ while $\boldsymbol{x}$ is the vector of unknown parameters. Similarly, $E_{\text{misfolded}} = E(\boldsymbol{s}_{\text{mis}}, \boldsymbol{x})$ is the energy of the misfolded sequence and $\boldsymbol{x}$ is the set of unknown parameters. Given a set of similar *decoy* structures to compare against, energy inequalities can be constructed for each one; ε is a small positive constant indicating that the energy of the native structure is lower than that of any of the decoy structures. The goal is to determine the unknown values of $\boldsymbol{x}$ which represents a linear combination of basis functions.

Since the energy function is a linear combination of basis functions, the values of $\boldsymbol{x}$ can be found by linear programming subject to the constraints arising from the energy inequalities for the decoy structures as defined above. The resulting models are very large, having hundreds or millions of constraints in a few hundred parameters (Wagner et al. 2004). Further, the models are very often infeasible, so a problem of identifying a maximum feasible subset of linear inequalities arises. The MAX FS solution helps define the region for the correct native structure. The techniques of Chap. 7 can be applied to this problem.

Given the scale of the resulting MAX FS problems, special large-scale solution techniques are needed. Meller et al. (2002) devised the interior-point heuristic of Sec. 7.7. Amaldi et al. (2005) tested a number of algorithms for solving the protein folding MAX FS problem. The exact Big-*M* MIP method (Sec. 7.1.1) was entirely unsuccessful on models in range of several hundred thousand inequalities, however the randomized thermal relaxation algorithms (Sec. 7.6) performed well, though there are no exact results for comparison.

11.3 Digital Video Broadcasting

As described by Amaldi et al. (2005), the digital video broadcasting planning problem results in models having large numbers of linear inequalities. Given m square areas representing test points for signal reception, and n transmitters, the goal is to determine the transmission power of each transmitter so that the signal reception at each test point is acceptable. Signals from different transmitters may arrive at the same test point with a delay; if this delay is too large, then the signals interfere. Hence the resulting model has a linear inequality for each test point of the form $\sum_{j=1}^{n} a_{ij}x_j \geq b_i$, where x_j is the unknown power of transmitter j, and a_{ij} is the strength of the signal arriving at test point i from transmitter j. a_{ij} is positive if the signal is useful, and negative if it interferes. b_i is the minimum signal strength needed at test point i to provide adequate signal strength with 95% probability. Each transmitter is also limited in transmission power: $x_j \leq p_{\max}$ for all j. Complete coverage of a large number of test points is not usually possible, so the problem becomes one of satisfying as many of the inequalities as possible, i.e. a MAX FS problem, so the methods of Chap. 7 are applicable.

Amaldi et al. (2005) formulate a discretized version of the problem for Italy using a few thousand transmitters that results in 55,000 inequalities. A version can also be formulated that is weighted by the population at each of the test points. They approach this problem using the randomized thermal relaxation (RTR) heuristic (Sec. 7.6) as well as the exact Big-*M* formulation (Sec. 7.1.1). The exact Big-*M* solution works well for the smaller models, but runs into difficulties and times out on the larger models. The RTR heuristic completes all solutions quickly (never taking more than 3 minutes of computation time on a PC). It is possible to assess the quality of the RTR solutions on the models completed by both methods within the time limits. The RTR solutions are in general not greatly worse than the exact Big-*M* solutions, and are even better in a couple of cases because of the numerical issues that arise in the Big-*M* method.

Amaldi et al. (2007) report on the use of the two-phase relaxation-based heuristic (Sec. 7.5) to solve the digital video broadcasting problem. The version of the algorithm using a linearization of the Big-*M* formulation as a first phase gives the best results, despite the fact that the second phase Big-*M* formulation times out frequently, hence returning only the incumbent solution available at time-out.

A similar problem formulation is likely applicable to cell phone towers.

11.4 Automated Test Assembly

Many types of tests (e.g. the Scholastic Aptitude Test) are automatically assembled subject to constraints, e.g. that certain categories of knowledge are tested a certain number of times, or that the word count is less than a prescribed limit, etc. Test questions are drawn from a pool of authorized questions. The problem of assembling a test from a bank of questions such that it meets all of the constraints is usually formulated as a binary program in linear constraints in which the binary variables indicate whether a particular question is to be included in the test or not (Huitzing et al. 2005). Questions are assigned an importance weight, and the objective is to maximize the total importance of the questions in the test. The test assembly formulation may be infeasible due to incompatible restrictions, or to a collection of questions that is too small for the desired test. Several researchers have addressed the issue of how to resolve the infeasibility.

Huitzing et al. (2005) approach the analysis of infeasibility in these problems in several ways, including IIS isolation (see Chap. 6), finding a maximum feasible subsystem (see Chap. 7), and finding the best approximation solution (see Chap. 8). Two categories of methods are applied. The first category provides a best approximate solution to the infeasible test assembly problem. Methods in this category include:

- A weighted elastic program. The objective is to minimize the sum of the weighted elastic variables (see Sec. 6.1.4). Huitzing et al. (2005) refer to this as *goal programming.*
- Multi-objective programming. An ordered application of the objective functions may be used, or a weighted combination of them. See Sec. 9.3.
- Greedy heuristics. Start the test assembly with a single item and successively add the next best item until a predetermined number of items has been added. The ordering of the items may be set by assigning weights to the violation of a constraint and choosing the next item as the one that minimizes the sum of the weights.

The second category of methods tries to determine the cause of the infeasibility so that it can be analyzed and repaired by the human test assembler. Various methods for isolating IISs are used:

- Standard deletion filtering applied to the LP-relaxation of the original binary test assembly problem. The deletion filter as implemented in Cplex 6.6 is used (termed RODA for *relaxed and ordered deletion algorithm*). See Sec. 6.1.2.
- A version of the deletion filter that randomizes the order of the constraints and respects the integrality constraints (termed IRDA for *integer and randomized deletion algorithm*).
- A sampling approach termed SCIS for *set covering and item sampling*. The sampling respects the binary nature of the variables, but otherwise the method is identical to that in Sec. 6.1.7.
- A method for successively relaxing bounds until feasibility is reached. This works by finding an IIS (by either RODA or IRDA), and then calculating a

measure for each member of the IIS to determine which one to relax. The process is repeated until the model is feasible. The measure for each member of the IIS is based on the amount it must be relaxed to make the IIS feasible relative to the size of its right hand side constant. A new LP/MIP must be solved for each constraint in the IIS to determine how much it must be relaxed to render the IIS feasible. Huitzing et al. (2005) refer to this process as the *IIS-solver*. It is similar to the methods of Sec. 7.2. This method can also be used to automatically provide an approximate solution to the infeasible problem.

Huitzing et al. (2005) evaluate the performance of the various methods on two examples derived from real test assembly problems.

Timminga (1998) further mentions a standard Big-M integer program for finding a minimum IIS set cover (see Sec. 7.1.1).

11.5 Buffer Overrun Detection

Computer buffer overruns are a serious problem. Unintentional buffer overruns may cause program failure while intentional buffer overruns are serious security vulnerabilities. Ganapathy et al. (2003) approach the problem of preventing possible buffer overruns by analyzing the source code for a computer program before it runs. Their analysis transforms the source code into a list of linear constraints, typically inequalities that relate to the size of the buffer (e.g. number of bytes) and to the current level of usage of the buffer, which can be analyzed for possible buffer overruns. Application of their technique to several commercial programs identified a number of previously-unknown buffer overruns. Infeasibility of the constraints indicates a modeling problem that must be corrected before the remainder of the overrun analysis can be applied (or may possibly indicate an overrun vulnerability). Ganapathy et al. (2003) carry out the infeasibility analysis via an elastic filter (see Sec. 6.1.4).

11.6 Customized Page Rankings

Page rankings are used by web search engines to sort a collection of web pages for presentation to the user. Ranking algorithms use web page features such as how many other pages point to a particular page, some measure of the “quality” of a page, the quality of the pages pointing to this page, etc. However the importance of a page is relative to the search at hand and the interests of the human searcher, hence customized page rankings are more suitable for particular fields of study. Tsoi et al. (2006) model user preferences on page rankings via linear constraints, e.g. that the rank of page A should be higher than the rank of page B, or that the rank of a page on a particular topic should be at least twice its general page ranking etc. Their page ranking algorithm results in a quadratic objective function subject to linear constraints.

The linear constraints may form an infeasible set. Tsoi et al. address the infeasibility in two ways. One approach is to solve an elastic program to minimize a weighted sum of the constraint violations (see Sec. 8.1.2). The second approach is a modified form of maximum feasible subset analysis (see Chap. 7). In the second approach, they first enumerate all of the maximum feasible subsets (an NP-hard problem in general), and then solve the linearly-constrained quadratic page ranking problem for each feasible subset. This approach is exercised on a small problem.

11.7 Backtracking in Tree-Structured Search

Tree-structured search arises in solving mixed-integer programs, in constraint programming, in solving the satisfiability problem (Sec. 4.1), and in many other contexts. When an infeasible node is reached in a tree-structured search, there is an opportunity to use the information about the infeasibility of the current node to decide which node to backtrack to, or to develop global constraints which will prevent branching to other nodes in the tree that have the same infeasibility. In both cases, the efficiency of the tree search is greatly improved. The efficiency relies on the ability to identify IISs.

Bruynooghe (1981) did some of the earliest work on *intelligent backtracking* in the context of constraint programs for combinatorial search problems. For such problems, each variable has a discrete set of possible values. Conflicts arise at a node in the search tree when the values of some of the variables have been fixed by the search tree and this fixing violates one or more constraints. The conflict set is then the set of fixed variables in the violated constraint. From the optimization perspective, this is similar to detecting infeasibility at a branch and bound node and identifying an infeasible subset that consists of a single functional constraint and the bounds on the variables it contains that have been altered by the branch and bound process. Bruynooghe defines a *minimal conflict set* similarly to an IIS: it is a conflict set for which every proper subset is not a conflict set. Bruynooghe suggests first finding all minimal conflict sets, and then ordering them from smallest to largest cardinality. Backtracking is then done on the variables in the chosen conflict set.

De Backer and Beringer (1991) find minimal conflict sets (i.e. IISs) in the subset of linear constraints that define the constraint program, again for the purposes of intelligent backtracking. They use a method similar to that of Thm. 6.16. Burg et al. (1994) present another method of finding minimal conflict sets in the subset of linear constraints in the model for the same purpose. Constraints are processed one at a time in constraint programming. Burg et al. maintain the current set of constraints in a special solved form achieved by Gaussian operations. The solved form appears to be similar to van Loon's (1981) form, and the minimal conflict sets appear to be isolated in the same manner. In both cases, the minimal conflict set directs the backtracking process to the nodes at which the members of the minimal conflict set were introduced so that another branch of the search can be explored.

Davey et al. (2002) introduce more efficient intelligent backtracking for binary linear programs consisting of inequalities. Each inequality of the form $\boldsymbol{a}_i \boldsymbol{x} \le b_i$ is a knapsack constraint, and a *knapsack-cover* is defined when a constraint is violated, of the form

$$\sum_{i \in C^+} a_i > b_i - \sum_{i \in N^- \setminus C^-} a_i \,.$$

N^- indexes the set of variables with negative coefficients (with C^- as a subset) and N^+ indexes the set of variables with positive coefficients (with C^+ as a subset). This equation states that given the current set of variable fixings (C^+, C^-), the constraint is violated no matter how the current unset variables are set. Based on this, a *knapsack-cover inequality* can be defined of the form

$$\sum_{i \in C^+} x_i - \sum_{i \in C^-} x_i \le \left| C^+ \right| - 1,$$

which states that not all of the variable settings in the knapsack-cover can hold simultaneously. The knapsack-cover inequality is used as a global constraint.

The efficiency of the method depends on finding small inconsistent sets with relatively little additional effort. Davey et al. find IISs using an approach based on identifying an improving ray in the dual cone of the unbounded dual problem associated with the infeasible primal problem. Their method is similar to that of Thm. 6.17. After the IIS is found, it is examined to see whether some of the constraints introduced by the branching process can be replaced by original constraints. For example, the IIS may include the constraint $x_1 + 5x_2 + 4x_3 + 2x_4 \le 8$, with the variables x_1, x_2, and x_3 all set to 1 by the branching process and the value of x_4 not yet set. This constraint is violated with these variable fixings, and is still violated if $x_1 = 1$ (fixed by the branching process) is replaced by the original constraint $x_1 \ge 0$. The resulting knapsack-cover inequality in this case expresses the fact that x_2 and x_3 cannot be set equal to 1 at the same time: $x_2 + x_3 \le 1$. Knapsack-cover inequalities involving fewer variables are more powerful and easier to check, hence the final examination of the functional constraint in an attempt to replace constraints introduced by the branching process with original variable bounds. This is done by a simple pass through the variables in the functional constraint whose values were set by the branching process, from smallest to largest coefficient, replacing the set value with the original bound as long as the functional constraint remains violated.

Davey et al. (2002) present empirical results for a variety of binary linear programs showing that their method solves binary linear programs faster than an unmodified branch and bound approach.

Fränzle and Herde (2005) describe a system for dealing with models that combines the Boolean satisfiability problem with linear inequalities. One aspect of the problem formulation uses Boolean variables to indicate whether or not a real-valued linear inequality is included in the model or not. There is thus a higher-level SAT problem in Boolean variables that controls various lower-level combinations of the linear inequalities. At some nodes of the search tree, the combination of linear inequalities may be infeasible. When this happens, an IIS is isolated and used to guide the backtracking in the SAT tree in the form of a conflict clause based on the IIS.

Codato and Fischetti (2006) use a similar approach for solving certain classes of MIPs. The model is separated into an all-binary master problem and slave continuous LP problems. If a slave LP problem is infeasible, an IIS analysis is used to generate a combinatorial cut constraint that is then added to the master binary problem to prevent recurrence of that particular infeasible combination of constraints. This approach greatly reduces the time to solve certain categories of MIPs as compared to a standard MIP solution method.

11.8 Piecewise Linear Model Estimation

Piecewise linear models are used to model a wide range of nonlinear phenomena. Such models typically result in a large set of linear equations, usually one for every data point, which is typically infeasible. Solving the MIN PFS problem for such a set of constraints (see Sec. 7.9) can be a very effective way of determining how to partition the model into linear segments. A feasible solution for each feasible partition then provides values for the model parameters.

Amaldi and Mattavelli (2002) show how piecewise model estimation problems can be converted to MIN PFS problems. First, to deal with noisy data, each equation i is replaced by a pair of inequalities defining a hyperslab of width 2ε, where ε is the noise tolerance: $\boldsymbol{a}_i\boldsymbol{x} \leq b_i+\varepsilon$ and $\boldsymbol{a}_i\boldsymbol{x} \geq b_i-\varepsilon$. The noise thresholds need not be identical over all equations since they may be subject to different physical effects. A MIN PFS solution finds a small set of hyperslabs that contains all of the original data points; see the example in Fig. 11.1.

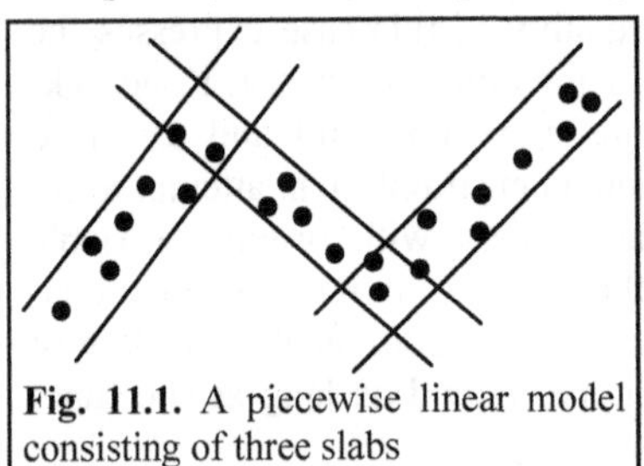

Fig. 11.1. A piecewise linear model consisting of three slabs

Amaldi and Mattavelli (2002) show how to apply heuristics for the solution of MIN PFS to line detection in digital images and to modeling of time series. For the line detection problem we are given a set of two-dimensional points p_i in the $x_1 \times x_2$ plane associated with contours extracted from the image. The goal is to detect line segments in this collection of contour points. Contour points lying on the same line will satisfy the constraint $a_1x_1 + a_2x_2 + a_3 = 0$. We thus construct one such constraint for each contour point $\boldsymbol{x}$. As described above, this equation is replaced by a pair of inequalities, which creates a hyperslab to allow for noise. Solving the MIN PFS problem will yield separate subsets of feasible inequalities. A solution for each feasible subset will yield the values of a_1, a_2, and a_3, defining the line segment. The problem can be further simplified by replacing a_3 by -1, which amounts to a scaling of each constraint; this simplification eliminates only solutions that pass exactly through the origin. Amaldi and Mattavelli test the greedy solution algorithm for the MIN PFS formulation (Sec. 7.9) by application to a number of line detection problems. The quality of the results returned is always at least as good as and sometimes better than a Hough transform.

Time series arise in many applications including various types of signal processing. These may be broken into linear submodels, as in threshold autoregressive models which choose the submodel at some time t by comparing the signal at time t-1 with predetermined thresholds. However it is difficult to choose the thresholds in advance. Such a piecewise linear model has the form

$$y_t = \sum_{j=1}^{n} x_j(\mathbf{y}_t) y_{t-j} + u_t$$

where $\mathbf{y}_t = (y_{t-1},\ldots,y_{t-n})$ are the known observations, and the coefficients of the current submodel $x_j(\mathbf{y}_t)$, $1 \leq j \leq n$, depend on the values of the observations in $\mathbf{y}_t$, i.e. where the point is located in the state space. $\{u_t\}$ is an independent and identically-distributed random sequence. The x_j coefficients and the partitions of the state space must both be estimated.

As Amaldi and Mattavelli (2002) show, this system can be represented by a set of linear equations of the form $\mathbf{Ax} = \mathbf{b}$ where

$$\mathbf{A} = \begin{bmatrix} y_n & y_{n-1} & \cdots & y_1 \\ y_{n+1} & y_n & \cdots & y_2 \\ \vdots & \vdots & \ddots & \vdots \\ y_{L-1} & y_{L-2} & \cdots & y_{L-n} \end{bmatrix} \text{ and } \mathbf{b} = \begin{bmatrix} y_{n+1} \\ y_{n+2} \\ \vdots \\ y_L \end{bmatrix}$$

for a sequence of observations $\{y_1,\ldots,y_L\}$. For anything beyond a simple model, this will be an infeasible system of linear equations, and the MIN PFS solution will define groupings of the vectors in the state space; the solution for each feasible subsystem will provide the parameters for the corresponding linear submodel. Noise is handled by replacing the equations by two oppositely-oriented inequalities so as to create slabs as we have seen above. With an appropriate choice of the error parameter ε for the slab separation, Amaldi and Mattavelli report very good results on a number of sample applications, as do Bemporad et al. (2005) for their related method.

11.9 Finding Sparse Solutions to Systems of Linear Equations

Given a feasible system of linear equations, what is the solution that has the smallest number of nonzeroes? The answer to this question can be important in numerous applications, e.g. signal processing. A number of special-purpose heuristics for solving this problem have been developed, under the general name of *basis pursuit*. However this problem can be cast as an instance of the MAX FS problem and hence solved by the many heuristics for this problem (see Chap. 7).

We are given a system of linear equations $\mathbf{Ax} = \mathbf{b}$ in which $m < n$ where m is the number of equations, n is the number of variables, and $\mathbf{b}$ is not empty. This is an underdetermined system, so there are multiple solutions. Defining $\|\mathbf{d}\|_0$ to be a norm giving the number of nonzeroes in some vector $\mathbf{d}$, the problem at hand is to solve $\min\{\|\mathbf{x}\|_0: \mathbf{Ax} = \mathbf{b}\}$. The most straightforward conversion of this problem to

MAX FS is to set up the infeasible system $\boldsymbol{Ax} = \boldsymbol{b}$, $\boldsymbol{x} = \boldsymbol{0}$ in which the $\boldsymbol{Ax} = \boldsymbol{b}$ constraints are strictly enforced, with only the constraints in $\boldsymbol{x} = \boldsymbol{0}$ removable in the search for the largest feasible subsystem. The largest feasible subsystem under these conditions corresponds directly with the smallest number of nonzeroes in the solution $\boldsymbol{x}$.

Amaldi and Kann (1998) show the equivalence of this problem and the MIN ULR problem. For an infeasible set of linear equations $\boldsymbol{Ax} = \boldsymbol{b}$, the MIN ULR problem is the same as finding $\boldsymbol{x}$ such that $\boldsymbol{Ax} + \boldsymbol{s} = \boldsymbol{b}$ has as few nonzero slacks in $\boldsymbol{s}$ as possible. Now imagine a matrix $\boldsymbol{D}$ that is orthogonal to $\boldsymbol{A}$, i.e. $\boldsymbol{DA} = \boldsymbol{0}$. Multiplying the system $\boldsymbol{Ax} + \boldsymbol{s} = \boldsymbol{b}$ by $\boldsymbol{D}$ yields the system $\boldsymbol{DAx} + \boldsymbol{Ds} = \boldsymbol{Db}$, but since $\boldsymbol{DA} = \boldsymbol{0}$ this reduces to $\boldsymbol{Ds} = \boldsymbol{Db}$. If $\boldsymbol{Ax} = \boldsymbol{b}$ is infeasible then $\boldsymbol{Db} \neq \boldsymbol{0}$ and there exists an $\boldsymbol{s}$ satisfying $\boldsymbol{Ds} = \boldsymbol{Db}$ with k nonzeroes if and only if there exists an $\boldsymbol{x}$ satisfying all but k equations of $\boldsymbol{Ax} = \boldsymbol{b}$.

A similar transformation is given by Jokar and Pfetsch (2007): using Gaussian elimination, transform the first m rows and columns of $\boldsymbol{Ax} = \boldsymbol{b}$ into a unit matrix, which is then simplified to the equivalent system $\boldsymbol{u} + \boldsymbol{Rv} = \boldsymbol{r}$, where $\boldsymbol{r}$ and $\boldsymbol{u}$ are $m \times 1$ vectors, $\boldsymbol{R}$ is an $m \times (n-m)$ matrix, and $\boldsymbol{v}$ is an $(n-m) \times 1$ vector. Setting $\boldsymbol{u} = \boldsymbol{0}$ and $\boldsymbol{v} = \boldsymbol{0}$ yields an infeasible system. Eliminating $\boldsymbol{u}$ reduces the infeasible system to $\boldsymbol{Rv} = \boldsymbol{r}$, $\boldsymbol{v} = \boldsymbol{0}$. A maximum feasible subsystem of this infeasible system yields a solution for the original system $\boldsymbol{Ax} = \boldsymbol{b}$ that has the fewest nonzeroes in the solution $\boldsymbol{x}$. The instance of the inequality-constrained MAX FS problem is $\boldsymbol{Rv} \leq \boldsymbol{r}$, $\boldsymbol{Rv} \geq \boldsymbol{r}$, $\boldsymbol{v} \leq \boldsymbol{0}$, $\boldsymbol{v} \geq \boldsymbol{0}$, which has $2n$ inequalities in n-m variables. If k inequalities are eliminated to render the system feasible, then a solution $\boldsymbol{v}$ to the remaining $2n$-k inequalities gives a solution $(\boldsymbol{u},\boldsymbol{v})$ for $\boldsymbol{u} + \boldsymbol{Rv} = \boldsymbol{r}$ that has at most k nonzeroes, where $\boldsymbol{u} = \boldsymbol{r}\text{-}\boldsymbol{Rv}$.

Jokar and Pfetsch (2007) study a variety of heuristic methods for solving their MAX FS transformation of the sparsest solution problem. Their experiments show that the best results are given by a heuristic method due to Mangasarian (1999) that is similar to the exponential approximation (Mangasarian 1996) given in Sec. 7.1.2. The exponential approximation gives better results than the special-purpose basis pursuit and orthogonal matching pursuit algorithms for this problem.

11.10 Various NP-Hard Problems

Several NP-hard problems can be reduced to MAX FS or its variants. Once converted, these problems can then be attacked via the methods of Chap. 7. One such problem is the MINIMUM FEEDBACK ARC SET problem.

Given a directed graph, the FEEDBACK ARC SET problem is to determine whether there is a subset S of the arcs such that $|S| \leq k$ and the directed graph that results when you remove the arcs in S is acyclic. Sankaran (1993) constructs a set of linear inequalities of the form $t_i - t_j \leq -1$ for every arc that connects some node i to some node j. He then goes on to prove that the graph can be made acyclic by removing at most k arcs if and only if the system of associated inequalities can be made feasible by deleting at most k constraints. The MINIMUM FEEDBACK ARC SET problem is to determine the smallest number of arcs to remove to render the graph acyclic; this is identical to the MIN ULR (or MAX FS) problem in the set of associated constraints.

Various other NP-hard problems reduce to MAX FS in binary variables. We do not have good heuristics for the solution of such problems yet. Two examples follow.

Given an undirected graph, the MAX INDEPENDENT SET problem is to find the largest set of nodes that are not pairwise adjacent (known as an independent set of nodes). Amaldi (2003) shows how to convert the MAX INDEPENDENT SET problem to MAX FS. Let there be a variable x_i for each node in the graph, which has a set of vertices v_i and set E of arcs (v_i,v_j). We can construct a constraint for each node of the form $x_i - \sum_{j:(v_i,v_j)\in E} x_j = 1$. The constraint is satisfied only if the node v_i represented by x_i is included in the independent set while all nodes directly connected to it via a single arc are excluded. A maximum feasible subset of this set of linear constraints corresponds to a maximum independent set.

Given m disjunctive clauses in n Boolean variables, the MAXSAT problem is to find a true/false assignment for the Boolean variables that satisfies the maximum number of clauses. Amaldi (2003) shows the conversion to MAX FS as follows. For each clause of the form $z_1 \vee \bar{z}_2 \vee ... z_n$ create a linear inequality of the form $x_1 + (1 - x_2) + ... x_n \geq 1$. The MAX FS solution then yields the largest subset of clauses that can be satisfied.

Given a set of m points on the surface of an n-dimensional unit sphere centred at the origin, the HEMISPHERE problem is to find the hemisphere that contains the largest number of points (this hemisphere may not be unique). If the points are given by n-vectors $\boldsymbol{x}_i$ for $i = 1$ to m, this amounts to finding an n-vector $\boldsymbol{a}$ such that $\boldsymbol{a}\boldsymbol{x}_i \geq 0$ for as many of the $\boldsymbol{x}_i$ vectors as possible (see Amaldi (2003)). Constructing one homogeneous inequality for each $\boldsymbol{x}_i$ vector converts this directly to the MAX FS problem.

12 Epilogue

Until recently, the main focus of algorithmic and computational work in optimization was firmly on finding optimum solutions. Issues of feasibility and infeasibility received much less attention, and even then mostly in the context of finding a feasible point as a preliminary step en route to the optimum. But the situation has changed markedly in the last twenty years or so. Computing power has increased dramatically and become inexpensive. As a consequence, optimization models have become larger and more complex, and hence more prone to modeling errors resulting in infeasibility, as well as difficulties in finding feasible solutions. New algorithms and computational methods have been developed in response to both problems. The algorithms for analyzing infeasibility have found unexpected applications outside of their original purpose. In addition, new application areas that pose challenging problems of feasibility and infeasibility have arisen. Examples include computational biology (e.g. Sec. 11.2) and medicine (e.g. Sec. 11.1), network security, and data analysis (Chap. 10), among others.

In short, the time is ripe for a summary of research on algorithms and computational methods related to feasibility and infeasibility in optimization, which is of course the point of this book. The field continues to be extremely active, with new developments almost daily, and much fascinating basic research remains to be done. Examples include ways to choose the best remedial actions in the face of infeasibility (delete constraints? shift constraints? reverse inequalities?), better infeasibility analysis for NLPs and MIPs, better feasibility-seeking methods for MIPs and mixed-integer nonlinear programs, etc. A particularly exciting development is the ongoing integration of ideas and methods from the mathematical programming and the constraint programming communities. This interaction continues to be extremely fruitful, resulting in powerful new hybrid methods; see for example the stimulating new book *Integrated Methods for Optimization* (Hooker 2007).

It is my hope that this book will serve as a valuable reference for researchers, practitioners and graduate students into the future.

References

Aardal K, Bixby RE, Hurkens CAJ, Lenstra AK, Smeltink JW (2000) Market Split and Basis Reduction: Towards a Solution of the Cornuéjols–Dawande Instances, INFORMS Journal on Computing 12:192–202.

Abadie J, Carpentier J (1969) Generalization of the Wolfe Reduced Gradient Method to the Case of Nonlinear Constraints, in Fletcher R (ed.), Optimization: 37–47, Academic Press, London.

Achterberg T (2007) Conflict Analysis in Mixed Integer Programming, Discrete Optimization 4:4–20.

Achterberg M, Berthold T (2005) Improving the Feasibility Pump, technical report 5–42, Konrad–Zuse–Zentrum für Informationstechnik Berlin.

Achterberg T, Koch T, Martin A (2006) MIPLIB 2003, Operations Research Letters 34:1–12.

Adem J, Gochet W (2006) Mathematical Programming Based Heuristics for Improving LP–Generated Classifiers for the Multiclass Supervised Classification Problem, European Journal of Operational Research 168:181–199.

Aggarwal C, Ahuja R, Hao J, Orlin JB (1998) Diagnosing Infeasibilities in Network Flow Problems, Mathematical Programming 81:263–280.

Agmon S (1954) The Relaxation Method for Linear Inequalities, Canadian Journal of Mathetmatics 6:382–392.

Amaldi E (1994) From Finding Maximum Feasible Subsystems of Linear Systems to Feedforward Neural Network Design, PhD thesis no. 1282, Département de Mathématiques, École Polytechnique Fédérale de Lausanne, Switzerland.

Amaldi E (2003) The maximum feasible subsystem problem and some applications, in Agnetis A, Di Pillo G, Eds., Modelli e Algoritmi per l'Ottimizzazione di Sistemi Complessi, Pitagora Editrice Bologna, 31–69.

Amaldi E, Belotti P, Hauser R (2005) Randomized relaxation methods for the maximum feasible subsystem problem, Proceedings of the 14th Integer Programming and Combinatorial Optimization conference (IPCO'05), Lecture Notes in Computer Science 3509, Springer–Verlag, 249–264.

Amaldi E, Bruglieri M, Casale G (2007) A Two–Phase Relaxation–Based Heuristic for the Maximum Feasible Subsystem Problem, Computers and Operations Research, to appear (available online now).

Amaldi E, Kann V (1995) The Complexity and Approximability of Finding Maximum Feasible Subsystems of Linear Relations, Theoretical Computer Science 147:181–210.

Amaldi E, Kann V (1998) On the Approximability of Minimizing Nonzero Variables or Unsatisfied Relations in Linear Systems, Theoretical Computer Science 209:237–260.

Amaldi A, Mattavelli M (2002) The MIN PFS Problem and Piecewise Linear Model Estimation, Discrete Applied Mathematics 118:115–143.

Amaldi E, Pfetsch M, Trotter Jr. L (1999) Structural and Algorithmic Properties of the Maximum Feasible Subsystem Problem, Proceedings of the Integer Programming and Combinatorial Optimization conference (IPCO'99), Lecture Notes in Computer Science 1610, Springer–Verlag, New York, NY, 45–59.

Amaral P (2001) Contribuições Para o Estudo de Sistemas Lineares Inconsistentes, PhD Disertation, Faculty of Science and Technology, UNL, Lisbon, Portugal (in Portuguese).

Amaral P, Barahona P (2005) Connections Between the Total Least Squares and the Correction of an Infeasible System of Linear Inequalities, Linear Algebra and its Applications 395:191–210.

Amaral P, Barahona P (2005a) A Framework for Optimal Correction of Inconsistent Linear Constraints, Constraints 10:67– 86.

Amaral P, Júdice J, Sherali HD (2006) A Reformulation–Linearization–Convexification Algorithm for Optimal Correction of an Inconsistent System of Linear Constraints, Computers and Operations Research, to appear (available online now).

Amarger RJ, Biegler LT, Grossmann IE (1992) An Automated Modelling and Reformulation System for Design Optimization, Computers and Chemical Engineering 16:623–636.

Andersen ED, Andersen KD (1995) Presolving in Linear Programming, Mathematical Programming 71:221–245.

Andersen ED, Gondzio J, Mészáros C, Xu X (1996) Implementation of interior point methods for large scale linear programming, in Terlaky T (ed.), Interior Point Methods in Mathematical Programming, Kluwer Academic Publishers: 189–252.

Atlihan MK, Schrage L (2006) Generalized Filtering Algorithms for Infeasibility Analysis, Computers and Operations Research, to appear (available online now).

Bailey J, Stuckey PJ (2005) Discovery of Minimal Unsatisfiable Subsets of Constraints Using Hitting Set Dualization, in Hermenegildo M (ed.), Proceedings of the International Conference of Practical Applications of Declarative Languages, Lecture Notes on Computer Science 3350:174–186, Springer–Verlag.

Bakker RR, Dikker F, Tempelman F, Wognum PM (1993) Diagnosing and solving over–determined constraint satisfaction problems, in Proceedings of IJCAI'93, 276–281.

Balas E, Ceria S, Dawande M, Margot F, Pataki G (2001) OCTANE: a New Heuristic for Pure 0–1 Programs, Operations Research 49:207–225.

Balas E, Martin C (1980) Pivot and Complement–a Heuristic for 0–1 Programming, Management Science 26:224–234.

Balas E, Martin C (1986) Pivot and Shift–a Heuristic for Mixed Integer Programming, GSIA Technical Report, Carnegie Mellon University.

Balas E, Ng SM (1989) On the set covering polytope: I. All the facets with coefficients in {0,1,2}, Mathematical Programming 43:57–69.

Balas E, Schmieta S, Wallace C (2004) Pivot and Shift–a Mixed Integer Programming Heuristic, Discrete Optimization 1:3–12.

Banerjee I, Ierapetritou MG (2005) Feasibilit Evaluation of Nonconvex Systems Using Shape Reconstruction Techniques, Industrial and Engineering Chemistry Research 44:3638–3647.

Bartak R (1999) Constraint Programming: In Pursuit of the Holy Grail, Proceedings of the 8th Annual Conference of Doctoral Students WDS'99.

Beale EML, Tomlin JA (1970) Special facilities in a general mathematical programming system for nonconvex problems using ordered sets of variables, Proceedings of the Fifth International Conference on Operational Research, Tavistock publication, London, 447–454.

Bemporad A, Garulli A, Paoletti S, Vicino A (2005) A Bounded–Error Approach to Piecewise Affine System Identification, IEEE Transactions on Automatic Control 50:1567–1580.

Benichou M, Gauthier JM, Girodet P, Hentges G, Ribiere G, Vincent O (1971). Experiments in mixed–integer linear programming, Mathematical Programming 1:76–94.

Bennett KP, Bredensteiner E (1997) A Parametric Optimization Method for Machine Learning, INFORMS Journal on Computing 9:311–318.

Berbee HCP, Boender CGE, Rinooy Kan AHG, Scheffer CL, Smith RL, and Telgen J (1987) Hit–and–Run Algorithms for the Identification of Nonredundant Linear Inequalities, Mathematical Programming 37:184–207.

Bertacco L, Fischetti M, Lodi A (2005) A Feasibility Pump Heuristic for General Mixed–Integer Problems, technical report OR–05–5, D.E.I.S. Operations Research Group, Università di Bologna.

Berthold T (2006) Primal Heuristics for Mixed Integer Programs, master's thesis, Technischen Universität Berlin.

Bixby R, Ceria S, McZeal CM, Savelsbergh MWP (1996) MIPLIB 3.0, World Wide Web http://www.caam.rice.edu/~bixby/miplib/miplib.html.

Boman EG (1999), Infeasibility and Negative Curvature in Optimization, PhD thesis, Scientific Computing and Computational Mathematics, Stanford University.

Bonami P, Cornuejols G, Lodi A, Margot F (2006) A Feasibility Pump for Mixed Integer Nonlinear Programs, IBM Research Report RC23862 (W0602– 029).

Boneh A (1983) PREDUCE–a Probabilistic Algorithm Identifying Redundancy by a Random Feasible Point Generator (RFPG) in Karwan, Lotfi, Telgen, Zionts (eds.), Lecture Notes in Economics and Mathematical Systems 206.

Bongartz I, Conn AR, Gould N, Toint PL (1995) CUTE: constrained and unconstrained testing environment, ACM Transactions on Mathematical Software 21:123–160. See http://www.sor.princeton.edu/~rvdb/ampl/nlmodels/cute/index.html for CUTE models in AMPL format.

Bonner & Moore (1979) RPMS (Refinery and Petrochemical Modeling System): a System Description, Bonner & Moore Management Science, Houston.

Boussemart F, Hemery F, Lecoutre C, Sais L (2004) Boosting systematic search by weighting constraints, in Proceedings of the 16th European Conference on Artificial Intelligence (ECAI'04), 146–150.

Bordetski AB, Kazarinov LS (1981) Determining the Committee of a System of Weighted Inequalities, Kibernetika 6:44–48.

Brearly AL, Mitra G, Williams (1975) Analysis of Mathematical Programming Problems Prior to Applying the Simplex Algorithm. Mathematical Programming 8:54–83.

Bremner D, Fukuda K, Rosta V (2006) Primal–Dual Algorithms for Data Depth, in Liu RY (ed.), Data Depth: Robust Multivariate Analysis, Computational Geometry and Applications, DIMACS series in Discrete Mathematics and Theoretical Computer Science 72:171–194.

Brown G, Graves G (1975) Elastic Programming: A New Approach to Large–Scale Mixed Integer Optimization, presented at ORSA/TIMS Conference, Las Vegas.

Bruni R (2005) On Exact Selection of Minimally Unsatisfiable Subformulae, Annals of Mathematics and Artificial Intelligence 43: 35–50.

Bruni R (2005a) Error Correction for Massive Data Sets, Optimization Methods and Software 20:295–314.

Bruni R, Reale A, Torelli R (2001) Optimization Techniques for Edit Validation and Data Imputation, Proceedings of Statistics Canada Symposium 2001, Ottawa.

Bruynooghe M (1981) Solving Combinatorial Search Problems by Intelligent Backtracking, Information Processing Letters 12:36–39.

Burg J, Lang SD, Hughes CE (1994) Finding Conflict Sets and Backtrack Points in CLP($\Re$), in Van Hentenryck P (ed.), Proceedings of the 11th International Logic Programming Conference, MIT Press, 323–338.

Byrne C, Censor Y (2001) Proximity Function Minimization Using Multiple Bregman Projections, with Applications to Split Feasibility and Kullback–Leibler Distance Minimization, Annals of Operations Research 105:77–98.

Caron RJ, Greenberg HJ, Holder AG (2002) Analytic centers and repelling inequalities, European Journal of Operational Research 143:268–290.

Carver WB (1921) Systems of Linear Inequalities, Annals of Mathematics 23, series 2:212–220.

Censor Y (2003) Mathematical Optimization for the Inverse Problem of Intensity Modulated Radiation Therapy, in Palta JR, Mackie TR (eds.), Intensity–Modulated Radiation Therapy: The State of The Art, American Association of Physicists in Medicine, Medical Physics Monograph No. 29, Medical Physics Publishing, Madison, Wisconsin, 25–49.

Censor Y (2006) Computational Acceleration of Projection Algorithms for the Linear Best Approximation Problem, Linear Algebra and its Applications 416:111–123.

Censor Y, Ben–Israel A, Xiao Y, Galvin JM (2006) On Linear Infeasibility Arising in Intensity–Modulated Radiation Therapy Inverse Planning, working paper, University of Haifa, Israel.

Censor Y, Elfving T (1982) New methods for linear inequalities, Linear Algebra and its Applications 42:199–211.

Censor Y, Elfving T, Herman, GT (2001) Averaging Strings of Sequential Iterations for Convex Feasibility Problems, in Butnariu D, Censor Y, Reich S, eds. Inherently Parallel Algorithms in Feasibility and Optimization and their Applications, Elsevier Science B.V., Amsterdam. 101–113.

Censor Y, Gordon D, and Gordon R (2001) Component Averaging: An Efficient Iterative Parallel Algorithm for Large and Sparse Unstructured Problems, Parallel Computing 27:777–808.

Censor Y, Zenios SA (1997) Parallel Optimization: Theory, Algorithms, and Applications. Oxford University Press, New York.

Chakravarti N (1994) Some Results Concerning Post–Infeasibility Analysis, European Journal of Operational Research 73:139–143.

Charnes A, Cooper WW (1961) Management Models and Industrial Applications of Linear Programming, John Wiley and Sons, New York.

Chen D (2007) A Branch and Cut Algorithm for the Halfspace Depth Problem, MCS thesis, the University of New Brunswick, Canada.

Chen XB, Kostreva MM (1999) Global Convergence Analysis of Algorithms for Finding Feasible Points in Norm–Relaxed MFD, Journal of Optimization Theory and Applications 100:287–309.

Chinneck JW (1990) MINOS(IIS) software, World Wide Web, http://www.sce.carleton.ca/faculty/chinneck/minosiis.html, 1990–2006.

Chinneck JW (1990a) Formulating Processing Networks: Viability Theory, Naval Research Logistics 37:245–261.

Chinneck JW (1990b) VIABLE1—Code for Identifying Nonviabilities in Processing Network Models, European Journal of Operational Research 44:119–120.

Chinneck JW (1992) Viability Analysis: A Formulation Aid for All Classes of Network Models, Naval Research Logistics 39:531–543.

Chinneck JW (1993) Netlib Repository of Infeasible LP Instances, World Wide Web http://www.netlib.org/lp/infeas/.

Chinneck, JW (1994), MINOS(IIS): Infeasibility Analysis Using MINOS, Computers and Operations Research 21:1–9.

Chinneck, JW (1995) Analyzing Infeasible Nonlinear Programs, Computational Optimization and Applications 4:167–179.

Chinneck JW (1996a) "Computer Codes for the Analysis of Infeasible Linear Programs", Journal of the Operational Research Society 47:61–72.
Chinneck JW (1996b) Localizing and Diagnosing Infeasibilities in Networks, ORSA Journal on Computing 8:55–62.
Chinneck, JW (1996c) An Effective Polynomial–Time Heuristic for the Minimum–Cardinality IIS Set–Covering Problem, Annals of Mathematics and Artificial Intelligence 17:127–144.
Chinneck JW (1997a) Feasibility and Viability, in Advances in Sensitivity Analysis and Parametric Programming, Gal T, Greenberg HJ (eds.), International Series in Operations Research and Management Science. 6:14–1 to 14–41, Kluwer Academic Publishers.
Chinneck JW (1997b), Finding a Useful Subset of Constraints for Analysis in an Infeasible Linear Program, INFORMS Journal on Computing 9:164–174.
Chinneck JW (1998) Improved linear classification via LP infeasibility analysis, Technical Report SCE–98–09, Department of Systems and Computer Engineering, Carleton University, Ottawa, Canada.
Chinneck JW (2001) Analyzing Mathematical Programs using MProbe, Annals of Operations Research 104:33–48.
Chinneck JW (2001a) Fast Heuristics for the Maximum Feasible Subsystem Problem, INFORMS Journal on Computing 13:210–223.
Chinneck JW (2002) Discovering the Characteristics of Mathematical Programs via Sampling", Optimization Methods and Software 17:319–352.
Chinneck JW (2002a) Guest Editor, Special Issue on The Merging of Mathematical Programming and Constraint Programming, INFORMS Journal on Computing 14.
Chinneck JW (2004) The Constraint Consensus Method for Finding Approximately Feasible Points in Nonlinear Programs, INFORMS Journal on Computing 16:255–265.
Chinneck JW, Dravnieks EW (1991) Locating Minimal Infeasible Constraint Sets in Linear Programs, ORSA Journal on Computing 3:157–168.
Chinneck JW, Michalowski M (1996) MOLP Formulation Assistance Using LP Infeasibility Analysis, in Tamiz M (ed.) Multi–Objective Programming and Goal Programming: Theories and Applications, Lecture Notes in Economics and Mathematical Systems, 432:87–106.
Chinneck JW, Saunders, MA (1995) MINOS(IIS) Version 4.2: Analyzing Infeasibilities in Linear Programs, European Journal of Operational Research 81:217–218.
Chvátal, V (1983) Linear Programming, W.H. Freeman and Company, New York.
Cimmino G (1938) Calcolo Approssimato per Soluzioni dei Sistemi di Equazioni Lineari. La Ricerca Scientifica XVI, Series II, Anno IX 1:326–333.
Codato G, Fischetti M (2004) Combinatorial Benders' Cuts, Proceedings of IPCO, Lecture Notes in Computer Science 3064:178–195.
Codato G, Fischetti M (2006) Combinatorial Benders' Cuts for Mixed–Integer Linear Programming, Operations Research 54:756–766.
Conrad J, Gomes CP, van Hoeve WJ, Sabharwal A, Suter J (2007) Connections in Networks: Hardness of Feasibility vs. Optimality, in Van Hentenryck P, Wolsey L (eds.), Integration of AI and OR Techniques in Constraint Programming for Combinatorial Optimization Problems, Proceedings of the 4th International Conference CPAIOR 2007, Springer, Lecture Notes in Computer Science 4510:16–28.
Cornuéjols G, Dawande M (1998) A Class of Hard Small 0–1 Programs, in Bixby RE, Boyd EA, Ríos–Mercado RZ (eds.), Integer Programming and Combinatorial Optimization, 6th International IPCO Conference, Lecture Notes in Computer Science 1412:284–293, Springer–Verlag, Berlin.
Crowder H, Johnson EL, Padberg M (1983) Solving Large–Scale Zero–One Linear Programming Problems, Operations Research 31: 803–834.

Dakin RJ (1965) A Tree Search Algorithm for Mixed Integer Programming Problems, Computer Journal 8:250–255.

Danna E, Rothberg E, Le Pape C (2005) Exploring Relaxation Induced Neighborhoods to Improve MIP Solutions, Mathematical Programming 102:71–90.

Dantzig GB, Thapa MN (1997) Linear Programming, 1: Introduction, Springer–Verlag, New York.

Dash Optimization (2006) XPRESS–Optimizer User Manual, Dash Optimization.

Davey B, Boland N, Stuckey PJ (2002) Efficient Intelligent Backtracking Using Linear Programming, INFORMS Journal on Computing 14:373–386.

Davis M, Logemann G, Loveland D (1962) A Machine Program for Theorem Proving, Communications of the ACM 5:394–397.

Davis M, Putnam H (1960) A Computing Procedure for Quantification Theory, Journal of the ACM 7:201–215.

Dax, A (2006) The l_1 Solution of Linear Inequalities, Computational Statistics and Data Analysis 50:40–60.

DeBacker B, Beringer H (1991) Intelligent Backtracking for CLP Languages, an Application to CLP(R), International Logic Programming Symposium, San Diego, 405–419.

De Backer B, Beringer H (1993) A CLP Language Handling Disjunctions of Linear Constraints, International Conference on Logic Programming, 550–563.

Debrosse CJ, Westerberg AW (1973) A Feasible–Point Algorithm for Structured Design Systems in Chemical Engineering, AIChE Journal 19:251–258.

Dechter R, Rossi F (2002) Constraint satisfaction, in Nadel L. (ed.), Encyclopedia of Cognitive Science, Nature Publishing Group, London, 2002.

de Givry S, Larrosa J, Meseguer P, Schiex T (2003) Solving Max–SAT as Weighted CSP, Principles and Practice of Constraint Programming CP 2003, Lecture Notes in Computer Science 2833:363–376.

De Pierro AR, Iusem AN (1985) A Simultaneous Projection Method for Linear Inequalities, Linear Algebra and its Applications 64:243–253.

de Siqueira N. JL, Puget JF (1988) Explanation–Based Generalisation of Failures. European Conference on Artificial Intelligence: 339–344.

Dravnieks EW (1989) Identifying Minimal Sets of Inconsistent Constraints in Linear Programs: Deletion, Squeeze and Sensitivity Filtering, MSc thesis, Systems and Computer Engineering, Carleton University.

Dravnieks EW, Chinneck JW (1997) Formulation Assistance for Global Optimization Problems, Computers and Operations Research 24:1151–1168.

Drud AS (1994) CONOPT–A Large Scale GRG Code, ORSA Journal on Computing 6:207–216.

Duran M, Grossmann IE (1986) An Outer–Approximation Algorithm for a Class of Mixed–Integer Nonlinear Programs, Mathematical Programming 36:307–339.

Dyer ME (1983) The Complexity of Vertex Enumeration Methods, Mathematics of Operations Research 8:381–402.

Eckstein J (1994) Parallel branch–and–bound algorithms for general mixed integer programming on the CM–5, SIAM Journal on Optimization 4:794–814.

Ellison EFD, Hajian M, Jones H, Levkovitz R, Maros I, Mitra G, Sayers D (1999) FortMP Manual, Numerical Algorithms Group and Brunel University.

Elwakeil OA, Arora JS (1995) Methods for Finding Feasible Points in Constrained Optimization, AIAA Journal 33:1715–1719.

Elwakeil OA, Arora JS (1996) Two algorithms for global optimization of general NLP problems, International Journal for Numerical Methods in Engineering 39:3305–3325.

Fan, K (1956) On Systems of Linear Inequalities, Annals of Mathematical Studies 38:99–156.

Feng, J (1999). Nonlinear Redundancy: Where is the Information? M.Sc., Mathematics, Department of Economics, Mathematics, and Statistics, University of Windsor, Canada.

Ferris MC, Dirkse SP, Meeraus A (2005) Mathematical programs with equilibrium constraints: Automatic reformulation and solution via constrained optimization, in Kehoe TJ, Srinivasan TN, Whalley J (eds.), Frontiers in Applied General Equilibrium Modeling, 67–93, Cambridge University Press.

Fischetti M, Glover F, Lodi A (2005) The Feasibility Pump, Mathematical Programming A 104:91–104.

Fischetti M, Lodi A (2003) Local Branching, Mathematical Programming B 98:23–49.

Ford LR, Fulkerson DR (1962) Flows in Networks, Princeton University Press, Princeton, NJ.

Fourer R, Gay DM, Kernighan BW (2003) AMPL: A Modeling Language for Mathematical Programming, Second Edition, Brooks/Cole, Pacific Grove, California, USA.

Fourer R, Orban D (2007) DrAmpl–A meta solver for optimization, technical report G–2007–10, GERAD, Montreal, Canada.

Fourier JBJ (1827) Analyse des Travaux de l'Académie Royale des Sciences Pendant l'Année 1824, Histoire de l'Académie Royale des Sciences de l'Institut de France 7:xlvii–lv.

Fränzle M, Herde C (2005) Efficient Proof Engines for Bounded Model Checking of Hybrid Systems, Electronic Notes in Theoretical Computer Science 133 :119–137.

Frean M (1992) A "Thermal" Perceptron Learning Rule, Neural Computation 4:946–957.

Freuder EC, Wallace RJ (1992) Partial Constraint Satisfaction, Artificial Intelligence 58: 21–70.

Frontline Systems Inc. (2007) World Wide Web http://www.solver.com/sdkplatformd.htm# Diagnosing%20Infeasibility.

Fukuda K, Rosta V (2005) Data Depth and Maximum Feasible Subsystems, in Avis D, Hertz A, Marcotte O (eds.), Graph Theory and Combinatorial Optimization, Springer 37–67.

Fulkerson DR (1959) A Network Flow Feasibility Theorem and Combinatorial Applications, Canadian Journal of Mathematics 11:440–451.

Fylstra D, Lason L, Watson J, Waren A (1998) Design and Use of the Microsoft Excel Solver, Interfaces 28:29–55.

Gale D (1957) A Theorem in Networks, Pacific Journal of Mathematics 7:1073–1082.

Ganapathy V, Jha S, Chandler D, Melski D, Vitek D (2003) Buffer Overrun Detection Using Linear Programming and Static Analysis, Proceedings of the 10th ACM Conference on Computer and Communications Security: 345–354.

Gauthier JM, Ribiere G (1977) Experiments in mixed–integer linear programming, Mathematical Programming 12:26–47.

Gertz M, Nocedal J, Sartenaer A (2004) A Starting–Point Strategy for Nonlinear Interior Methods, Applied Mathematics Letters 17:945–952.

Gill PE, Murray W, Saunders MA (2005) SNOPT : an SQP Algorithm for Large–Scale Constrained Optimization, SIAM Review 47:99–131.

Gleeson J, Ryan J (1990) Identifying Minimally Infeasible Subsystems of Inequalities, ORSA Journal on Computing 2:61–63.

Glover F (1968) Surrogate Constraints, Operations Research 16:741–749.

Glover F (1990) Tabu Search: A Tutorial, Interfaces, 20:74–94.

Glover F (2003) Tutorial on Surrogate Constraint Approaches for Optimization in Graphs, Journal of Heuristics 9:175–227.

Glover F, Laguna M, Martí R (2000) Fundamentals of Scatter Search and Path Relinking, Control and Cybernetics 29:653–684.
Glover F, Laguna M, Martí R (2003) Scatter search and path relinking: Advances and applications", in Glover FW, Kochenberger GA (eds.), Handbook of Metaheuristics, International Series in Operations Research & Management Science 57:1–36, Kluwer Academic Publishers, Boston.
Glover F, Laguna M, Martí R (2004) New Ideas and Applications of Scatter Search and Path Relinking, in New Optimization Technologies in Engineering, Onwubolu GC, Babu BV (eds.), Studies in Fuzziness and Soft Computing 141:367–384, Springer.
Goyal V, Ierapetritou MG (2003) Framework for Evaluating the Feasibility/Operability of Nonconvex Processes, American Institute of Chemical Engineering Journal 49:1233–1240.
Grant M (2004) Disciplined Convex Optimization, PhD thesis, Electrical Engineering, Stanford University.
Grant M, Boyd S, Ye Y (2006) Disciplined Convex Programming, in Liberti L, Maculan N (eds.) Global Optimization: From Theory to Implementation, Nonconvex Optimization and its Applications 84:155–210, Springer.
Greenberg HJ (1978) Pivot Selection Tactics, in Greenberg HJ (ed.), Design and Implementation of Optimization Software, Sijthoff and Noordhoff:143–174.
Greenberg HJ (1981a) The Scope of Computer–Assisted Analysis and Model Simplification, In: Greenberg HJ, Maybee JS (eds.), Computer–Assisted Analysis and Model Simplification, Academic Press, New York: 17–26.
Greenberg HJ (1981b) Implementation Aspects of Model Management: A Focus on Computer–Assisted Analysis, In: Greenberg, HJ, Maybee JS (eds.), Computer–Assisted Analysis and Model Simplification, Academic Press, New York: 455–479.
Greenberg HJ (1983) A Computer–Assisted Analysis System for Linear Programming Models, ACM Transactions on Mathematical Software 9:18–56.
Greenberg HJ (1987a) Computer–Assisted Analysis for Diagnosing Infeasible or Unbounded Linear Programs, Mathematical Programming Studies 31:79–97.
Greenberg HJ (1987b) Diagnosing Infeasibility in Min–cost Network Flow Problems; Part I: Dual Infeasibility, IMA Journal of Mathematics in Management 1:99–109.
Greenberg HJ (1987c) The Development of an Intelligent Mathematical Programming System, WORMSC Proceedings, Washington, D.C., November.
Greenberg HJ (1988) Diagnosing Infeasibility in Min–cost Network Flow Problems; Part II: Primal Infeasibility, IMA Journal of Mathematics in Management 2:39–50.
Greenberg HJ (1991) An Industrial Consortium for the Development of an Intelligent Mathematical Programming System, Interfaces 20:88–93.
Greenberg HJ (1992) An Empirical Analysis of Infeasibility Diagnosis for Instances of Linear Programming Blending Models, IMA Journal of Mathematics in Business & Industry 4:163–210.
Greenberg HJ (1993) How to Analyze the Results of Linear Programs–Part 3: Infeasibility Diagnosis, Interfaces 23:120–139.
Greenberg HJ (1993a) A Computer–Assisted Analysis System for Mathematical Programming Models and Solutions: A User's Guide for ANALYZE, Kluwer Academic Publishers, Boston.
Greenberg HJ (1996a) Consistency, Redundancy, and Implied Equalities in Linear Systems, Annals of Mathematics and Artificial Intelligence 17:37–83.
Greenberg HJ (1996b) A bibliography for the development of an intelligent mathematical programming system, Annals of Operations Research 65:55–90
Greenberg HJ (2003) Mathematical Programming Glossary Supplement: Tolerances, World Wide Web http://glossary.computing.society.informs.org/notes/tolerances.pdf.

Greenberg HJ (2003a) Mathematical Programming Glossary Supplement: Convex Cones, Sets, and Functions, World Wide Web http://glossary.computing.society.informs.org/notes/convexity.pdf.

Greenberg HJ, Murphy FH (1991) Approaches to Diagnosing Infeasibility for Linear Programs, ORSA Journal on Computing 3:253–261.

Greenberg HJ, Pierskalla WP (1971) A Review of Quasi–Convex Functions, Operations Research 19:1553–1570.

Guieu O, Chinneck JW (1999) Analyzing Infeasible Mixed–Integer and Integer Linear Programs", INFORMS Journal on Computing 11:63–77.

Gupta P, Vlach M, Bhatia D (2004) Fuzzy Approximation to an Infeasible Generalized Linear Complementarity Problem, Fuzzy Sets and Systems 146:221–233.

Han SP (1980) Least–squares solution of linear inequalities. Technical Report 2141, Mathematics Research Center, University of Wisconsin–Madison.

Heath D, Kasif S, Salzburg S (1993) Learning Oblique Decision Trees, Proceedings of the 13th International Conference on Artificial Intelligence, Chambery, France, Morgan Kaufmann, San Mateo, CA, 1002–1007.

Hemery F, Lecoutre C, Sais L, Boussemart F (2006) Extracting MUCs from Constraint Networks, in Proceedings of the 17th European Conference on Artificial Intelligence (ECAI'2006), 113–117.

Holder A (2004) Radiotherapy Treatment Design and Linear Programming, Operations Research and Health Care: A Handbook of Methods and Applications, Brandeau ML, Sainfort F, Pierskalla WP (eds.), Chap. 29, Kluwer Academic Publishers.

Holder A (2006) Mathematical Programming Glossary, INFORMS Computing Society, World Wide Web, http://glossary.computing.society.informs.org/.

Holmström K, Göran AO, Edvall MM (2006) User's Guide for Tomlab/Xa V14, Tomlab Optimization Inc.

Holzbaur C, Menezes F, Barahona P (1996) Defeasibility in CLP(Q) Through Generalized Slack Variables, Principles and Practice of Constraint Programming–CP 96, Lecture Notes in Computer Science 1118:209–223.

Hoffman AJ (1960) Some Recent Applications of the Theory of Linear Inequalities to Extremal Combinatorial Analysis, Proceedings of Symposia on Applied Mathematics 10.

Hooker JN (2007) Integrated Methods for Optimization, Springer Science+Business Media LLC, New York.

Huitzing HA, Veldkamp BP, Verschoor AJ (2005) Infeasibility in Automated Test Assembly Models: A Comparison Study of Different Methods, Journal of Educational Measurement 42:223–243.

Ibrahim W, Chinneck JW (2005) Improving Solver Success in Reaching Feasibility for Sets of Nonlinear Constraints, Computers and Operations Research, to appear (available online at www.sciencedirect.com).

Ignizio JP, Cavalier TM (1994) Linear Programming, Prentice Hall, Englewood Cliffs.

Ilog (2006) Cplex software, World Wide Web http://www.ilog.com/products/cplex/.

John E, Yildirim EA (2006) Implementation of warm–start strategies in interior–point methods for linear programming in fixed dimension, Computational Optimization and Applications, to appear.

Johnson EL, Nemhauser GL, Savelsbergh MWP (2000) Progress in Linear Programming–Based Algorithms for Integer Programming: An Exposition, INFORMS Journal on Computing 12:2–23.

Jokar S, Pfetsch ME (2007) Exact and Approximate Sparse Solutions of Underdetermined Linear Equations, Konrad–Zuse–Zentrum für Informationstechnik Berlin, technical report 07–05.

Jones DR, Schonlau M, Welch WJ (1998) Efficient global optimization of expensive black–box functions, Journal of Global Optimization 13:455–492.

Juloski AL, Heemels WPMH, Ferrari–Trecate G, Vidal R, Paoletti S, Niessen JHG (2005) Comparison of Four Procedures for the Identification of Hybrid Systems, Lecture Notes in Computer Science 3414:354–369, Springer–Verlag, Berlin.

Junker U (2001) Quickxplain: Conflict detection for arbitrary constraint propagation algorithms, in IJCAI–2001 Workshop on Modeling and Solving Problems with Constraints, 75–82.

Kaczmarz S (1937) Angenäherte Auflösung von Systemen Linearer Gleichungen, Bulletin de l'Académie Polonaise des Sciences et Lettres, A35:355–357.

Kirkpatrick S., Gelatt Jr. CD, Vecchi MP (1983) Optimization by Simulated Annealing, Science 220:671–680.

Koene J (1982) Minimal Cost Flow in Processing Networks, a Primal Approach, CWI Tract 4.

Kumar V (1992) Algorithms for Constraint–Satisfaction Problems: a Survey, AI Magazine, Spring 1992:32–44.

Kurator WG, O'Neill RP (1980) PERUSE: An Interactive System for Mathematical Programs, ACM Transactions on Mathematial Software 6:489–509.

Lagoudakis MG, Littman ML (2001) Learning to Select Branching Rules in the DPLL Procedure for Satisfiability, Electronic Notes in Discrete Mathematics 9, LICS 2001 Workshop on Theory and Applications of Satisfiability Testing (SAT 2001), Boston, MA, June 14–15, 2001.

Laguna M, Martí R (2005) Experimental Testing of Advanced Scatter Search Designs for Global Optimization of Multimodal Functions, Journal of Global Optimization 33:235–255.

Land AH, Doig AG (1960) An Automatic Method for Solving Discrete Programming Problems, Econometrica 28:497–520.

Lasdon LS (1970) Optimization Theory for Large Systems, Macmillan Company, New York.

Lasdon L, Plummer J (2006) Multistart Algorithms for Seeking Feasibility, Computers and Operations Research, to appear (available online now).

Lasdon L, Plummer J, Ugray Z, Bussieck M (2004) Improved filters and randomized drivers for multi–start global optimization, McCombs School of Business Research Paper Series No. IROM–06–06, University of Texas at Austin.

Lasdon L, Waren AD (1978) Generalized Reduced Gradient Software for Linearly and Nonlinearly Constrained Problems, in Greenberg HJ (ed.), Design and Implementation of Optimization Software, Sijthoff and Noordhoff.

Lawrence CT, Tits AL (2001) A Computationally Efficient Feasible Sequential Quadratic Programming Algorithm, SIAM Journal on Optimization 11:1092–1118.

Lee EK, Gallagher RJ, Zaider M (1999) Planning Implants of Radionuclides for the Treatment of Prostate Cancer: An Application of Mixed Integer Programming, OPTIMA Mathematical Programming Society Newsletter 61:1–7.

León T, Liern V (2001) A Fuzzy Method to Repair Infeasibility in Linearly Constrained Problems, Fuzzy Sets and Systems 122:237–243.

Liffiton MH, Sakallah KA (2005) On Finding All Minimally Unsatisfiable Subformulas, Proceedings of the 8th International Conference on Theory and Applications of Satisfiability Testing (SAT–2005):173–186, June.

Lim J, Ferris MC, Shepard DM, Wright SJ, Earl MA (2006) An Optimization Framework for Conformal Radiation Treatment Planning, INFORMS Journal On Computing, to appear.

Linderoth JT, Savelsbergh MWP (1999) A computational study of search strategies for Mixed Integer Programming, INFORMS Journal on Computing 11:173–187.

Lingo Systems Inc. (2007) LINGO, World Wide Web http://www.lindo.com/products/lingo/lingom.html.

Lustig IJ, Puget JF (2001) Program Does Not Equal Program: Constraint Programming and its Relationship to Mathematical Programming, Interfaces 31:29–53.

MacLeod M (2006) Multistart Constraint Consensus for Seeking Feasibility in Nonlinear Programs, MASc thesis, Systems and Computer Engineering, Carleton University, Ottawa, Canada.

MacLeod M, Chinneck JW (2007) Multistart Constraint Consensus for Seeking Feasibility in Nonlinear Programs, technical report, Systems and Computer Engineering, Carleton University.

Main RA (1993a) Infeasibility Analysis Using CLAUDIA–I, BP Oil International, Oil Technology Centre, technical report.

Main RA (1993b) Infeasibility Analysis Using CLAUDIA–II. BP Oil International, Oil Technology Centre, technical report.

Mammen DL, Hogg T (1997) A New Look at the Easy–Hard–Easy Pattern of Combinatorial Search Difficulty, Journal of Artificial Intelligence Research 7:47–66.

Mangasarian OL (1993) Mathematical Programming in Neural Networks, ORSA Journal on Computing 5:349–360.

Mangasarian OL (1994) Misclassification Minimization, Journal of Global Optimization 5:309–323.

Mangasarian OL (1996) Machine Learning via Polyhedral Concave Minimization, Applied Mathematics and Parallel Computing, in Fischer H, Riedmueller B, Schaeffler S (eds.), Physical–Verlag:175–188.

Mangasarian OL (1999) Minimum–Support Solutions of Polyhedral Concave Programs, Optimization 45:149–162.

McCarl B (1998) Repairing Misbehaving Mathematical Programming Models: Concepts and a GAMS–Based Approach, Interfaces 28:124–138.

Meller J, Wagner M, Elber R (2002) Maximum Feasibility Guideline to the Design and Analysis of Protein Folding Potentials, Journal of Computational Chemistry 23:111–118.

Meneses CN, Pardalos PM, Resende MGC (2005) GRASP for nonlinear optimization, Technical Report TD–6DUTRG, AT&T Labs Research, Florham Park, NJ, June.

Meseguer P, Bouhmala N, Bouzoubaa T, Irgens M, Sánchez M (2003) Current Approaches for Solving Over–Constrained Problems, Constraints 8:9–39.

Mészáros Cs, Suhl UH (2003) Advanced Preprocessing Techniques for Linear and Quadratic Programming, Operations Research Spectrum 25:575–595.

Michalewicz Z, Logan TD, Swaminathan S (1994) Evolutionary Operators for Continuous Convex Parameter Spaces, in Proceedings of the 3rd Annual Conference on Evolutionary Programming, Sebald AV, Fogel LJ (eds.), World Scientific Publishing, River Edge, NJ, 84–97.

Michalewicz Z, Nazhiyath G (1995) Genocop III: a co–evolutionary algorithm for numerical optimization problems with nonlinear constraints, IEEE International Conference on Evolutionary Computation 1995, 2:647–651.

Michalowski W, Szapiro T (1992) A Bi–reference Procedure for Interactive Multiple Criteria Programming, Operations Research 40:247–258.

Miguel I (2001) Dynamic Flexible Constraint Satisfaction and Its Application to AI Planning, PhD Thesis, University of Edinburgh.

Mitchell D, Selman B, Levesque H (1992) Hard and Easy Distributions of SAT Problems, Proceedings of the 10th Annual Conference on Artificial Intelligence AAAI–92:459–465.

Mitra G, Tamiz M (1988) FortLP Reference Manual, NAG Ltd.

Motzkin TS (1936) Beiträge zur Theorie der linearen Ungleichungen, Ph.D. thesis, Azriel, Jerusalem.

Motzkin TS, Schoenberg JJ (1954) The Relaxation Method for Linear Inequalities, Canadian Journal of Mathematics 6:393–404.

Murtagh BA, Saunders MA (1987) MINOS 5.1 User's Guide, technical report SOL 83–20R, Systems Optimization Laboratory, Department of Operations Research, Stanford University.

Murthy S, Kasif S, Salzberg S(1994) A System for induction of oblique decision trees, Journal of Artificial Intelligence Research 2:1–32.

Murty KG (1983) Linear Programming, John Wiley & Sons, New York.

Murty KG, Kabadi SN, Chandrasekaran R (2000) Infeasibility Analysis for Linear Systems, a Survey, Arabian Journal of Science and Technology 25:3–18.

Nadel A (2002) Backtrack Search Algorithms for Propositional Logic Satisfiability: Review and Innovations, master's thesis, Hebrew University of Jerusalem.

Nazareth JL (1987) Computer Solution of Linear Programs, Oxford University Press, New York.

Newman DJ, Hettich S, Blake CL, Merz CJ (1998) UCI Repository of machine learning databases [http://www.ics.uci.edu/~mlearn/MLRepository.html]. Irvine, CA: University of California, Department of Information and Computer Science.

Nemhauser GL, Savelsbergh MWP, Sigismondi GC (1994) MINTO: a Mixed INTeger Optimizer, Operations Research Letters 15:47–58.

Nemhauser GL, Wolsey LA (1988) Integer and Combinatorial Optimization, Wiley–Interscience Series in Discrete Mathematics and Optimization, John Wiley & Sons, New York.

Obuchowska WT (1998) Infeasibility Analysis for Systems of Quadratic Convex Inequalities, European Journal of Operational Research 107:633–643.

Obuchowska WT (1999) On Infeasibility of Systems of Convex Analytic Inequalities, Journal of Mathematical Analysis and Applications 234:223–245.

Ordonez F, Freund RM (2003) Computational Experience and the Explanatory Value of Condition Measures for Linear Optimization, SIAM Journal on Optimization 14:307–333.

Padberg M (1999) Linear Optimization and Extensions, 2nd edition, Springer–Verlag.

Pannell DJ (1997) Introduction to Practical Linear Programming, John Wiley and Sons Inc., New York.

Pardalos PM (1994) On the Passage from Local to Global in Optimization, in Birge JR, Murty KG (eds.), Mathematical Programming: State of the Art 1994, The University of Michigan.

Parker M (1995) A Set Covering Approach to Infeasibility Analysis of Linear Programming Problems and Related Issues, PhD thesis, University of Colorado at Denver.

Parker M, Ryan J (1996) Finding the Minimum Weight IIS Cover of an Infeasible System of Linear Inequalities, Annals of Mathematics and Artificial Intelligence 17:107–126.

Patel J, Chinneck JW (2006) Active–Constraint Variable Ordering for Faster Feasibility of Mixed Integer Linear Programs, Mathematical Programming, to appear.

Petit T, Regin JC, Bessiere C (2000) Meta–Constraints on Violations for Over Constrained Problems, Proceedings of the 12th IEEE International Conference on Tools with Artificial Intelligence 2000 (ICTAI 2000):358–365

Pfetsch ME (2002) The Maximum Feasible Subsystem Problem and Vertex–Facet Incidences of Polyhedra, PhD thesis, Dept. of Mathematics, Technischen Universität Berlin.

Pfetsch ME (2005) Branch–and–Cut for the Maximum Feasible Subsystem Problem, ZIB Report 05–46, Konrad–Zuse–Zentrum für Informationstechnik Berlin.

Pintér JD (1998) Continuous global optimization: An introduction to models, solution approaches, tests and applications, Interactive Transactions of ORMS 2, World Wide Web http://catt.bus.okstate.edu/itorms/pinter/.

Popescu E (2001) Use of the Interior–Point Method fort Correcting and Solving Inconsistent Linear Inequality Systems, Analele Stiinţifice ale Universităţii "Ovidius" Constanţa, Seria Matematică 9:65–72.

Press WH, Teukolsky SA, Vetterling WT, Flannery BP (1992) Numerical Recipes in C: The Art of Scientific Computing, Second Edition, Cambridge University Press, Cambridge.

Rardin RL (1998) Optimization in Operations Research, Prentice Hall, Upper Saddle River, New Jersey.

Renegar J (1994) Some Perturbation Theory for Linear Programming, Mathematical Programming 65:73–91.

Resende MGC, Ribeiro CC (2003a) Greedy randomized adaptive search procedures, in Glover FW, Kochenberger G (eds.), Handbook of Metaheuristics, International Series in Operations Research and Management Science 57:219–249, Kluwer Academic Publishers, Boston.

Resende MGC, Ribeiro CC (2003b) GRASP with path–relinking: Recent advances and applications, Technical Report TD–5TU726, AT&T Labs Research, December.

Riera–Ledesma J, Salazar–Gonzalez JJ (2007) A Branch–and–Cut Algorithm for the Continuous Error Localization Problem in Data Cleaning, Computers and Operations Research 34:2790–2804.

Rosenthal R (2007) GAMS — A User's Guide, GAMS Development Corporation, Washington, D.C., World Wide Web http://www.gams.com/docs/gams/GAMSUsersGuide.pdf.

Roodman GM (1979) Post–Infeasibility Analysis in Linear Programming, Management Science 25:916–922.

Russell S, Norvig P (2002) Artificial Intelligence, A Modern Approach, Second Edition, Prentice Hall.

Sadegh P (1999) A Maximum Feasible Subset Algorithm with Application to Radiation Therapy, Proceedings of the American Control Conference, San Diego, California: 405–408.

Sahinidis NV (1996) BARON: A general purpose global optimization software package, Journal of Global Optimization 8:201–205.

Sahindis NV (2000) BARON: Brand and Reduce Optimization Navigator User's Manual. Version 4.0.

Sandholm T, Shields R (2006) Nogood Learning for Mixed Integer Programming, CMU Computer Science Department technical report CMU–CS–06–155.

Sankaran JK (1993) A Note on Resolving Infeasibility in Linear Programs by Constraint Relaxation, Operations Research Letters 13:19–20.

Savelsbergh MWP (1994) Preprocessing and Probing Techniques for Mixed Integer Programming Problems, ORSA Journal on Computing 6:445–454.

Sepulveda AE, Epstein L (1996) The repulsion algorithm, a new multistart method for global optimization, Structural and Multidisciplinary Optimization 11:145–152.

Schrage L (1991) LINDO: An Optimization Modeling System 4th edition, The Scientific Press, San Francisco.

Scharge L (1997) Optimization Modeling with LINDO, Duxbury Press.

Sherali HD, Tuncbilek CH (1992) A Global Optimization Algorithm for Polynomial Programming Problems using a Reformulation–Linearization Technique, Journal of Global Optimization 2:101–112.

Smith S, Lasdon L (1992) Solving Large Sparse Nonlinear Programs Using GRG, ORSA Journal on Computing 4:2–15.

Steuer RE (1986) Multiple Criteria Optimization: Theory, Computation and Application, Wiley, New York.

Steuer RE, Schuler AT (1981) Interactive Multiple Objective Linear Programming Applied to Multiple Use Forestry Planning, publication FWS–1–81, School of Forestry and Wildlife Resources, Virginia Polytechnic Institute and State University.

Tamiz M, Mardle SJ, Jones DF (1995) Resolving Inconsistency in Infeasible Linear Programmes, technical report, School of Mathematical Studies, University of Portsmouth, U.K.

Tamiz M, Mardle SJ, Jones DF (1996) Detecting IIS in Infeasible Linear Programmes using Techniques from Goal Programming, Computers and Operations Research 23:113–119.

Timminga E (1998) Solving Infeasibility in Computerized Test Assembly, Applied Psychological Measurement 22:280–291.

Tsoi AC, Hagenbucher M, Scarselli F (2006) Computing Customized Page Ranks, ACM Transactions on Internet Technology 6:381–414.

Tu W, Mayne RW (2002a) An approach to multi–start clustering for global optimization with non–linear constraints, International Journal for Numerical Methods in Engineering 53:2253–2269.

Tu W, Mayne RW (2002b) Studies of multi–start clustering for global optimization, International Journal for Numerical Methods in Engineering 53:2239–2252.

Ugray Z, Lasdon L, Plummer J, Glover F, Kelly J, Martí R (2006) Scatter Search and Local NLP Solvers: A Multistart Framework for Global Optimization, INFORMS Journal on Computing, to appear.

Van Hentenryck P (1999) The OPL Optimization Programming Language, MIT Press, Cambridge, Massachusetts.

van Loon J (1981) Irreducibly Inconsistent Systems of Linear Inequalities, European Journal of Operations Research 8:283–288.

Vatolin AA (1992) An LP–Based Algorithm for the Correction of Inconsistent Linear Equation and Inequality Systems, Optimization 24:157–164.

Vera JR (1998) On the Complexity of Linear Programming Under Finite Precision Arithmetic, Mathematical Programming 80:91–123.

Wagner M, Meller J, Elber R (2004) Large–Scale Linear Programming Techniques for the Design of Protein Folding Potentials, Mathematical Programming 101:301–318.

Williams HP (1978) Model Building in Mathematical Programming, John Wiley and Sons, Chichester.

Winston WL, Venkataramanan M (2003) Introduction to mathematical programming, 4th edition. Brooks/Cole, Pacific Grove, USA.

Wolfe P (1965) The Composite Simplex Method, SIAM Review 7:42–54.

Wright, SJ (1997) Primal–Dual Interior–Point Methods, SIAM Publications.

Wu X, Barbará D (2002) Learning Missing Values from Summary Constraints, ACM SIGKDD Explortions Newsletter 4:21–30.

Xiao Y, Censor Y, Michalski D, Galvin J (2003) The Least–Intensity Feasible Solution for Aperture–Based Inverse Planning in Radiation Therapy. Annals of Operations Research 119:183–203.

Yang J (2006) Infeasibility Resolution Based on Goal Programming, Computers and Operations Research, to appear (available online now).

Yarnold PR, Soltysik RC (2004) Optimal Data Analysis: a Guidebook with Software for Windows, American Psychological Association.

Yildirim EA, Wright SJ (2002) Warm–Start Strategies in Interior–Point Methods for Linear Programming, SIAM Journal on Optimization 12:782–810.

You Z (1993) Localization and Diagnosis of Structural Problems in Petri Net Models, MSc thesis, Systems and Computer Engineering, Carleton University, Ottawa, Canada.

Zionts S, Wallenius J (1983) An Interactive Multiple Objective Linear Programming Method for a Class of Underlying Nonlinear Utility Functions, Management Science 29:519–529.

Index

A
Active constraints method, 37–42
Additive adaptive grouping, 104
Additive method, 98–101, 104, 106, 109–110, 119, 124, 126, 129, 132, 134–137, 140–143, 144, 149, 153, 154, 155–157, 165–167
Additive/deletion method, 109, 110, 120, 137, 140–143
Additive/sensitivity method, 119, 120
Adjusting the constraint matrix, 206–208
Advanced start, 17, 18, 23, 109, 112, 149, 156, 170, 175, 176
Affine-scaling method, 65, 199
Algorithm 2(k), 174, 175
Algorithm 3(k), 175
Alldiff, 46
Almost convex region effect, 57, 58
Alpha-shape technique, 60
Altering constraints, 197
AMPL, 54, 76, 96, 180, 230
ANALYZE, 90, 127, 128, 130, 131
Analyzing infeasibility, 7, 25, 89–209, 213
Approximating LP, 205, 206
Arc consistency, 47, 195
Artificial variables, 11–13, 16, 17, 101, 114, 115, 124, 167, 168, 198, 201, 202
Automated test assembly, 212, 238–239
Average direction-based constraint consensus, 69

B
Back jumping, 195
Backtracking, 43, 46–49, 118, 131, 156, 190–192, 212, 240, 241
BARON, 87
Basis pursuit, 243, 244
Basis reduction, 43, 44
Best approximation problem, 209
Big-M method, 11, 13, 181, 230, 237
Bilinear relationship, 163
Binary constraint, 47
Binary integer program, 193
Binary program, 25, 27–29, 130, 238, 242
Bootstrapping, 52, 59, 60, 62, 85, 87, 88, 144, 184
Bound tightening, 10, 75, 76, 95, 96, 130, 131
Branch and bound, 8, 17, 23–24, 44, 130–133, 138–144, 167, 195, 208, 212, 230, 232, 240–241
Branch and cut, 23, 24, 28, 30, 130, 180, 230, 232
Branch down, 24
Branch up, 24
Branching variable selection, 24, 37, 139, 140
Bregman distance, 200
Buffer overrun detection, 239

C
Candidate variable, 15, 23, 24, 37–40
Check inequality, 205
CLAUDIA, 128, 129
Closed hemisphere problem, 231
Column protection, 122, 125
Combinatorial Bender's cuts, 167, 180, 183
Combining methods, 118
Complementary inequalities, 189–192
Complete inconsistency, 232
Component averaging, 20, 21, 66
Composite objective, 18
Concave function, 55–57
Conflict analysis, 42–43
Conflict constraints, 42, 43
Conflict-directed backtracking, 48
Conflict refiner, 129, 143
Conflict set, 43, 48, 143, 156, 220, 240
Consensus vector, 4, 19–21, 66, 67, 69, 71, 81
Constrained region, 57, 58, 60, 62
Constraint consensus, 19, 65–73, 77, 80, 81, 83, 84

Constraint effectiveness, 59, 111
Constraint learning, 42, 48
Constraint logic programming, 45, 156
Constraint programming, 42, 45–50, 96, 98, 154, 156, 157, 193, 195, 212, 240
Constraint propagation, 46, 47, 96
Constraint satisfaction problem, 45, 46, 48, 49, 156, 157, 193, 195
Constraint sensitivity, 173, 174
Constraint shifting, 197–199, 202, 204–205
Constraint violation, 2, 3, 19, 23, 51, 52, 58, 80, 148, 149, 167, 168, 172–174, 199, 240
Constructive method, 156
Control row, 128, 129
Control sequence, 19, 20
Convex function, 9, 54–56
Convexify, 8, 9
Convex region effect, 57–60, 62
Convex sampling enclosure, 58–60, 75
Convex set, 19, 36, 51, 54, 56–58, 206
Cplex, 32, 34, 37, 40–42, 113, 129, 143, 168, 169, 177, 178, 183, 199, 230, 238
Crash start, 11, 15, 16
Crossover from an infeasible basis, 16, 17
CSADT, 229
CUTE, 64, 77

D

Data analysis, 227–233
Data classification, 227, 231
Data depth, 211, 227, 231, 232
Davis-Putnam-Logemann-Loveland algorithm, 49
DBavg constraint consensus, 69
DBbound constraint consensus, 71, 72
DBmax constraint consensus, 70, 71
Debug command, 129
Degree heuristic, 46
Deletion filter, 43, 97–98, 104–110, 112, 118–122, 123–126, 128–129, 132–134 137, 140–143, 144–146, 149–153, 154, 156–157, 160, 167, 238
Deletion/sensitivity filter, 118–122, 125, 129, 150, 160
Depth first binary search filter (DFBS), 105–107, 109, 142
Destructive method, 156
Diagnosis of over determined constraint satisfaction problems (DOC), 156
Dichotomic method, 156
Digital video broadcasting, 180, 181, 183, 212, 235, 237
Disciplined convex optimization, 58
Discriminant analysis, 183
Distance to ill-posedness, 208
Dom/Wdeg, 156
Dr. AMPL, 54
Duality gap, 5
Dubious constraint, 133, 134, 137, 141
Dynamic reordering additive method, 100, 136, 141

E

Easy-hard-easy pattern, 7
Efficient global optimization (EGO), 78
Elastic filter, 101–104, 108, 112, 120, 122, 125, 126, 128, 129, 144, 239
Elastic programming, 102, 105, 172, 198, 199
Elastic SINF, 169–171, 175
Elastic variable, 98, 101, 102, 114, 115, 122, 167, 168, 171–174, 178, 198–200, 202, 204, 205, 238
Enforcing a constraint, 102, 125
Equilibrium constraints, 162, 163
Errors in massive data sets, 211, 232, 233
Exact penalty function, 80
Extreme aspiration level, 220–222, 225

F

Fail-first heuristic, 46
FDfar algorithm, 68, 69
FDnear algorithm, 68, 69
Feasibility-distance based constraint consensus, 68
Feasibility distance tolerance, 67–70, 72, 84
Feasibility pump, 30–36
Feasibility vector, 4, 19–21, 66–72, 83, 84
Feasible sequential quadratic programming, 51
Forcing substructure, 213
FortMP, 15–17
Forward checking, 47, 195
Frequency-based heuristic, 165
Frobenius norm, 207, 208
Frontline systems solvers, 130, 154
Full elastic program, 168, 172, 173, 199
Function tolerance test, 3
Fuzzy sets, 202

G
GAMS, 8
Generalized binary search filter (GBS), 107–109, 142
Generalized network, 214, 215
Generalized reduced gradient algorithm (GRG), 51, 65
Genetic algorithm, 60, 78
Global optimization, 52, 78, 79, 87–88, 153, 154, 162, 208
Goal programming, 98, 199, 202, 203, 238
GPIIS, 98
Gradient projection, 66
GRASP, 78
Greedy unit propagation (GUP), 50
Grouping constraints, 149, 157
Grow method, 160
Guard constraints, 145, 146
Guide codes, 121, 122, 129, 221, 223–225
Guiding the isolation, 120–122, 143

H
Half space depth, 231
Hard constraint, 195, 217, 218, 220, 223, 225
Hard objective, 217, 220, 223
Hemisphere problem, 231, 245
Hill-climbing, 48
Hit-and-run methods, 59–62, 75
Hyperslab, 189, 242

I
IIS cover, 159–161, 164, 166–171, 175–178, 184–190, 227, 228, 231, 232
IIS enumeration, 164, 165
IIS pivoting, 117, 157
Implied equality, 59, 111, 152
Incumbent solution, 7, 24, 25, 180, 230, 237
Infeasible due to mathematical error, 145, 146
Infeasible in the ordinary sense, 145, 146
Infeasible network LP, 127, 128
Infeasible subset, 48, 88–90, 93, 132–139, 141–143, 154, 240
Infeasible-path interior point methods, 11
Initial point placement, 63–65, 76, 77
Integer and randomized deletion algorithm (IRDA), 238
Integer infeasibility, 4, 5, 26, 32
Intelligent backtracking, 43, 48, 156, 240, 241
Intelligent Mathematical Programming System (IMPS), 90
Intensity-Modulated Radiation Therapy (IMRT), 235, 236
Interaction analysis, 217, 218, 223
Interior point heuristic, 183, 184, 237
Interior point methods, 11, 18, 65, 112, 118
Irreducible infeasible subset of constraints (IIS), 42, 48, 85–90, 93, 94, 96–106
Isolating infeasibility, 93–157

J
Jeroslaw-Wang branching heuristic, 49, 50

K
k-consistency, 47
killing constraint, 152, 153
knapsack cover, 241

L
L1 norm, 31, 203, 204, 206, 207
Least constraining value heuristic, 47
Least-squares, 199, 200
LINDO, 129, 143, 154, 185, 213
Linear program, 2, 11–21, 45, 90, 93–95, 101, 112–130, 159–209, 213, 216–226, 228–232, 235–236
Linear program with equilibrium constraints (LPEC), 162, 179, 180, 229
l-infinity norm, 199
LINGO, 143, 154
Local search, 25, 46, 48, 78, 79, 183
Location depth, 231
Logarithmic barrier function, 61, 62
Logic programming, 45, 155, 156
Logical reduction, 95, 96, 127
LP. *See* Linear program
LP relaxation, 23–28, 30–33, 37, 38, 131, 132, 134, 138, 140, 141, 166, 167, 238
LSGRG(MIS), 147, 149, 154

M

Machine learning, 160, 162, 180, 181, 227, 228
Maintaining Arc Consistency (MAC), 47
Mams branching heuristic, 49, 50
Marker point, 81–84
Market split problems, 43, 44
Max independent set problem, 245
Maximum feasibility guideline algorithm, 184
Maximum feasible subsystem problem (MAX FS), 89, 159–195, 198, 211, 212, 227, 228, 231, 232, 236, 237, 240, 243–245
Maximum satisfiability problem (MAXSAT), 49, 187, 189, 193, 245
MAXO branching heuristic, 49, 50
Measuring infeasibility, 2–5
Method of feasible directions, 65
Min-conflicts heuristic, 48
Minimal conflict set, 48, 156, 240
Minimal cover, 172, 187
Minimal infeasible system, 189
Minimal intractable subsystem (MIS), 90, 145, 147
Minimal unbounded set of variables, 213
Minimal unsatisfiable core, 156
Minimally unsatisfiable subformula (MUS), 154, 155
Minimum feedback arc set problem, 244
Minimum misclassification cardinality, 227
Minimum number of feasible partitions problem (MIN PFS), 189–193
Minimum remaining values heuristic, 46
Minimum unsatisfied linear relation problem (MIN ULR), 159
Minimum-cardinality IIS set-covering problem (MIN IIS COVER), 159–161, 164, 227, 228, 231
Minimum-weight IIS cover, 171
MINLP. *See* Mixed-integer nonlinear program
MINOS(IIS), 121, 122, 124, 128–130
MINOS, 18, 63
MINTO, 131
MIP. *See* Mixed-integer linear program
MISMIN, 229
Mixed-integer linear program (MIP or MILP), 2, 23–44, 96, 102, 104–105, 130–143, 157, 161, 166–167, 179–180, 183, 236, 237, 242
Mixed-integer nonlinear program (MINLP), 34–36
Mixed-integer program (MIP), *See* Mixed-integer linear program
MOMS branching heuristic, 49, 50
Most violated constraint control, 20
Movement tolerance, 67–70, 72
MProbe, 55, 56, 59, 61, 73, 75, 76, 90, 111
MSNLP, 79, 80, 83
Multi-objective program, 195, 202, 203, 238
Multiple constraint, 86, 202
Multiple-objective linear program (MOLP), 216–226
Multiplicative adaptive grouping, 104, 105
Multistart constraint consensus, 80–84
Multistart methods, 73, 77–79, 87
Mutually incompatible constraints (MIC), 129

N

Necessary constraint, 62, 199
Netlib, 112, 113, 166, 175
Neural networks, 227–231
Node selection, 24, 46, 140
Nogood learning, 42, 48
Nonbasic variable, 13, 14, 16, 17, 25–27, 116
Nonconserving processing network, 214, 215
Nonconvex region effect, 57–59
Nonlinear program (NLP), 2, 34, 51, 144, 204
Nonlinear range cutting, 75
Nonlinear turnabout, 86
Nontermination, 130, 132
Nonviability, 214–216
NP-hard, 159, 161, 212, 240, 244–245
Nucleus box, 74, 75
Number of infeasibilities (NINF), 3, 14, 67, 167

O

Objective interference, 221–223, 225
Oblique projection, 20
OC1, 229
OCTANE, 28–30
Open hemisphere problem, 231

OPL, 45
Optimal data analysis, 231
Optimality gap, 32, 41
OptQuest, 78
Orthogonal projection, 4, 19–21, 66
OSL, 129
Overconstrained problem, 193, 195
Overlapped IISs, 119, 185

P
Page ranking, 239, 240
Partial constraint satisfaction, 193–195
Partial inconsistency, 232
Path-consistent, 47
Penalty function, 51–53, 80, 145
PERUSE, 90
Petri net, 216
Phase 1 algorithm, 11, 12
Phase 1 heuristics, 167–169
Phase 2 algorithm, 174, 175, 177, 229
Piecewise affine autoregressive exogenous (PWARX), 193
Piecewise linear model estimation, 242, 243
Pivot-and-complement, 25–28
Pivot-and-shift, 25–28
Pivoting methods, 116–118
Pointer, 82–84
Posynomial function, 9
PREDUCE, 55
Preference constraint, 46
Presolving, 47, 75, 95, 96
Primal-dual interior point methods, 18
Prime analytic centre, 61
Processing node, 214, 216
PROFLOW, 129
Projection methods, 19–21
Propagation, 47, 50
Protein folding, 183, 236, 237
Pseudo-costs, 37
Pure network, 214
Pure processing network, 214, 215
Purify algorithm, 17
Push algorithm, 17

Q
Quadratic program, 14, 151–153, 204
Quadratically-constrained quadratic program (QCP, QCQP), 14, 150, 151, 153

R
Radiation treatment planning, 235, 236
Random sampling, 58, 73
Random walks, 48
Randomized standard heuristic, 63–65
Randomized thermal relaxation algorithm (RTR), 181–183
Ratio equation, 214
Reciprocal filter, 113
Refinery and Petrochemical Modeling System, 90
REFORM, 8
Reformulation, 8–10
Relaxation method, 21
Relaxation parameter, 20
Relaxation-induced neighbourhood search (RINS), 25
Relaxed and ordered deletion algorithm (RODA), 238
Remotest set control, 20, 69
Reverse deletion filter, 124–126, 156
Root node, 24, 32, 38
Row aggregation, 130

S
Sampling box, 73, 74, 76–78, 83
Sampling enclosure, 55, 57–60
Sampling methods, 110, 111
Satisfiability problem (SAT), 49, 240, 241
Scaling, 3, 4, 9, 71, 162, 242
Scatter search, 78
Selective unit propagation (SUP), 50
Sensitivity filter, 114–116, 119, 126, 129, 138, 160
Sequential linear approximation, 163
Sequential projection, 19, 181
Sequential quadratic programming (SQP), 204, 205
Set covering, 110, 112, 164–166, 238
Set covering and item sampling (SCIS), 238
Shifting constraints, 197–206
Simple phase 1, 168, 170
Simulated annealing, 48, 182
Simultaneous projection, 20
SINF-reduction heuristic, 169–178
Single-start methods, 73, 77
Small-cardinality IIS, 116
Soft constraint, 195, 217–220, 223
Soft objective, 217, 220, 223

Space-covering global optimizers, 153
Spanning line segment, 59, 63, 75
Sparse solutions to systems of linear equations, 243, 244
Special ordered sets, 37
Stand-and-hit algorithm, 59
Standard elastic program, 168–170
Standard heuristic, 63–65, 77, 83
Stochastic local search, 48
Strictly complementary partition, 118
Strong branching, 37
Structural infeasibility, 87
Structural relationships, 214
Successive overrelaxation, 17
Sufficient constraint, 134, 185
Sum of the infeasibilities (SINF), 3, 14, 15, 18, 167, 182
Superbasic variable, 17
Surface of maximal intersection, 85
Surrogate constraints, 37

T
Tabu search, 48
Tightening variable bounds, 9, 73–76
Tolerances, 3, 5, 90, 94, 144
Tree-structured search, 212, 240
True MOLP, 219, 223–225
Tukey depth, 231
Tukey median, 231
Two-phase method, 11, 13, 180, 181
Two-phase relaxation-based heuristic, 179–181
Type 1 pivot, 25–28
Type 2 pivot, 26–28
Type 3 pivot, 26–28

U
Unary constraint, 46
Unbounded linear program, 213–216
Undecidable point, 192–194
Underdetermined system, 243
Unit basis, 15
Unit propagation (UP), 50
Useful isolation, 122–127

V
Viability, 213–215
Voting heuristics, 20
Voxel, 235, 236

W
Warm start, 17, 18
Wcore, 156, 157
Weighted random multistart, 81
Witness nodes, 128

X
XA, 130
Xpress, 28, 34, 129

Z
Zooming and domain elimination, 79

Early Titles in the
INTERNATIONAL SERIES IN
OPERATIONS RESEARCH & MANAGEMENT SCIENCE
Frederick S. Hillier, Series Editor, *Stanford University*

Saigal/ *A MODERN APPROACH TO LINEAR PROGRAMMING*
Nagurney/ *PROJECTED DYNAMICAL SYSTEMS & VARIATIONAL INEQUALITIES WITH APPLICATIONS*
Padberg & Rijal/ *LOCATION, SCHEDULING, DESIGN AND INTEGER PROGRAMMING*
Vanderbei/ *LINEAR PROGRAMMING*
Jaiswal/ *MILITARY OPERATIONS RESEARCH*
Gal & Greenberg/ *ADVANCES IN SENSITIVITY ANALYSIS & PARAMETRIC PROGRAMMING*
Prabhu/ *FOUNDATIONS OF QUEUEING THEORY*
Fang, Rajasekera & Tsao/ *ENTROPY OPTIMIZATION & MATHEMATICAL PROGRAMMING*
Yu/ *OR IN THE AIRLINE INDUSTRY*
Ho & Tang/ *PRODUCT VARIETY MANAGEMENT*
El-Taha & Stidham/ *SAMPLE-PATH ANALYSIS OF QUEUEING SYSTEMS*
Miettinen/ *NONLINEAR MULTIOBJECTIVE OPTIMIZATION*
Chao & Huntington/ *DESIGNING COMPETITIVE ELECTRICITY MARKETS*
Weglarz/ *PROJECT SCHEDULING: RECENT TRENDS & RESULTS*
Sahin & Polatoglu/ *QUALITY, WARRANTY AND PREVENTIVE MAINTENANCE*
Tavares/ *ADVANCES MODELS FOR PROJECT MANAGEMENT*
Tayur, Ganeshan & Magazine/ *QUANTITATIVE MODELS FOR SUPPLY CHAIN MANAGEMENT*
Weyant, J./ *ENERGY AND ENVIRONMENTAL POLICY MODELING*
Shanthikumar, J.G. & Sumita, U./ *APPLIED PROBABILITY AND STOCHASTIC PROCESSES*
Liu, B. & Esogbue, A.O./ *DECISION CRITERIA AND OPTIMAL INVENTORY PROCESSES*
Gal, T., Stewart, T.J., Hanne, T. / *MULTICRITERIA DECISION MAKING: Advances in MCDM Models, Algorithms, Theory, and Applications*
Fox, B.L. / *STRATEGIES FOR QUASI-MONTE CARLO*
Hall, R.W. / *HANDBOOK OF TRANSPORTATION SCIENCE*
Grassman, W.K. / *COMPUTATIONAL PROBABILITY*
Pomerol, J-C. & Barba-Romero, S. / *MULTICRITERION DECISION IN MANAGEMENT*
Axsäter, S. / *INVENTORY CONTROL*
Wolkowicz, H., Saigal, R., & Vandenberghe, L. / *HANDBOOK OF SEMI-DEFINITE PROGRAMMING: Theory, Algorithms, and Applications*
Hobbs, B.F. & Meier, P. / *ENERGY DECISIONS AND THE ENVIRONMENT: A Guide to the Use of Multicriteria Methods*
Dar-El, E. / *HUMAN LEARNING: From Learning Curves to Learning Organizations*
Armstrong, J.S. / *PRINCIPLES OF FORECASTING: A Handbook for Researchers and Practitioners*
Balsamo, S., Personé, V., & Onvural, R./ *ANALYSIS OF QUEUEING NETWORKS WITH BLOCKING*
Bouyssou, D. et al. / *EVALUATION AND DECISION MODELS: A Critical Perspective*
Hanne, T. / *INTELLIGENT STRATEGIES FOR META MULTIPLE CRITERIA DECISION MAKING*
Saaty, T. & Vargas, L. / *MODELS, METHODS, CONCEPTS and APPLICATIONS OF THE ANALYTIC HIERARCHY PROCESS*
Chatterjee, K. & Samuelson, W. / *GAME THEORY AND BUSINESS APPLICATIONS*
Hobbs, B. et al. / *THE NEXT GENERATION OF ELECTRIC POWER UNIT COMMITMENT MODELS*
Vanderbei, R.J. / *LINEAR PROGRAMMING: Foundations and Extensions, 2nd Ed.*
Kimms, A. / *MATHEMATICAL PROGRAMMING AND FINANCIAL OBJECTIVES FOR SCHEDULING PROJECTS*
Baptiste, P., Le Pape, C. & Nuijten, W. / *CONSTRAINT-BASED SCHEDULING*
Feinberg, E. & Shwartz, A. / *HANDBOOK OF MARKOV DECISION PROCESSES: Methods and Applications*
Ramík, J. & Vlach, M. / *GENERALIZED CONCAVITY IN FUZZY OPTIMIZATION AND DECISION ANALYSIS*
Song, J. & Yao, D. / *SUPPLY CHAIN STRUCTURES: Coordination, Information and Optimization*
Kozan, E. & Ohuchi, A. / *OPERATIONS RESEARCH/ MANAGEMENT SCIENCE AT WORK*

Early Titles in the
INTERNATIONAL SERIES IN
OPERATIONS RESEARCH & MANAGEMENT SCIENCE
(Continued)

Bouyssou et al. / *AIDING DECISIONS WITH MULTIPLE CRITERIA: Essays in Honor of Bernard Roy*
Cox, Louis Anthony, Jr. / *RISK ANALYSIS: Foundations, Models and Methods*
Dror, M., L'Ecuyer, P. & Szidarovszky, F. / *MODELING UNCERTAINTY: An Examination of Stochastic Theory, Methods, and Applications*
Dokuchaev, N. / *DYNAMIC PORTFOLIO STRATEGIES: Quantitative Methods and Empirical Rules for Incomplete Information*
Sarker, R., Mohammadian, M. & Yao, X. / *EVOLUTIONARY OPTIMIZATION*
Demeulemeester, R. & Herroelen, W. / *PROJECT SCHEDULING: A Research Handbook*
Gazis, D.C. / *TRAFFIC THEORY*
Zhu/ *QUANTITATIVE MODELS FOR PERFORMANCE EVALUATION AND BENCHMARKING*
Ehrgott & Gandibleux/ *MULTIPLE CRITERIA OPTIMIZATION: State of the Art Annotated Bibliographical Surveys*
Bienstock/ *Potential Function Methods for Approx. Solving Linear Programming Problems*
Matsatsinis & Siskos/ *INTELLIGENT SUPPORT SYSTEMS FOR MARKETING DECISIONS*
Alpern & Gal/ *THE THEORY OF SEARCH GAMES AND RENDEZVOUS*
Hall/*HANDBOOK OF TRANSPORTATION SCIENCE - 2nd Ed.*
Glover & Kochenberger/ *HANDBOOK OF METAHEURISTICS*
Graves & Ringuest/ *MODELS AND METHODS FOR PROJECT SELECTION: Concepts from Management Science, Finance and Information Technology*
Hassin & Haviv/ *TO QUEUE OR NOT TO QUEUE: Equilibrium Behavior in Queueing Systems*
Gershwin et al/ *ANALYSIS & MODELING OF MANUFACTURING SYSTEMS*
Maros/ *COMPUTATIONAL TECHNIQUES OF THE SIMPLEX METHOD*
Harrison, Lee & Neale/ *THE PRACTICE OF SUPPLY CHAIN MANAGEMENT: Where Theory and Application Converge*
Shanthikumar, Yao & Zijm/ *STOCHASTIC MODELING AND OPTIMIZATION OF MANUFACTURING SYSTEMS AND SUPPLY CHAINS*
Nabrzyski, Schopf & Węglarz/ *GRID RESOURCE MANAGEMENT: State of the Art and Future Trends*
Thissen & Herder/ *CRITICAL INFRASTRUCTURES: State of the Art in Research and Application*
Carlsson, Fedrizzi, & Fullér/ *FUZZY LOGIC IN MANAGEMENT*
Soyer, Mazzuchi & Singpurwalla/ *MATHEMATICAL RELIABILITY: An Expository Perspective*
Chakravarty & Eliashberg/ *MANAGING BUSINESS INTERFACES: Marketing, Engineering, and Manufacturing Perspectives*
Talluri & van Ryzin/ *THE THEORY AND PRACTICE OF REVENUE MANAGEMENT*
Kavadias & Loch/*PROJECT SELECTION UNDER UNCERTAINTY: Dynamically Allocating Resources to Maximize Value*
Brandeau, Sainfort & Pierskalla/ *OPERATIONS RESEARCH AND HEALTH CARE: A Handbook of Methods and Applications*
Cooper, Seiford & Zhu/ *HANDBOOK OF DATA ENVELOPMENT ANALYSIS: Models and Methods*
Luenberger/ *LINEAR AND NONLINEAR PROGRAMMING, 2nd Ed.*
Sherbrooke/ *OPTIMAL INVENTORY MODELING OF SYSTEMS: Multi-Echelon Techniques, Second Edition*
Chu, Leung, Hui & Cheung/ *4th PARTY CYBER LOGISTICS FOR AIR CARGO*
Simchi-Levi, Wu & Shen/ *HANDBOOK OF QUANTITATIVE SUPPLY CHAIN ANALYSIS: Modeling in the E-Business Era*
Gass & Assad/ *AN ANNOTATED TIMELINE OF OPERATIONS RESEARCH: An Informal History*
Greenberg/ *TUTORIALS ON EMERGING METHODOLOGIES AND APPLICATIONS IN OPERATIONS RESEARCH*
Weber/ *UNCERTAINTY IN THE ELECTRIC POWER INDUSTRY: Methods and Models for Decision Support*
Figueira, Greco & Ehrgott/ *MULTIPLE CRITERIA DECISION ANALYSIS: State of the Art Surveys*

Early Titles in the
INTERNATIONAL SERIES IN
OPERATIONS RESEARCH & MANAGEMENT SCIENCE
(Continued)

Reveliotis/ *REAL-TIME MANAGEMENT OF RESOURCE ALLOCATIONS SYSTEMS: A Discrete Event Systems Approach*
Kall & Mayer/ *STOCHASTIC LINEAR PROGRAMMING: Models, Theory, and Computation*
Sethi, Yan & Zhang/ *INVENTORY AND SUPPLY CHAIN MANAGEMENT WITH FORECAST UPDATES*
Cox/ *QUANTITATIVE HEALTH RISK ANALYSIS METHODS: Modeling the Human Health Impacts of Antibiotics Used in Food Animals*
Ching & Ng/ *MARKOV CHAINS: Models, Algorithms and Applications*
Li & Sun/ *NONLINEAR INTEGER PROGRAMMING*
Kaliszewski/ *SOFT COMPUTING FOR COMPLEX MULTIPLE CRITERIA DECISION MAKING*

***** *A list of the more recent publications in the series is at the front of the book* *